Virtual and Augmented Reality Applications in the Automobile Industry

Devaraj E.
CMR University, India

IGI Global
Scientific Publishing
Publishing Tomorrow's Research Today

Vice President of Editorial — Melissa Wagner
Director of Acquisitions — Mikaela Felty
Director of Book Development — Jocelynn Hessler
Production Manager — Mike Brehm
Cover Design — Phillip Shickler

Published in the United States of America by
IGI Global Scientific Publishing
701 East Chocolate Avenue
Hershey, PA, 17033, USA
Tel: 717-533-8845
Fax: 717-533-7115
Website: https://www.igi-global.com E-mail: cust@igi-global.com

Library of Congress Cataloging-in-Publication Data

Names: E, Devaraj, 1987- editor.
Title: Virtual and augmented reality applications in the automobile industry / edited by Devaraj E.
Description: Hershey, PA : IGI Global, [2025] | Includes bibliographical references and index. | Summary: "This book aims to bridge the gap between theoretical advancements and practical applications, providing a comprehensive resource that not only highlights current developments but also fosters further research and innovation. Advancing the State of Knowledge By compiling cutting-edge research and case studies, this book seeks to advance the state of knowledge regarding VR and AR applications in the automotive industry"-- Provided by publisher.
Identifiers: LCCN 2024047138 (print) | LCCN 2024047139 (ebook) | ISBN 9798369372876 (hardcover) | ISBN 9798369372883 (paperback) | ISBN 9798369372890 (ebook)
Subjects: LCSH: Automobile industry and trade. | Automobiles--Technological innovations.
Classification: LCC HD9710.A2 V56 2025 (print) | LCC HD9710.A2 (ebook) | DDC 338.4/76292--dc23/eng/20250107
LC record available at https://lccn.loc.gov/2024047138
LC ebook record available at https://lccn.loc.gov/2024047139
British Cataloguing in Publication Data
A Cataloguing in Publication record for this book is available from the British Library.

Table of Contents

Detailed Table of Contents

Machine Leanring (ML) and Artificial Intelligence (AI) can change the game of transport with Intelligent Transportation Systems (ITS). The connected graphs observe architecture and enable from the ML and AI across domains to change the performing method of connectivity and make available better traffic performance, security enhancement, and mobility optimization. The big data from sensors, vehicles, and infrastructure are processed by machine learning algorithms to provide data-driven, real-time decisions, predictive analysis, and traffic flow optimization for smooth flow, less jam, and fewer emissions. Resulting in the smart cities intelligent mobility, with the rise of AI and ML in ITS, the smart motors are being transformed into adaptive devices that can change as per evolving traffic scenarios. It explores how AI and ML may be integrated with smart cities and ITS to make these connected, efficient, and sustainable and examines current and emerging applications of ITS with a future outlook.

Virtual prototyping has evolved to the automotive industry and allows for a more speedy and accurate car development process at a fraction of the previous costs. The application of virtual prototyping and utilization of computer aided design, simulations and digital twin has been discussed in this chapter. These approaches reduce development time and costs as well as improve design performance over traditional use of physical models. These tests add value to vehicle performance and safety so there are few tests such as crash, aerodynamics, and virtual assembly are discussed. It also introduces some recent tendencies on virtual prototyping development including Artificial Intelligence, Machine Learning and Virtual Reality. It identifies various challenges in applying these technologies and examines future prospects and implications in introducing them in automobile industry respectively.

A major contributor to accidents is alcohol, and shift workers, business drivers, and truck drivers are the most susceptible to dozing off while operating a motor vehicle. Along with putting lives in danger, these problems put a heavy financial strain on business owners, who are held responsible for these kinds of accidents. In this research, a mobile application-integrated adaptive driver and business owner alert system for tracking driving behaviors is presented. The system uses advanced sensors to detect drunkenness, tiredness, and seat belt usage, triggering a buzzer when risky behaviors are detected. The system uses GPS and GSM modules to communicate the car's location to the designated individual, deactivating the motor if critical thresholds are crossed.This system, controlled by a Cortex M3 microprocessor, proactively addresses major crash causes, reducing accidents, injuries, fatalities, and property damage. The virtual and augmented reality-related future trends in vehicle safety with Auto Alert have also been explained.

S. Pushpa, Department of Computer Science and Engineering, St. Peter's Institute of Higher Education and Research, Chennai, India
R. M. Sathiyamoorthy, Department of Mechanical Engineering, Easwari Engineering College, Ramapuram, India
S. Anita, Department of Science and Humanities (General Engineering), R.M.K. Engineering College, Kavaraipettai, India
R. Sathishkannan, Department of Computer Science and Engineering, Bannari Amman Institute of Technology, Erode, India
M. Renuga, Department of Electronics and Communication and Engineering, Prince Shri Venkateshwara Padmavathy Engineering College, India
S. Boopathi, Muthayammal Engineering College, Namakkal, India

This chapter explores an overview of using virtual reality (VR) to train technicians and factory personnel in the automotive industry. It also shows how immersive virtual reality technologies can simulate real environments, providing a safe, controlled area for the skills to be developed and practiced on. VR enables workers to practice how to operate complex machinery, perform repair procedures, and work on assembly lines without risking injury to themselves or damage to equipment—often at a fraction of the cost of training the old-fashioned way. This can allow industrial corporations within automotive design and restoration to boost their performance and security. The chapter continues explaining how useful it is to implement VR to engage employees with the application, where it may even enhance learning, reduce training time, and increase the general output quality of the workers.

Virtual Reality (VR) and Augmented Reality (AR) Technologies in Intelligent Transport System.Presently, VR and AR technologies play a significant role in enhancing the transport system. Their applications include traffic simulation, route planning, and driver training. Furthermore, AR-enhanced mappings and navigation systems also boosted the efficiency of commuters through trusty Return to Base. Applications of virtual reality allow engineers to create planning for the smart infrastructure by embedding future transport projects within the built environment. Out of these technologies, semiconductor and wireless technologies form the foundation for vehicle-to-everything (V2X) communication, the cornerstone of intelligent traffic management systems through seamless data exchange. The fusion of VR with superbusiness forms of AR and ITS then leads into adaptive and interactive and smart transportation systems which were also sustainable transport solutions to urban mobility challenges.

The incorporation of Virtual reality (VR) and Augmented reality (AR) technologies in the automobile industry has made a major leap forward in vehicle design, manufacturing, marketing, and upkeep. The conventional approach involves physical prototypes, manual quality assurance, and static training resources, making them very time consuming and expensive with minimal efficiency. Current trends show that VR and AR are becoming increasingly accepted in the automobile industry due to their ability to increase precision, reduce costs, speed up processes, or improve interaction with users. Virtual models can be developed and tested much faster in the design and prototyping stages using VR. Manufacturing accuracy can be increased by AR through improved quality control and reduced error incidence. This chapter gives a large explanation of how VR and AR are applied in the automotive sector. It stresses how the utilization of VR will affect design works alongside prototyping while at the same time outlining AR's contribution to manufacturing processes.

This chapter explores the foundations of Virtual Reality (VR) technology and its significant implications and applications in the automotive industry. The objective is to examine how VR is reshaping automotive design, manufacturing, and customer engagement, while identifying the opportunities and challenges it presents. A multi-method approach is used, combining a literature review, case study analysis, and expert interviews to assess the current state of VR in the automotive sector. Key findings reveal that VR enhances the design process, optimizes manufacturing practices, and improves customer experience through virtual showrooms and test drives. However, challenges such as high implementation costs and the need for standardized VR platforms persist. The chapter concludes that VR is a transformative technology with the potential to drive future innovations, yet its widespread adoption depends on overcoming technical and economic barriers. Further research is needed to explore the integration of VR with emerging technologies such as AI and AR in automotive applications.

Augmented Reality (AR) technology has emerged as a disruptive force with the potential to redefine human interaction with the digital world. This chapter provides a comprehensive overview of the foundational aspects of AR, focusing on its underlying principles, key components, and transformative applications. At its essence, AR technology seamlessly blends virtual elements with real-world environments, offering users enriched experiences through contextual information and interactive overlays. It also dissects the core components of AR systems, including sensors, displays, tracking mechanisms, and rendering algorithms. Rendering algorithms, such as occlusion handling and real-time object tracking, contribute to creating convincing AR experiences by generating realistic virtual objects and integrating them seamlessly into the real world. Additionally, it scrutinizes the theoretical frameworks supporting AR, including computer vision, spatial computing, and human-computer interaction paradigms, to clarify the cognitive and perceptual mechanisms underlying AR interfaces.

This chapter focuses on the integration of VR and AR technologies with automobile companies, especially through what transformative impacts they bring along for the design and process of automobile manufacturing. Therefore, with the use of these VR and AR technologies, the car manufacturers are facilitated in producing immersive, interactive experiences to streamline the designs thus producing better visions, prototypings, and testing cars as may be needed. These technologies allow designers and engineers, along with many other stakeholders, to provide collaborative working with real-time access to virtual environments that allow exploring and refining of concepts. Apart from this, AR increases efficiency in manufacturing by speeding up on-site assembly as well as reducing errors. This particular chapter includes case studies demonstrating the successful application of both VR and AR to improve workflow productivity, increase precision, as well as fuel innovation in auto-making.

M. K. Srilekha, Department of Electronics and Communication Engineering, SRM Institute of Science and Technology, Kattankulathur, India
A. Maria Jossy, Department of Electronics and Communication Engineering, SRM Institute of Science and Technology, Kattankulathur, India
S. Hema Priyadarshini, Department of Medical Electronics Engineering, Dayananda Sagar College of Engineering, Bangalore, India
G. V. Krishna Pradeep, Department of Mechanical Engineering, Aditya University, Surampalem, India
S. Devi, Science and Humanities, Prince Shri Venkateshwara Padmavathy Engineering College, Chennai, India
Arun Chokkalingam, Sona College of Technology, India

Augmented Reality (AR)-based HUD (Heads-Up Display) systems have changed the face of driver assistance in vehicles, which significantly improves safety, situational awareness, and driving efficiency. These systems project key information directly onto the windshield, superimposing navigation directions, speed, and collision avoidance alerts in a way that merges the real and digital worlds. AR-HUDs have solved some key problems with a society of current transportation by taking away driver distraction and making reaction times better. Utilizing advanced sensors, machine learning, and real-time data processing, these displays can dynamically adapt to changing conditions and ultimately provide a driver-specific experience. This chapter discusses the mechanism, the design principle, and the technology behind the AR-HUDs with a better effect on accident reduction, increasing the safety on the road overall. It includes current limitations, advancements on the way, and how these are being examined for autonomous and connected vehicles.

Sunil Sharma, Lovely Professional University, India

This chapter provides a comprehensive examination of Augmented Reality (AR) in manufacturing, covering its introduction, types, implementation strategies, advantages, challenges, applications, and methodologies for implementation. We present a brief overview of AR technology in the context of Industry 4.0 and its significance in manufacturing. It highlights the potential of AR to bridge the gap between the physical and digital worlds. We then discuss the various types of AR technologies relevant to manufacturing along with the working of AR technology.

The numerous benefits of adopting AR in manufacturing and the systematic methodology for implementing AR in manufacturing environments is presented. Five applications of AR in manufacturing from real industries have been presented. The advantages, common challenges and barriers to AR implementation in manufacturing are discussed. This research provides a comprehensive roadmap for leveraging AR technology in manufacturing, offering insights into its potential, challenges, and practical implementation strategies.

N. Venkateswaran, Department of Master of Business Administration, Panimalar Engineering College, Chennai, India
A. Srinivasa Reddy, Department of Computer Science and Engineering (Data Science), CVR College of Engineering, Hyderabad, India
B. Uma Maheswari, Department of Computer Science and Engineering, St. Joseph's College of Engineering, Chennai, India
P. Suthanthira Devi, Department of Data Science and Business Systems, School of Computing, SRM Institute of Science and Technology, Kattankulathur, India
M. Aishwarya, Department of Science and Humanities, Prince Shri Venkateshwara Padmavathy Engineering College, Chennai, India
J. K. Periasamy, Department of Computer Science and Engineering, Sri Sai Ram Engineering College, Chennai, India

This chapter explores the use of Augmented Reality (AR) and Virtual Reality (VR) for the development of immersive showrooms, where vehicles, city-wide progressive mobility, smart city development, etc. can be personalized and experienced. AR-VR tech can generate immersed experiences such as real-time customizing, interactive test drives, and even position the client's own vehicle into his area to study how it will function on his street, for example. This all reduces the need for physical showrooms, speeds up decision-making processes, and promotes sustainable practices. Additionally, the chapter scopes the AR-VR potential in enabling urban mobility solutions by delivering real-time scenarios simulation to optimize vehicle designs and enhance the sustainability targets. By bringing together cutting-edge technology with responsive user experiences, it sets a new bar for smart city infrastructure and next-gen mobility systems.

The Autonomous Vehicle is an urban mobility sector user-generative disruptor. The chapter describes the core technology that allows AVs to function—sensory systems, machine learning, and the artificial intelligence that orchestrates the autonomous process itself. Finally, how can ACT align itself with urban systems, not just at pragmatic levels, such as what the unavoidable safety issues end up being and in terms of legislation, but with ACT itself considered an infrastructure? It will also consider how A.V.s may affect traffic patterns, as well as environmental sustainability and equity. This chapter elaborates on the myriad systems that a city employs to ensure that a fleet of vehicles does what it is designed to do, arguing that many details of this framework are what innovators will need to build if they hope to succeed. And it also lays out the immediate benefits the innovations provide to the city itself.

Virtual and augmented reality (VR/AR) technologies are revolutionizing the automobile industry by enhancing design processes, improving manufacturing precision, and enriching customer experiences. Despite these benefits, significant challenges and limitations hinder widespread adoption. This paper explores these obstacles, including technical constraints, high costs, and integration difficulties, while also discussing potential solutions and future trends that could mitigate these issues and pave the way for broader implementation.

This survey paper aims to discuss the areas of enhancement of virtual reality (VR) and augmented reality (AR) in the automotive industry: the basic principles, practice, problems, and points of improvement. It also emphasizes the application of VR in design improvement, prototyping and manufacturing, and sales presenting extraordinary experience and increased efficacy. Applications of AR in repair and maintenance, navigations, and driver assistance are analyzed with reference to enhancements of diagnostics, trainings and safety. Some of the issues that stem from the implementation of the two technologies include; technicalities, costs and acceptability by users. Finally, the paper outlines a future outlook touching on future trends and how VR and AR need constant enhancement and deployment in the automotive industry.

Air intake management has a tremendous effect on an automobile engine. It can affect everything from combustion to fuel efficiency to engine power. Using a fine mesh, we proposed an optimized air intake design and performed flow simulation analysis of air intake distribution uniformity in order to overcome the disadvantage of turbulence in the inlet. The fine mesh location, as well as priority between filtering and flow, leads us to create a new intake manifold fine mesh design. The effect of the fine mesh on air velocity, pressure distribution, and turbulence is examined using computational fluid dynamics (CFD) simulations. Experimental validation in the form of flow bench tests and engine dynamometer runs compares the proposed system to conventional intake designs. The results demonstrate improved flow consistency, lower loss of intake manifold pressure, and potential benefits in overall engine efficiency and power output. This offers advancements in automotive engineering, such as improving engine performance and fuel economy, which could help cut emissions.

Preface

The automotive industry stands on the brink of a technological revolution, where innovation is not a luxury but a necessity. As manufacturers, engineers, marketers, and service professionals strive to meet evolving consumer expectations and remain competitive, the adoption of transformative technologies has become paramount. Among these, Virtual Reality (VR) and Augmented Reality (AR) are redefining the very foundations of how vehicles are designed, built, marketed, and maintained.

Virtual and Augmented Reality Applications in the Automobile Industry is a comprehensive exploration of the growing impact of VR and AR across the automotive value chain. This book aims to bridge the divide between theoretical advancement and real-world implementation, offering insights that serve both academic researchers and industry practitioners.

Through detailed chapters and case studies, readers will discover how VR is reshaping the design and engineering processes by enabling immersive 3D prototyping and simulation, drastically reducing development time and cost. AR, on the other hand, is enhancing manufacturing workflows and operational efficiency by overlaying critical digital information onto the physical environment, revolutionizing assembly lines and maintenance procedures.

The influence of VR and AR extends beyond production into the realms of sales, marketing, and customer experience. From virtual showrooms and immersive test drives to AR-enhanced user manuals and remote diagnostics, these technologies are personalizing the buyer journey and optimizing post-purchase support.

This book also delves into the integration of VR and AR with other emerging technologies such as artificial intelligence, machine learning, and the Internet of Things (IoT). The convergence of these innovations promises to unlock even greater potential—enabling autonomous driving simulations, predictive maintenance, and smarter vehicle systems.

This book presents a comprehensive exploration of emerging technologies that are transforming the automotive and transportation sectors, with a particular focus on Artificial Intelligence (AI), Machine Learning (ML), Virtual Reality (VR), Augmented Reality (AR), and intelligent systems.

Chapter 1 delves into how AI and ML are revolutionizing Intelligent Transportation Systems (ITS), using big data analytics to improve traffic flow, reduce emissions, and create adaptive mobility solutions within smart cities.
Chapter 2 introduces virtual prototyping as a game-changer in automotive development, emphasizing the role of CAD, simulations, and digital twins in reducing design costs and development cycles while enhancing vehicle safety and performance.
Chapter 3 presents an adaptive driver monitoring system designed to prevent accidents caused by drowsy or impaired driving. This solution leverages sensor technology, mobile integration, and alert systems to enhance road safety for both individuals and businesses.
Chapter 4 focuses on the application of VR in training automotive technicians and assembly line workers, showing how immersive environments can safely and cost-effectively improve skills and reduce errors in manufacturing.
Chapter 5 highlights how AR and VR contribute to ITS through real-time simulations, smart navigation, and infrastructure planning. It also discusses the underlying wireless and semiconductor technologies enabling vehicle-to-everything (V2X) communication.
Chapter 6 explores how VR and AR are being applied in vehicle design, quality control, and maintenance. These technologies streamline workflows, enhance precision, and offer immersive design and manufacturing experiences.
Chapter 7 provides a structured examination of VR's impact on the automotive industry, from design optimization to customer engagement through virtual showrooms. It also outlines current limitations and future integration with AI and AR.
Chapter 8 offers a technical deep dive into the foundations of AR, explaining its core components and the computational frameworks behind immersive digital overlays that interact with physical environments.
Chapter 9 discusses collaborative design and real-time prototyping through AR/VR technologies. It includes case studies that show their practical impact on improving manufacturing efficiency and fostering innovation in automotive workflows.
Chapter 10 details the development and impact of AR-based Heads-Up Displays (HUDs), which provide real-time navigation and safety alerts directly on windshields, significantly enhancing driver awareness and road safety.

Chapter 11 analyzes the strategic implementation of AR in industrial manufacturing. It presents real-world applications, discusses challenges such as integration and cost, and provides a roadmap for adoption in Industry 4.0 environments.

Chapter 12 explores immersive showrooms powered by AR and VR, demonstrating their potential to personalize vehicle experiences, promote sustainable urban mobility, and support smart city infrastructure planning.

Chapter 13 examines the foundational technologies and societal impacts of Autonomous Vehicles (AVs), addressing how AVs interact with urban systems, influence traffic patterns, and align with sustainability goals.

Chapter 14 reviews the current limitations of VR/AR adoption in the automotive sector, including cost and technical hurdles, while proposing solutions and forecasting future advancements.

Chapter 15 provides a critical survey of VR and AR applications in automotive design, diagnostics, and driver support. It outlines practical benefits, persistent challenges, and the continuous need for innovation and improvement.

Chapter 16 shifts focus to automotive engineering, introducing a new air intake manifold design optimized using computational fluid dynamics (CFD). This chapter shows how airflow improvements can lead to better fuel efficiency and engine performance.

In addition to highlighting the benefits, this work addresses the challenges that accompany the adoption of immersive technologies, including technical limitations, cost concerns, and organizational readiness. By doing so, it offers a balanced and forward-looking perspective that encourages further exploration and dialogue.

A key strength of this volume lies in its interdisciplinary approach, combining insights from engineering, computer science, manufacturing, marketing, and user experience design. The result is a rich and diverse examination of how immersive technologies can be leveraged across multiple domains.

Ultimately, *Virtual and Augmented Reality Applications in the Automobile Industry* is not only a record of current developments but a catalyst for future research. It is our hope that this book will inspire readers to engage with these powerful technologies, drive innovation in their respective fields, and contribute to shaping the future of mobility.

E. Devaraj
CMR University, India

Chapter 1

Leveraging Machine Learning and Artificial Intelligence for Enhanced Connectivity in Vehicles: Intelligent Transportation Systems

Santosh Naduvinamani
https://orcid.org/0009-0006-7076-9570
Department of Artificial Intelligence and Data Science, SG Balekundri Institute of Technology, Belagavi, India

R. Bhuvana
Department of Biomedical Engineering, GKM College of Engineering and Technology, Chennai, India

Rupali Atul Saraf
https://orcid.org/0009-0002-7609-4333
Department of MCA, MKSSS K.B. Joshi Institute of Information Technology, Pune, India

M. N. Babitha
Department of Computer Science and Engineering, Sri Siddhartha Institute of Technology, Sri Siddhartha Academy of Higher Education, Tumakuru, India

Indira Ghosh Dastidar
https://orcid.org/0009-0008-4621-0034
Department of Automation and Robotics, Dr. D.Y. Patil Institute of Technology, Pune, India

DOI: 10.4018/979-8-3693-7287-6.ch001

ABSTRACT

Machine Leanring (ML) and Artificial Intelligence (AI) can change the game of transport with Intelligent Transportation Systems (ITS). The connected graphs observe architecture and enable from the ML and AI across domains to change the performing method of connectivity and make available better traffic performance, security enhancement, and mobility optimization. The big data from sensors, vehicles, and infrastructure are processed by machine learning algorithms to provide data-driven, real-time decisions, predictive analysis, and traffic flow optimization for smooth flow, less jam, and fewer emissions. Resulting in the smart cities intelligent mobility, with the rise of AI and ML in ITS, the smart motors are being transformed into adaptive devices that can change as per evolving traffic scenarios. It explores how AI and ML may be integrated with smart cities and ITS to make these connected, efficient, and sustainable and examines current and emerging applications of ITS with a future outlook.

INTRODUCTION

Transportation systems are one of those areas that are also passing through a new stage of evolution induced by technology, mainly the adoption of machine learning and artificial intelligence in the so-called Intelligent Transportation Systems (ITS). A deeper evaluation of these technologies for solutions of more effective, safe, and sustainable traffic paradigms. ITS stands for Intelligent Transport Technologies, which are the modern telecommunications networks, sensor devices, data analytics, and artificial intelligence technologies that have recently optimized transport infrastructure and mobile-based advancements. In fact, they are constantly evaluating the real-time state of the roadways, using multi-terabyte data to make immediate and autonomous decisions en route for optimal traffic flow and congestion mitigation while also predicting potential crashes(Dash et al., 2018).

ML is the part of the AI that deals with the capacity of the system to understand from the data and further enhance its accuracy. For ITS, ML algorithms will combine historical data and real-time feeds from a wide range of sources—from traffic cameras to roadway surface sensors to vehicle GPS logs to social media." Once data is collected, it is then analyzed by ML (machine learning) algorithms to recognize patterns and to predict future occurrences and even decisions about traffic flow, accident probability zones, and time taken to reach a particular destination. Based on historical data, for example, ML is used to predict heavy traffic conditions in a sensitive area, so the system can make changes to, for example, traffic lights or suggest the driver different routes to save travel time and fuel consumption(Boukerche et al., 2020).

AI systems in ITS are meant to replicate human modes of thinking and beliefs to take crucial decisions, make real-time changes to evolving situations, and address complex problems independently. Well, one of the most impactful things is the area where AI is making an impact is through the optimization of the traffic signals. Traditional traffic phase signals operate based upon a fixed timer that runs irrespective of actual traffic conditions, resulting in non-optimal control of intersections and intolerable waits. AI programs are dynamically counting vehicles on the road, enabling them to change the timings of traffic lights based on real time as opposed to existing traffic light systems balancing out traffic based on static routing with little or no traffic congestion, as they are new AI-driven systems that will balance the traffic out. Such AI systems anticipate traffic patterns and adapt signal timings according to data from sources including traffic cameras, sensors, and vehicle counts(Gangwani & Gangwani, 2021).

Intelligent Transportation System (ITS) Utilizing Artificial Intelligence (AI) and Machine Learning (ML) – The 1st Contribution The vehicles essentially share V2V communication in which vehicles can technically share the location of the vehicle, direction of movement, and speed of the vehicle with other running vehicles on the same road to minimize vehicular collisions and assist traffic management. If a vehicle detects that it's about to strike something or that the traffic ahead has suddenly slowed, it can send a heads-up to surrounding vehicles and tell them what to do—either stop or switch lanes. On the same track, V2I (Vehicle to Infrastructure) communication allows cars to connect with devices on the road, such as traffic lights, toll booths, and parking meters, thus optimizing road usage and also providing the driver with immediate useful information(Ang et al., 2022).

ITS also makes use of AI and ML to improve the safety of the roads. Traffic accidents are yet another area where AI and ML can have enormous potential to make a big impact, killing and injuring millions of people around the world in just one single year. AI systems are trained to study a mix of historical accident records, live traffic conditions, weather patterns, and even driver behavior to determine how likely an accident is to occur in a location. AI algorithms, for instance, are capable of implementing automatic alerts, automatic speed limit reduction, and traffic signal changes to prevent delay/kickbacks. Moreover, AI plus machine learning used in self-driving cars can minimize human error or typical mistakes that are responsible for most of the road traffic accidents. Autonomous vehicles rely on a combination of sensors, cameras, and machine learning algorithms to navigate safely on the road, deciding what action to take based on the surroundings while driving(Bharadiya, 2023).

This should not be confused with an entirely automated public transit system and visionary statements about 100% AI and ML in ITS. A.I. can be used by agencies to deter demand surges for public transportation, refine routes, and customize schedules using real-time traffic feeds. AI algorithms could interpret the latest data from buses, trains, and ride-hailing services and determine the best routes to send such vehicles or at least deploy them during high post-reduction delays. In addition, there is lower energy consumption and little energy focused on implementing electric buses and trains, besides additional sustainable goals(Deepika & Pandove, 2024).

In the recent literature of ITS, the pie charts for AI and ML for topics on smart cities have maybe been some of the hottest papers. It is quite simple: the density of a city makes it harder to control the mobility within the city. AI & ML is helping ITS to disrupt and therefore may evolve into smarter urban centers with better transportation networks, lesser congestion, and lesser emissions. Using data from several sources, such as traffic detectors, smart parking systems, and weather stations, AI and ML algorithms can also develop a detailed model of urban mobility and offer suggestions on how to optimize the movement of people and goods across the city. These can include measures such as redirecting traffic, suggesting other means of getting to a destination, or even moving public transport facilities in response to demand(Sreelekha & Midhunchakkaravarthy, 2024).

Thus, AI and ML become breakthrough technologies for the next-gen connective smart transportation system in ITS. The transportation network, on the globe, changed via these technologies: decision-making integrated in real-time, vehicle-to-vehicle networking, optimal trajectory for traffic, and multiplexing transport of movement and safety. M2M (Machine to Machine) is a tightly knit and, at the same time, many devices and infrastructure are providing real-time input for routing and switching applications. Here we have the possibility of AI and ML in transport, which could be ubiquitous. These technologies combine to enhance an excellent traveling experience while simultaneously developing smart, resilient cities that can more effectively house increasing populations.

Objectives

The study aims to evaluate the impact of machine-learning- and artificial-intelligence-based systems on the effectiveness and efficacy of intelligent transfer systems. The initiative aims to probe how such technologies improve monitoring of traffic, facilitate decisions in real time, and allow action to be taken to alleviate congestion. Finally, the chapter states that the scope also includes AI and ML needs that relate to safety, sustainable capital, and vehicle-to-vehicle/infrastructure networking. Therefore, in the wake of these existing applications and future predictions, the chapter will necessarily focus on the transformative implications of AI and ML on mobility in cities and their implications for the future of the transport ecosystem.

BACKGROUND: AI AND ML IN ITS

AI and ML have revolutionized traditional transport systems; they now learn from data, respond to situational changes, and take reasoned decisions. This section describes the fundamental concepts behind AI and ML, how such techniques can be applied to ITS, and how they can provide connectivity and efficiency through a variety of data sources(Nikitas et al., 2020; Ullah et al., 2020).

Definitions and Key Concepts

Artificial Intelligence (AI)—the act of creating machines capable of performing tasks that normally require human intelligence (reasoning, learning, decision-making, problem-solving). AI technologies facilitate the autonomous absorption of data related to transportation, traffic, and travel—ultimately, ensuring that operations run as smoothly as possible in real time. ML, or machine learning, is one of the most spoken categories of artificial intelligence, which is a computer's ability to learn from data and make decisions without being explicitly programmed. ML algorithms discover patterns, relationships, and trends through iterative training, and these algorithms then use that information to make predictions or decisions.

Supervised Learning—Predicts the outcome from a set of data. Unsupervised Learning—Identifies the hidden pattern/intrinsic structures from unlabeled data. Deep learning spans a broader array of use cases, the utilization of neural nets to digest large sets of data, making it well-suited for offering insight such as detecting frames from video from highway traffic cameras, while reinforcement learning excels at use cases where new actions are learned and optimized for use in application-based uses like adapting reward structures to environmental states to learn optimal triggers such as the timing of traffic signal lights.

Types of AI and ML Algorithms in ITS

Various AI and ML algorithms have been used in ITS to address various transportation problems. The type of the algorithm is determined based on the data and the required objective. The classification algorithms, such as decision trees, SVMs, and logistic regression, classify data such as, for example, road segments. Regression algorithms (used to predict continuous outcomes (travel time, fuel consumption), linear regression, and neural networks. Some common examples are K-means and hierarchical clustering, which cluster similar data points within one cluster. One of the deep learning representations that is well suited to working with high-dimensional data (such as images or video streams from traffic cameras) is the Convolutional Neural Network (CNN) model. Reinforcement learning is used in dynamic systems, such as adaptive real-time traffic signal control systems that use algorithms, such as Q-learning or deep Q-networks (DQN), that adapt to their environment by learning strategies through progressive interactions with the environment.

Data Sources and Their Integration

Integration of heterogeneous data sources is essential for effective AI and ML in intelligent transportation systems (ITS). Read Also: ITS: Intelligent Solutions for Urban Mobility DrainTransportation systems generate massive amounts of data from a large number of points, which, when intelligently processed, can lead to enhanced decision-making and relationships. Modern traffic management systems face a myriad of data sources, including traffic sensors, closed-circuit television and traffic cameras, GPS data from vehicles, V2X communication data from vehicles, weather data, and social media platforms. These systems are responsible for analyzing vehicle counts, speed, and lane occupancy, applying computer vision algorithms to process video feeds and real-time location data in order to find ways to optimize routes and the flow of traffic. They also take weather conditions into account to modify traffic estimates and ensure things are safe in inclement weather. Social media such as Twitter and Waze give real-time updates about traffic and accidents. These two data types can be integrated into a single model by using advanced data fusion techniques that combine data from heterogeneous sources. This enables prompt and data-driven insights that enhance the efficiency and responsiveness of transport systems.

ENHANCING TRAFFIC MANAGEMENT WITH AI AND ML

The integration of Artificial Intelligence (AI) and Machine Learning (ML) in Intelligent Transportation Systems (ITS) improves the efficiency of traffic management systems through real-time data processing, predictive capabilities, and dynamic decision-making. Static systems and manual interventions are being replaced by intelligent, data-driven techniques that optimize traffic flow, minimize congestion, and improve urban mobility(Olugbade et al., 2022; Yuan et al., 2022). This section delves into the above three critical traffic management use cases of AI and ML: real-time traffic analysis and prediction, adaptive traffic signal control systems, and dynamic route optimization. Processes involved in enhancing traffic management with AI and ML as shown in Figure 1.

Figure 1. Processes involved in enhancing traffic management with AI and ML

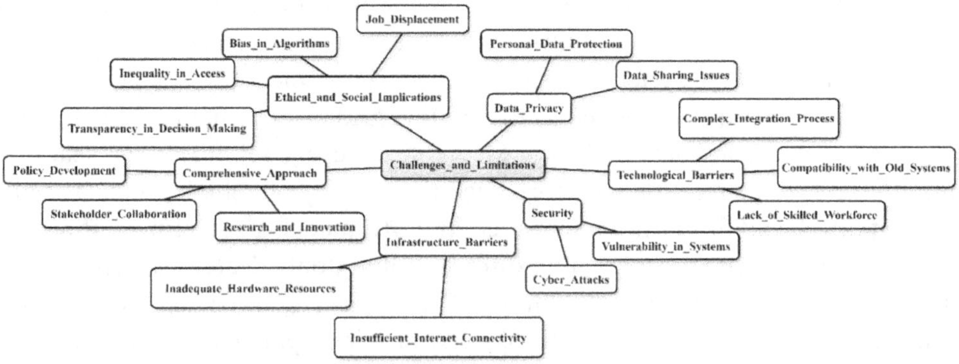

Real-Time Traffic Analysis and Prediction

One of the most intriguing areas where AI and ML can significantly play a role in traffic management is analyzing live data, predicting traffic flow, and adjusting traffic parameters as needed. SAP Customization: Enormous amounts of data are being generated everywhere from sensors built into the roadways, GPS data from vehicles that drive over them, and feeds that come off traffic cameras, and so AI and ML algorithms work with all that information to look for patterns and anomalies. ITS does this by means of predictive models that enable it to predict traffic conditions with high accuracy so that problems can be managed before they become serious. Supervised ML algorithms (e.g., regression models, neural networks) based on previous and current traffic flow data, weather conditions, and special events can predict congestion hotspots. Transportation planners can preemptively respond to such predictions by optimizing traffic signals, re-routing hedging vehicles, or dispatching traffic enforcement units to hotspots. It can also make possible real-time information analysis of differences in traffic density and speed of flow, thus providing drivers with instant updates of travel times and alternative routes to reduce delays while maximizing opportunities for exploration and efficiency(Boukerche et al., 2020; Dash et al., 2018).

Adaptive Traffic Signal Control Systems

Traditional traffic signal systems are optimized for static timers, which are not suitable for variable traffic volumes. This challenge is mitigated with our AI-powered adaptive traffic signal control systems, which compute and adaptively set a dynamically optimal configuration of signal green time by analyzing the real-time traffic condition. These systems learn how to configure the traffic signals in a way that maximizes conditions conducive to avoiding gridlock by interacting with their environment and adapting to changing conditions using reinforcement learning algorithms. The intersection's sensors track the number of vehicles, their speeds, and wait times, and this information is fed into ML models to determine how long the signal should stay green in either direction. It not only decreases waiting time and congestion, but also makes fuel more efficient as unwanted idling of trucks is eliminated. More sophisticated traffic systems also incorporate predictive features, so they can "see" surges in traffic and then adjust signals ahead of time. That means establishing, for example, adaptive systems that offer more green light time to heavily trafficked routes and fewer delays to less congested streets during peak travel times or following large events. Adaptive traffic management with V2I systems is also extremely flexible, as real-time traffic data enables vehicles to communicate directly with traffic signals. Signals can also prioritize emergency vehicles or public transportation, further improving mobility and safety.

Dynamic Route Optimization

AI and ML for Dynamic Route Optimization Process. They draw on real-time data from GPS devices, traffic cameras, and weather stations to propose the most efficient routes for drivers. They are constantly refreshed as conditions change—whether through accidents, road closures, or congestion. Specifically, clustering algorithms are employed to create clusters of vehicles that have similar destination points or alternative routes. AI-based systems for auto-routing and fixing traffic problems AI systems can always find out what would be the pinch points and how to adjust the traffic moving accordingly to get the optimum traveling experience. This technique reduces travel time, fuel consumption, and operational expenditures for ride-sharing and logistics companies. The smart routing systems can also provide options for multimodal transport to reduce urban congestion and lessen the environmental footprint.

Traffic Management: Real-time analysis and reactive traffic management enabled by AI and ML systems Such applications enable the authorities to anticipate congestion, to streamline the movement of cars at intersections, and to advise on travel times. This leads to the heights of scintillating and responsive transport systems that enhance mobility and reduce safety concerns, facilitating sustainable urbanization. As the populations swell, and with traffic demand continuously increasing, AI and ML will have a big role in traffic management and tackling contemporary urban mobility issues.

VEHICLE-TO-VEHICLE (V2V) AND VEHICLE-TO-INFRASTRUCTURE (V2I) COMMUNICATION

One of the technologies—the connected vehicles, particularly Vehicle-to-Vehicle (V2V) and Vehicle-to-Infrastructure (V2I) communication—pays special attention to ITS. This technology allows for the communication of real-time information between vehicles and infrastructure, leading to optimized traffic flow, increased safety, and the deployment of smart infrastructure. Artificial Intelligence (AI) is the brain behind all these communication technologies, resulting in enormous computation for processing huge data in real time and providing actionable insights(Deep & Verma, 2023; Dikshit et al., 2023; Makanadar & Shahane, 2023). It also looks at ways AI will enhance traffic flow and help prevent crashes. Figure 2 depicts the communication between vehicles (V2V) and infrastructure (V2I).

Figure 2. Vehicle-to-Vehicle (V2V) and Vehicle-to-Infrastructure (V2I) communication

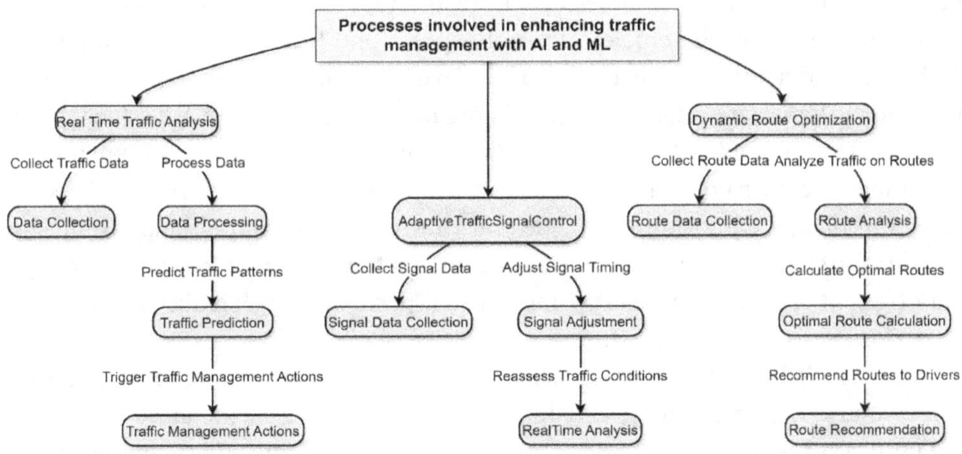

Role of AI in Enabling Communication

From AI-based vehicle-to-vehicle (V2V) and vehicle-to-infrastructure (V2I) communication to innovative data processing and data interpretation between connected cars and road signs, the use of AI in transportation networks is transforming them. AI algorithms continuously analyze the data to identify patterns, predict risks, and facilitate decision-making. In V2V communication, lane changes, emergency brake events, and road obstructions are shared, and AI analyzes this data, enabling vehicles to predict threats and advise fundamental safety further actions. V2I (vehicle-to-infrastructure) communication is where an AI hears data from things around it that govern its environment, such as traffic signals, cameras, and weather observation systems(Deepika & Pandove, 2024; Hossain et al., 2024; Zhu et al., 2023). This means that AI algorithms are able to analyze traffic camera data to ensure they monitor congestion levels and communicate them to the vehicles so they can modify their routes. By merging the data from many sources of information, AI gives vehicles a broad understanding of their surroundings that informs decisions about when to slow down in icy conditions or to divert from construction sites.

Improving Traffic Flow and Collision Avoidance

The communication of V2V and V2I for improved traffic flow and safety on the road depends on the vehicles and infrastructures interacting with each other. Information about each vehicle's position, speed, and heading can be shared with all the others that are in the network to help avoid traffic and accident risks, so AI-powered systems can greatly help in this regard. We can develop V2V technology that lets cars in mixed traffic share information with each other to help synchronize movement and to trigger adjustments if a slow down, stall, or error is detected. Using this data, AI algorithms provide real-time recommendations and instant suggestions for human drivers or automated systems to implement, enabling the rapid and efficient responses necessary for optimal performance(Jagatheesaperumal et al., 2024; Prakash et al., 2024).

V2V and V2I powered by AI, which keep the vehicle moving, avoiding side-swipes and other impending collisions. Accident-prone zones can be pinpointed and collision scenarios predicted through predictive analytics that are powered by machine learning models. Another example includes AI algorithms that can monitor vehicle paths at traffic signals and issue warnings in the event of possible collisions. In autonomous vehicles, AI systems can literally control the braking and steering in order to prevent an accident when human input is not present. AI System Analysis for Collision Avoidance with High Accuracy and Speed.

Infrastructure welcomes changes to achieve improved traffic management by adapting to the dynamic state, i.e., with the help of communication between vehicle and infrastructure. They aggregate real-time data collected from connected vehicles and AI-powered traffic signals that help optimize timing. Moreover, at peak travel times, AI algorithms provide priority to the green light along busy routes, thus preventing jams on adjacent roads. AI also helps to control multiple vehicles at intersections to simplify gridlocks and increase transportation efficiency in terms of lower trip time, less fuel usage, and lower emissions.

Applications in Smart Infrastructure

It is onward for Advanced Safety and Automation in V2V and V2I communication, which is widely used in the enhancement of traffic management and safety to further develop smarter infrastructure. Driving Urban IoT: Connected vehicles of AI assistant-powered V2I systems can collect data from other connected vehicles—optimizing smart parking systems, time spent in search of parking spots, and enhancing urban congestion by providing real-time performance and availability information. Artificial intelligence is an essential part of smart infrastructure, especially in the connection of renewable energy sources with transportation infrastructures. It can optimize energy usage by coordinating interactions between electric vehicles and charging stations, lowering grid stress. Dynamic pricing is one of the utility implementations of V2I communication, while it can also reserve charging slots and distribute power flexibly for e-vehicle owners. AI can also play a role in smart tolling systems, charging vehicles by location and travel distance automatically, cutting down on delays.

Besides these, the AI-based V2I systems are essential for disaster management and emergency response. In addition, they permit the flow of real-time information to emergency services during natural catastrophes and major accidents. AI goes through the data and understands what areas are affected, estimates early evacuation routes, and helps in arranging other resources as well. For example, the DSRC sent to and from vehicles directly to V2I systems can give priority to emergency vehicles by providing them green traffic lights and the quickest paths there. AI is revolutionizing transport by enabling V2V and V2I communication systems to enhance connectivity, improve traffic flow, and aid the development of smart infrastructure. These innovations work in tandem with real-time data analytics, communication, and intelligent decision-making capabilities; together, they contribute to enhanced safety, efficiency, and sustainability across KSA's transportation systems. They address urban mobility challenges and lay the groundwork for connected and autonomous transportation, enabling smarter and more resilient cities.

AI AND ML IN ROAD SAFETY

Artificial Intelligence and Machine Learning have revolutionized road safety by providing innovative solutions to reduce accidents, save lives, and improve driving experiences. Traditional methods, such as traffic signs and physical barriers, have been replaced by dynamic, data-driven strategies. This section explores three critical applications of AI and ML in road safety: predictive analytics for accident prevention, AI-powered driver assistance systems, and the role of autonomous vehicles in improving safety standards(Amani et al., 2020; Atitallah et al., 2020; Bibri et al., 2020). Figure 3 depicts the significant impact of AI and ML on road safety.

Figure 3. AI and ML impact road safety

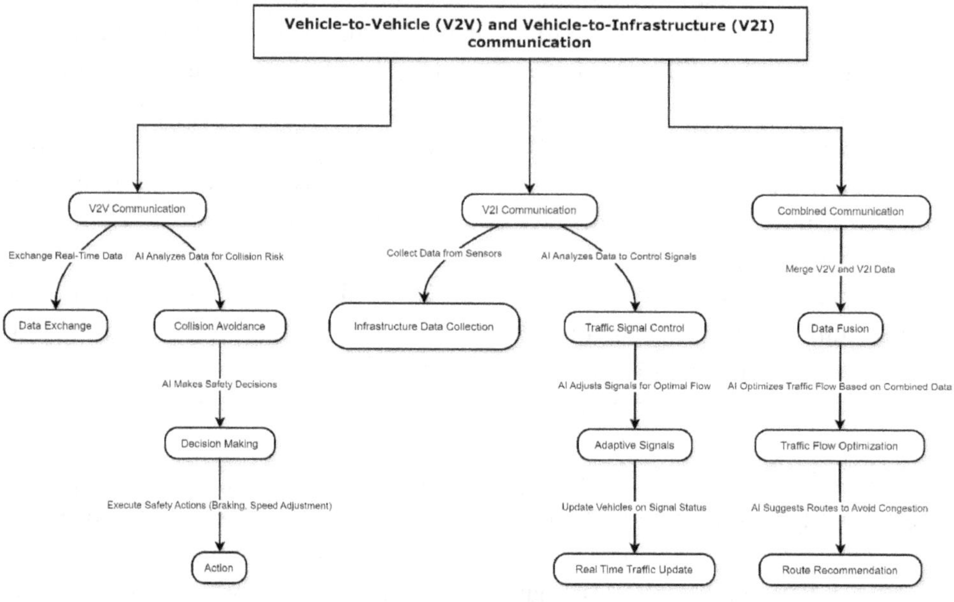

Predictive Analytics for Accident Prevention

AI- and ML-driven predictive analytics are changing road safety by leveraging historical accident data, traffic patterns, and weather. They facilitate intervention by making road environment improvements, reallocating enforcement resources, or optimizing traffic signal timings. Similar to decaying corpses, in the shape of decision trees or neural networks, machine heuristics can accommodate enormous datasets and spot hidden correlations between variable combinations, like vehicle speed, road quality, and accident figures. AI-based systems can also make dose predictions on the sections of road towards which it is predicted there will be a high volume of accidents, factoring in things like curves, visibility, or heavy traffic.

Predictive analytics provides drivers with real-time alerts of potential danger—from sudden slowdowns to icy roads—using AI models. They can also prevent collisions by simply notifying the drivers, who can then make informed decisions. Fleet management in the case of predictive analytics is another application in which AI systems analyze driver behavior, including evaluation of the vehicle itself for identifying any driver behavior or situations that may lead to an accident. Things like heavy braking, full-throttle acceleration, or tire wear can be tracked, and corrective measures can be taken by the companies before the incident actually occurs.

AI-Powered Driver Assistance Systems

One of the critical applications of AI can be found in Advanced Driver Assistance Systems (ADAS), which have made driving safer than ever before by continuously monitoring the driver, the outside environment, and the vehicle itself by providing real-time assistance & automated functionalities aimed at accident prevention. They utilize artificial intelligence (AI) and machine learning (ML) to constantly monitor the surroundings of the vehicle, recognize driving conditions, and execute important functions, reducing the burden on the driver as well as human errors, which contribute to road accidents. These include lane departure alerts, adaptive cruise control, automatic emergency braking, and blind-spot detection. Lane-keeping assist systems rely on cameras that have been trained with the help of artificial intelligence to detect lane markings and warn the driver if he or she drifts out of his or her lane. Adaptive cruise control changes vehicle speed based on the behavior of traffic in vehicles ahead of and behind it, and automatic emergency braking systems use cameras and sensors to identify obstructions and apply brakes automatically if a driver does not respond in time. Using sensors, the AI management monitored blind spots, warning the driver when another car was nearby. With the presence of these systems minimizing the instances of rear-end collisions, making travel safer as a whole(Koppula et al., 2024; Olaniyi et al., 2023).

Contributions of Autonomous Vehicles to Safety

AVs Will Transform the Safety Landscape on Our Roads — Human Error is The Leading Cause of All Traffic Accidents Automating real-world driving, AVs need to combine their sensory inputs and situational analysis with intelligence to plan their actions with a high degree of precision in reference to their environment. They can ensure consistent driving behavior, especially avoiding hazards due to over-speed, reckless driving, or a sluggish response time AVs can instantly monitor the traffic in real time, respond to weather conditions at the earliest, and maintain a comfortable distance between other vehicles. They also have fail-safe systems in case of emergencies — like pulling over if the system malfunctions. AVs would be a game changer for the safety of our roads.

AVs also communicate with each other and with infrastructure, which could enhance road safety by cooperating and acting in concert. This link-up reduces the potential for chain-reaction accidents and traffic jams, providing transit systems with extra safety. AVs will also be key to countering the trend toward safer streets for vulnerable road users—i.e., pedestrians and cyclists. AVs use AI algorithms to detect nearby objects and predict their movement, enabling the AV to react and glide out of the way when the need arises. This function is ideal for dense urban environments where the number of vulnerable road users is high. Thus, AVs likewise assist in forming a safe and effective transportation system.

OPTIMIZING PUBLIC TRANSPORTATION SYSTEMS

Public transportation remains a core aspect of smart urban mobility, with urban planners and policymakers actively working to amplify this service for better mobility, reduced congestion, and lower environmental impacts. These systems are being transformed into emerging, refined networks of efficient, data-driven systems by integrating artificial intelligence and machine learning. More than AI & ML can improve public transit by increasing its availability, dependability, and sustainability through accurate demand forecasting, dynamic routing, dynamic scheduling, and energy-efficient operations(Rosca et al., 2024; Sreelekha & Midhunchakkaravarthy, 2024). These transformations are leading to advances in demand prediction, real-time scheduling, and the efficiency and sustainability of energy systems. The process of optimizing public transportation systems is depicted in Figure 4.

Figure 4. Process of optimizing public transportation systems

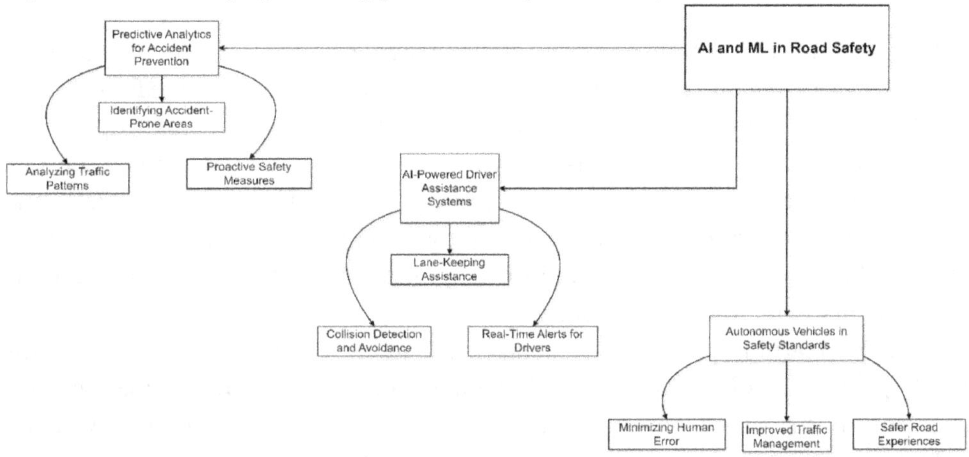

Demand Prediction and Route Optimization

Passenger demand prediction plays a key role in the effective operation of systems in a discrete manner to determine passengers per future instant based on the continuous analysis of historical data, demographics, social events, and weather in areas in which air transportation, road transportation, or boat transportation can be applied in realizing such systems built on AI or ML. All data is pushed into a data warehouse, and the time series analysis and neural network predictive models identify many trends and seasonal variations to help operators prepare for peak times of day and desired routes. For big events, AI systems may predict more ridership and change service levels accordingly.

In public transport, route optimization plays a significant role, which is achieved using AI and ML. AI algorithms can analyze traffic data, passenger flow, and road conditions to determine effective routes for buses, trains, and other transit vehicles. Dynamic route optimization allows fleet logistics to skip congested areas and keep running in a timely manner, improving passenger experience. When they propose a route, they can include in the balance between efficiency and increased environmental degradation such factors as fuel consumption, emissions, and travel time. These clustering algorithms can be utilized in order to determine destinations that are proximal to one another so that routing can be optimized in order to minimize distance and maximize coverage, thus reducing overall operation costs as well as ensuring underserved areas remain within reach in terms of transportation.

Scheduling with Real-Time Adjustments

Conventional public transport has a static schedule, which is not suitable for real-time changes in demand or traffic conditions. AI and ML pave the path for dynamic scheduling, as they enable transit systems to make real-time adjustments to operations in response to unanticipated events. Using data from GPS trackers, traffic sensors, and passenger counters, AI-powered systems can monitor the status of vehicles and detect any deviations from the schedule and make real-time adjustments on the fly, whether that means sending out additional vehicles, rerouting buses, or extending service hours in the midst of emergencies.

Machine learning-based fleet management models use historic and real-time data to predict vehicle availability and maintenance needs. Predictive maintenance systems utilized by transit agencies are used to monitor the health of transit vehicles and notify the enterprise of an issue before it results in a breakdown, minimizing agency downtime. AI algorithms can also optimize vehicle allocation by tabulating both fixed seating and expected ridership so that buses or trains are neither over- nor underloaded. This encourages operational efficiency, with smaller vehicles serving low-demand routes and larger ones servicing high-density corridors. Real-time scheduling features can enhance passenger experience with transit services. AI-led crowd management solutions help monitor passenger density in real time, enabling staggered boarding and additional staffing during peak hours, making travel safer and more comfortable amidst the turbulence.

Energy Efficiency and Sustainability

One area where AI & ML are critical is sustainability in public transport. It also helps optimize energy consumption in electric trains and buses by analyzing data from energy meters, weather sensors, and passenger flows using AI algorithms. It helps ensure that vehicles run efficiently while keeping occupants comfortable by predicting the energy demand and changing operational settings accordingly. AI helps minimize environmental impact by reducing fuel consumption or emissions using route optimization in the system that selects routes with fewer stops and congestion. It also schedule-syncs with renewable energy availability, for example, charging electric buses when grid demand is low or renewable generation is high. This reduces operating expenses and encourages the integration of renewable energy sources into public transport systems.

AI and ML are fueling the adoption of alternative energy solutions such as electric and hydrogen-powered vehicles, revolutionizing deployment and infrastructure needs. Data analysis at scale: AI algorithms analyze tons of historical travel patterns to recommend the ideal location of the charging station based on proximity to customers while optimizing the utilization of infrastructure. ML also predicts battery and range limitations, alleviating range anxiety. AI-enhanced transportation systems enable modal shifts—public transit instead of personal vehicles, for instance—resulting in reduced emissions and greater reliability. Anticipated Mobility Shift: With advanced multimodal integration, there would be a seamless transition between shared mobility options, facilitating a sustainable transportation ecosystem.

Demand prediction, dynamic routing, and optimized energy-efficient operation are reinventing the way public transport systems work using AI and ML. Through real-time route optimization and reduced energy consumption, these data points contribute to ensuring passengers have the best experience possible while also being kinder to the environment. In the years to come, as these technologies continue to mature, their values will be integrated to shape the future of transportation.

Figure 5. Role of AI and ML in smart city developments

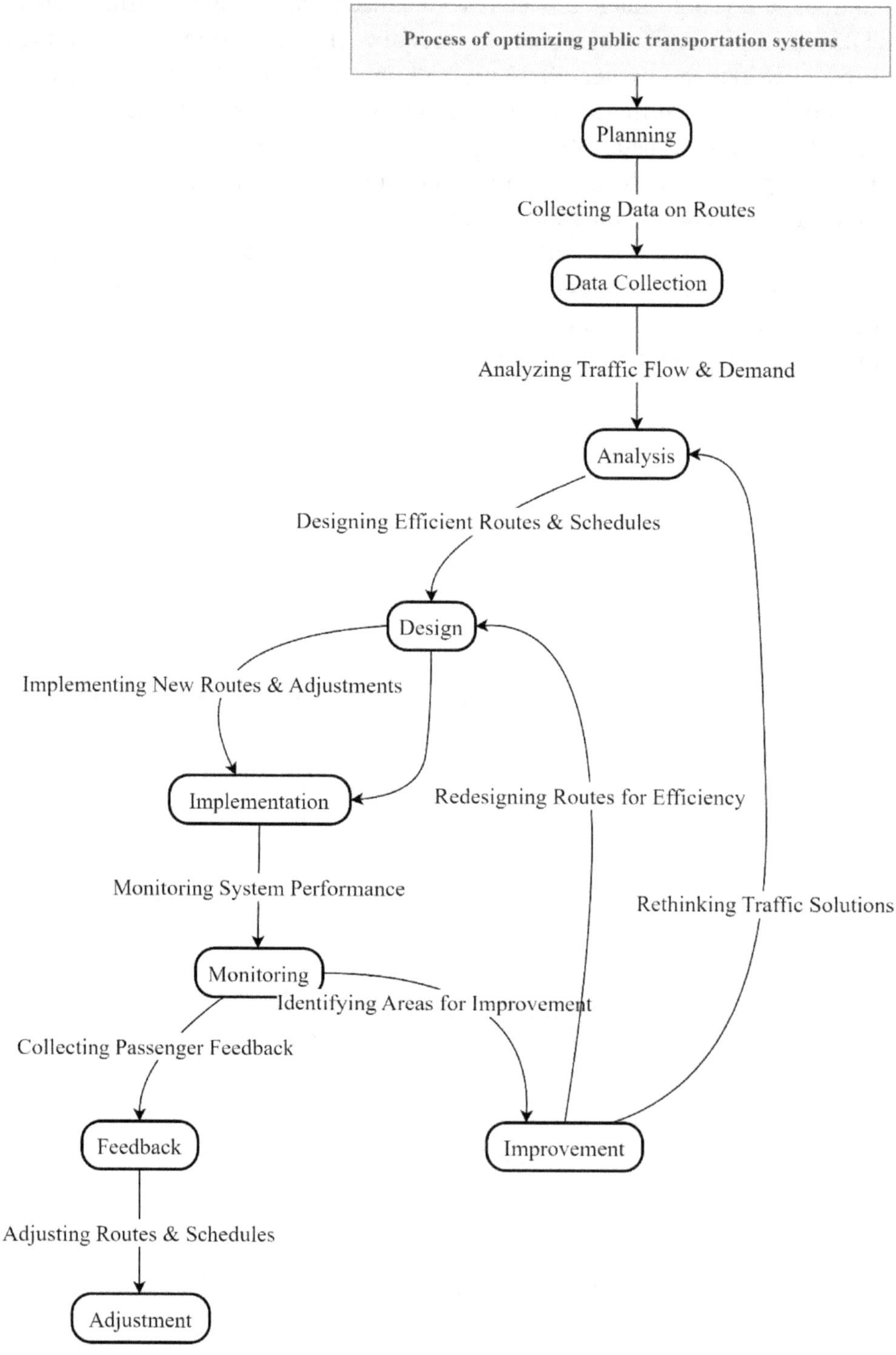

AI AND ML IN SMART CITIES

AI and ML are revolutionizing smart cities by transforming urban ecosystems into efficient, sustainable, and highly connected ones. They help manage congestion, optimize transportation systems, and mitigate environmental impacts. These technologies are crucial in enhancing the quality of life for residents, reducing congestion, integrating multi-modal transportation systems, and achieving environmental benefits through emission reduction(Boopathi, 2024; Molakatala et al., 2024). The figure 5 illustrates the significant role of AI and ML in the development of smart cities.

Managing Urban Mobility and Reducing Congestion

AI and ML technologies are being used to improve urban mobility by monitoring and managing traffic flow in real time. These systems, based on data from traffic cameras, GPS devices, and connected vehicles, can predict congestion hotspots and implement measures to prevent gridlock. This enables authorities to proactively address potential bottlenecks in urban development. AI algorithms can improve traffic efficiency by dynamically adjusting signal timings based on real-time traffic flow. This reduces waiting times, fuel consumption, and delays. AI can prioritize high-traffic routes during peak hours and minimize delays during off-peak times. Additionally, AI-powered incident detection systems can identify accidents in real-time. AI aids in smart parking systems, reducing congestion in urban areas. These systems use AI-driven sensors and ML algorithms to provide real-time updates on parking availability, guiding drivers to the nearest vacant spots. Predictive models analyze historical data to anticipate demand and optimize space utilization.

Integration of Multi-Modal Transportation Systems

Featured contentAI and ML significantly alter multimodal transportation systems by enhancing the integration of multiple forms of transportation. The post transforming multi-modal transportation systems appeared first on Macmillan. Thereby enabling public and shared mobility and reducing dependence on private vehicles. In smart cities, these systems fed with AI can develop centralized platforms, giving citizens real-time insight into transit schedules, routes, and availability—making journey planning a much less painful process. Mobility-as-a-Service (MaaS) platformsIndiaStays uses AI to integrate data from various transportation providers into one interface, providing everything from real-time navigational information to ticket bookings and payment processing. These platforms allow users to integrate multiple transit modes into one trip. Machine learning algorithms interpret user behavior to provide tailored suggestions, recommending optimal routes based on travel history, road congestion, and environmental parameters. Demand prediction models that are AI-powered enable transit authorities to use resources more intelligently and deploy additional buses or trains in higher-demand areas. The importance of AI and ML integration in transportation systems extends to shared mobility services such as ride-sharing and bike-sharing, where the routes are optimized for each service, resulting in a lower wait period. This lessens private vehicle ownership, encourages ridesharing, and leads to less crowded city streets and sustainable modes of transportation.

Environmental Benefits and Emission Reduction

AI and ML play an important key role in smart cities in minimizing environmental pitfalls and optimizing sustainability. Transportation is one of the leading producers of greenhouse gases, and AI-laid optimization systems can alleviate emissions generated by it. This is particularly relevant to public transit fleets and logistics firms, for whom improving fuel efficiency comes with lower costs and, by extension, lower emissions as well. Electric vehicle (EV) adoption: AI is enhancing the adoption of electric vehicles (EVs) by driving smart deployments in the smart urban transport ecosystem. A battery optimization model that maximizes the battery to minimize production costs is based on machine learning data on travel patterns, charging station energy, and grid capacity. AI-driven smart energy management systems enable smart charging at off-peak hours or at times when renewable energy is being produced, thus reducing the carbon footprint of urban transport. AI also enables citywide sustainability initiatives involving real-time monitoring of air quality and emissions, informing policymakers and enabling proactive intervention—like adding green zones and promoting active transport.

AI and ML are revolutionizing urban transportation systems, optimizing electric buses and train systems as well as changing how these transportation methods emit carbon and use energy. They are revolutionizing smart cities with congestion management, multi-modal integration, and energy optimization. Thus, these AI-enabled solutions prove increasingly essential as cities worldwide embrace new ways to make their transportation networks more robust and linked—to drive toward global sustainability targets. Not only do these innovations enhance the quality of life for the residents, but they also establish entirely new benchmarks for innovation and sustainability in urban development.

CHALLENGES AND LIMITATIONS

The integration of AI and ML in transportation and smart city systems presents significant opportunities but also presents challenges such as data privacy, security, technological and infrastructure barriers, and ethical and social implications, necessitating a comprehensive approach(Boopathi et al., 2021; Gopal et al., 2024). The integration of AI and ML in transportation and smart city systems is characterized by challenges and limitations, as illustrated in Figure 6.

Figure 6. Challenges and limitations of integrating AI and ML in transportation and smart city systems

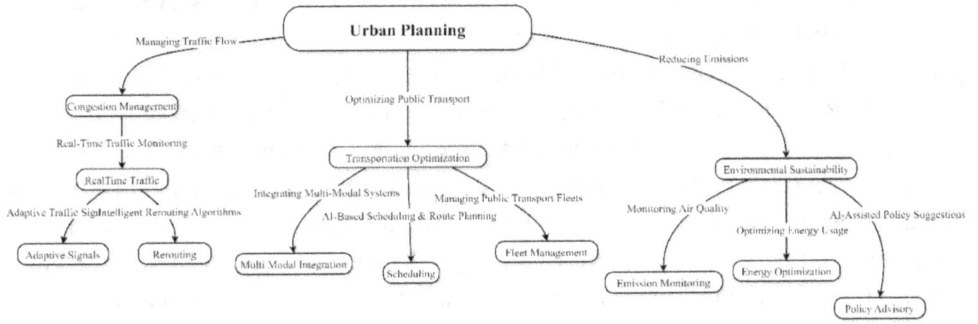

Data Privacy and Security Concerns

Earning trust in AI and ML solutions is one of the greatest challenges. Their systems are perceived as cornerstones of intelligent transportation systems and smart city concepts, which depend on a vast amount of data, even personal data, such as user location histories and user action patterns. But if these data sources aren't managed properly, they can become the target of hacks, which puts users at risk of identity theft, financial fraud, or surveillance. Users can get exploited and be exploited for situations that they wouldn't want to participate in, as there isn't good regulation. Transparent data governance policies, along with robust cybersecurity measures, are crucial to ensuring privacy and security.

Technological and Infrastructure Barriers

AI and ML technologies for transportation and urban systems are limited, probably due to a lack of infrastructure, computational abilities, and costs. So there are no high-speed internet connections in cities, particularly in developing nations, advanced sensors, and continuing power sources. In order to achieve the optimal usage of AI in supply chain processes, legacy systems need a whole lot of modernization in order to be looped into a new age AI-driven alternative; this will, however, be complicated and costly. Processing data and taking decisions in real-time requires huge computing power and advanced hardware. The high cost of data inhibits widespread adoption, particularly in areas without the financial infrastructure to support it.

Ethical and Social Implications

The paper presents ethical and social implications of AI and ML in transportation and urban management contexts. It may favor certain groups over others and thus lead to differential access to services. AI does have the potential to make many industries more efficient and make life more accessible, but as with all things, more of something isn't always better; AI is set to permeate human society, and with it come challenges, particularly in the workforce, where human workers may be faced with the threat of displacement due to automation replacing humans in driving and logistics at a greater rate than, in the current climate, society can invest in workforce retraining and upskilling programs that the displaced can access. Examples include ethical dilemmas related to making decisions not just in safety scenarios but also in decision-making scenarios, for instance, for autonomous foundation models. "The AI programming of a self-driving vehicle will have to adjudicate complex moral dilemmas like whether to prioritize the safety of the passengers in the vehicle or nearby pedestrians, but such moral complexity is not necessarily amenable to algorithmic codification, muddying the waters over accountability and liability." Conclusively, AI, along with ML, is going to transform personal mobility and urban systems, but several key obstacles like data privacy, technical constraints, and ethical concerns must be overcome by all stakeholders, including the government, industries, and researchers.

CONCLUSION

Hence, AI and ML play a central role in revolutionizing Intelligent Transportation Systems (ITS), allowing cities to gain insight into, analyze, and respond to the challenges of urban mobility and transport. By supporting intelligent data analysis, decision-making processes, and intelligent technologies, it has greatly enhanced traffic management, vehicle-vehicle communication, improvements in public transport, and road safety. In addition to better traffic flow and reduced congestion as well as lesser air and climate impacts, AI and ML have also played a critical role in shaping sustainable urban mobility.

But this chapter has also been a case study of the challenges and limitations that must be overcome if these technologies are to be able to realize their full potential. Addressing challenges associated with data privacy, security, infrastructure limitations, and ethical considerations requires proactive measures and collaboration between governments, researchers, and industry stakeholders. Overcoming these barriers will be critical to building inclusive, trustworthy, and resilient systems. Emerging fields such as edge computing, digital twins, and the advance of autonomous and connected vehicles will inevitably reshape the fabric of Intelligent Transportation Systems (ITS) in the upcoming decades. Aligning with sustainability objectives, AI and ML can provide a platform for newer and greener innovations. Using these technologies to deliver enhanced safety, access, and environmental benefits to help make life better for everyone, we are opening up a new era of transportation.

REFERENCES

Amani, M., Ghorbanian, A., Ahmadi, S. A., Kakooei, M., Moghimi, A., Mirmazloumi, S. M., Moghaddam, S. H. A., Mahdavi, S., Ghahremanloo, M., Parsian, S., Wu, Q., & Brisco, B. (2020). Google earth engine cloud computing platform for remote sensing big data applications: A comprehensive review. *IEEE Journal of Selected Topics in Applied Earth Observations and Remote Sensing*, *13*, 5326–5350. DOI: 10.1109/JSTARS.2020.3021052

Ang, K. L.-M., Seng, J. K. P., Ngharamike, E., & Ijemaru, G. K. (2022). Emerging technologies for smart cities' transportation: Geo-information, data analytics and machine learning approaches. *ISPRS International Journal of Geo-Information*, *11*(2), 85. DOI: 10.3390/ijgi11020085

Atitallah, S. B., Driss, M., Boulila, W., & Ghézala, H. B. (2020). Leveraging Deep Learning and IoT big data analytics to support the smart cities development: Review and future directions. *Computer Science Review*, *38*, 100303. DOI: 10.1016/j.cosrev.2020.100303

Bharadiya, J. (2023). Artificial intelligence in transportation systems a critical review. *American Journal of Computing and Engineering*, *6*(1), 34–45. DOI: 10.47672/ajce.1487

Bibri, S. E., Krogstie, J., & Kärrholm, M. (2020). Compact city planning and development: Emerging practices and strategies for achieving the goals of sustainability. *Developments in the Built Environment*, *4*, 100021. DOI: 10.1016/j.dibe.2020.100021

Boopathi, S. (2024). Advancements in Machine Learning and AI for Intelligent Systems in Drone Applications for Smart City Developments. In *Futuristic e-Governance Security With Deep Learning Applications* (pp. 15–45). IGI Global. DOI: 10.4018/978-1-6684-9596-4.ch002

Boopathi, S., Gavaskar, T., Dogga, A. D., Mahendran, R. K., Kumar, A., Kathiresan, G., N., V., Ganesan, M., Ishwarya, K. R., Ramana, G. V., & others. (2021, August 4). *Emergency medicine delivery transportation using unmanned aerial vehicle (Patent Grant)*.

Boukerche, A., Tao, Y., & Sun, P. (2020). Artificial intelligence-based vehicular traffic flow prediction methods for supporting intelligent transportation systems. *Computer Networks*, *182*, 107484. DOI: 10.1016/j.comnet.2020.107484

Dash, B., Sharma, P., & Ansari, M. (2018). *A Data-Driven AI Framework to Improve Urban Mobility and Traffic Congestion in Smart Cities*. ed.

Deep, G., & Verma, J. (2023). Embracing the future: AI and ML transforming urban environments in smart cities. *J. Artif. Intell*, *5*(0), 57–73. DOI: 10.32604/jai.2023.043329

Deepika, & Pandove, G. (2024). A Comparison of ML Models for Predicting Congestion in Urban Cities. International Journal of Intelligent Transportation Systems Research, 22(1), 171-188.

Dikshit, S., Atiq, A., Shahid, M., Dwivedi, V., & Thusu, A. (2023). The use of artificial intelligence to optimize the routing of vehicles and reduce traffic congestion in urban areas. *EAI Endorsed Transactions on Energy Web, 10.*

Gangwani, D., & Gangwani, P. (2021). Applications of machine learning and artificial intelligence in intelligent transportation system: A review. *Applications of Artificial Intelligence and Machine Learning: Select Proceedings of ICAAAIML 2020*, 203–216.

Gopal, M., Lurdhumary, J., Bathrinath, S., Priya, A. P., Sarojwal, A., & Boopathi, S. (2024). Energy Harvesting and Smart Highways for Sustainable Transportation Infrastructure: Revolutionizing Roads Using Nanotechnology. In *Principles and Applications in Speed Sensing and Energy Harvesting for Smart Roads* (pp. 136–165). IGI Global. DOI: 10.4018/978-1-6684-9214-7.ch005

Hossain, M., Khalid, H., Rao, A. P., Lootah, M., Al-Mohammedi, S. S. K., & Majeed, S. R. (2024). Comprehensive Review of AI, IoT, and ML in Enhancing Urban Mobility and Reducing Carbon Footprints. *2024 Third International Conference on Sustainable Mobility Applications, Renewables and Technology (SMART)*, 1–6. DOI: 10.1109/SMART63170.2024.10815521

Jagatheesaperumal, S. K., Bibri, S. E., Huang, J., Rajapandian, J., & Parthiban, B. (2024). Artificial intelligence of things for smart cities: Advanced solutions for enhancing transportation safety. *Computational Urban Science*, *4*(1), 10. DOI: 10.1007/s43762-024-00120-6

Koppula, V. K., Birundadevi, M., Ramprathap, K. P. P., Nageswaran, A., & Balajee, R. (2024). Climate Change and Food Security Management–based Urban Health Care Systems Using Artificial Intelligence Techniques. *Remote Sensing in Earth Systems Sciences*, *7*(4), 328–336. DOI: 10.1007/s41976-024-00125-z

Makanadar, A., & Shahane, S. (2023). Urban Mobility: Leveraging AI, Machine Learning, and Data Analytics for Smart Transportation Planning-A case study on New York City. *2023 14th International Conference on Computing Communication and Networking Technologies (ICCCNT)*, 1–6.

Molakatala, N., Kumar, D. A., Patil, U., Mhatre, P. J., Sambathkumar, M., & Boopathi, S. (2024). Integrating 5G and IoT Technologies in Developing Smart City Communication Networks. In *Blockchain-Based Solutions for Accessibility in Smart Cities* (pp. 147–170). IGI Global. DOI: 10.4018/979-8-3693-3402-7.ch006

Nikitas, A., Michalakopoulou, K., Njoya, E. T., & Karampatzakis, D. (2020). Artificial intelligence, transport and the smart city: Definitions and dimensions of a new mobility era. *Sustainability (Basel)*, *12*(7), 2789. DOI: 10.3390/su12072789

Olaniyi, O. O., Okunleye, O. J., & Olabanji, S. O. (2023). Advancing data-driven decision-making in smart cities through big data analytics: A comprehensive review of existing literature. *Current Journal of Applied Science and Technology*, *42*(25), 10–18. DOI: 10.9734/cjast/2023/v42i254181

Olugbade, S., Ojo, S., Imoize, A. L., Isabona, J., & Alaba, M. O. (2022). A review of artificial intelligence and machine learning for incident detectors in road transport systems. *Mathematical & Computational Applications*, *27*(5), 77. DOI: 10.3390/mca27050077

Prakash, J., Murali, L., Manikandan, N., Nagaprasad, N., & Ramaswamy, K. (2024). RETRACTED ARTICLE: A vehicular network based intelligent transport system for smart cities using machine learning algorithms. *Scientific Reports*, *14*(1), 468. DOI: 10.1038/s41598-023-50906-7

Rosca, C.-M., Stancu, A., Neculaiu, C.-F., & Gortoescu, I.-A. (2024). Designing and Implementing a Public Urban Transport Scheduling System Based on Artificial Intelligence for Smart Cities. *Applied Sciences (Basel, Switzerland)*, *14*(19), 8861. DOI: 10.3390/app14198861

Sreelekha, M., & Midhunchakkaravarthy, . (2024). Intelligent Transportation System for Sustainable and Efficient Urban Mobility: Machine Learning Approach for Traffic Flow Prediction. *International Conference on Multi-Strategy Learning Environment*, 399–412. DOI: 10.1007/978-981-97-1488-9_30

Ullah, Z., Al-Turjman, F., Mostarda, L., & Gagliardi, R. (2020). Applications of artificial intelligence and machine learning in smart cities. *Computer Communications*, *154*, 313–323. DOI: 10.1016/j.comcom.2020.02.069

Yuan, T., da Rocha Neto, W., Rothenberg, C. E., Obraczka, K., Barakat, C., & Turletti, T. (2022). Machine learning for next-generation intelligent transportation systems: A survey. *Transactions on Emerging Telecommunications Technologies*, *33*(4), e4427. DOI: 10.1002/ett.4427

Zhu, L., Chen, C., Wang, H., Yu, F. R., & Tang, T. (2023). Machine learning in urban rail transit systems: A survey. *IEEE Transactions on Intelligent Transportation Systems*.

Chapter 2
Virtual Prototyping in the Automotive Industry

Shankar Kumar
Department of Automobile Engineering, Birla Institute of Technology, Mesra, India

J. Ananth
https://orcid.org/0000-0003-4197-5595
Department of Marine Engineering, AMET University (Deemed), Chennai, India

N. Manikandan
Department of Data Science and Business Systems, SRM Institute of Science and Technology, Kattankulathur, India

K. Karthik
https://orcid.org/0009-0001-8253-7260
Department of Electronics and Communication Engineering, Velammal College of Engineering and Technology, Madurai, India

P. Vijayaraghavan
Department of Mechanical Engineering, Prince Shri Venkateshwara Padmavathy Engineering College, Chennai, India

M. Sudhakar
https://orcid.org/0009-0002-4269-5480
Department of Mechanical Engineering, Sri Sai Ram Engineering College, Chennai, India

ABSTRACT

Virtual prototyping has evolved to the automotive industry and allows for a more speedy and accurate car development process at a fraction of the previous costs. The application of virtual prototyping and utilization of computer aided design, simulations and digital twin has been discussed in this chapter. These approaches reduce development time and costs as well as improve design performance over traditional use of physical models. These tests add value to vehicle performance and safety so there are few tests such as crash, aerodynamics, and virtual assembly are discussed. It also introduces some recent tendencies on virtual prototyping de-

DOI: 10.4018/979-8-3693-7287-6.ch002

velopment including Artificial Intelligence, Machine Learning and Virtual Reality. It identifies various challenges in applying these technologies and examines future prospects and implications in introducing them in automobile industry respectively.

INTRODUCTION

Virtual prototyping has become one of the most progressive methodologies in the automotive industry and has significantly impacted the manner in which automotive products are conceived, built and assembled. The pressures of increasing interest in aesthetic and unique designs, higher safety standards, the need for shorter production cycles, and concerns for costs have compelled the industry to go digital. Virtual prototyping which incorporates CAD, simulation aids and digital twin solution is a very effective solution to the above challenges since it allows manufacturers to construct virtual models of vehicles before physical models are developed. This chapter focuses on defining virtual prototyping, its best practises and possible repercussions on the automotive industry, as well as its possibilities for the future(Firu et al., 2021).

The automotive industry like many others has always been innovative because of stiff market competition and also change in consumer preferences. The conventional approaches to prototyping while useful in the past have been becoming less useful in handling the modern vehicles development. Physical prototype is always expensive, take a lot of time and go through many cycles before being optimized hence leading to longer development periods and huge costs. Virtual prototyping resolves these iwoke by allowing engineers to model, analyse and optimize the designs from a virtual platform. With high fidelity models and millennium simulations the actual performance of a particular vehicle can be simulated so that certain prematerial design defects can be corrected at the earliest. This not only helps to cut expenses but also bring the time-to-market to a lower level(Besnea et al., 2019).

The most apparent benefit of virtual prototyping is that the model can reveal the effects of various conditions with a good deal of precision. For instance, crash testing, which is part of automotive safety, can be run on computers through the help of softwares. What these simulations do is to model the impact of collision on the structures of automobiles and this enables engineers to obtain the best designs for safety of passengers. Correspondingly, such characteristics as aerodynamic characteristics, which have an impact on fuel consumption and car stability, can also be examined and improved through the use of even a simulated wind tunnel. Through these tools, manufacturers are able to produce designs that are very close to the target with minimal dependency on physical prototype testing(Henriques & Winkler, 2021).

The second significant use of virtual prototyping is within manufacturing and assembly functions of a business. Through virtual assembly simulations engineers can analyze whether it is possible to assemble a certain part of a vehicle and decide whether it is worthwhile, if there are going to be problems in the production line. This means that the process from design to manufacturing is seamless and this minimises the occurrence of costly setbacks due to non convergence between design and manufacturing. In addition, the advancement of the digital twin which is a real time model of a physical system has also improved virtual prototyping. Digital twins receive real-time information from physical systems to drive performance evaluation, predict maintenance, and introduce design enhancements concurrently(Gromova, 2019).

The usage of the virtual prototyping process in the realization of car models is not without its fair share of problems. The use of these technologies entails high costs in terms of software, hardware, and most importantly highly trained personnel. Other emerging issues include data management since generation and analysis of voluminous data require strong systems that can guarantee the right results. Furthermore, resistance to change is a barrier where organizational stakeholders are averse to change from conventional work practices to embracing of technology remedies. Nonetheless, decision-makers face early challenges as they implement this innovation, such as increased initial costs and longer development cycles; but in the long run, its blessings such as cost reduction, enhanced product quality, and shorter concept-to-production cycle, make it a critical element in the automotive design and manufacturing of the future(Yildiz et al., 2021).

New dimensions in virtual prototyping models are expected to take the automotive business to a whole new next level. AI and ML are also being implemented in systems used for virtual prototyping thus improving the virtual prototyping design's efficiency. These technologies can maintain large datasets and can find out patterns and suggest improvements to the design which can more than enhance the decision making. Virtual reality (VR) and augmented reality (AR) are also used significantly in enhancing the utility of the virtual prototyping services. Due to the natural manner in which they engage the users, VR and AR enable engineers, designers, and other stakeholders to be engaged with the virtual models in a way that helps in enhancing collaborative training of the projects and designs(Iqbal et al., 2020).

A special word is to be said about the application of 3D printing and additive manufacturing technologies in virtual prototyping. These fit with virtual prototyping by allowing the production of physical prototypes from virtual models at a very short time. This enables the assessment of virtual simulations and comes up with a concrete product at the end. Also, the use of virtual prototypes offers the chance to design sustainable products in the auto industry, and sustainability is now a key factor to consider. Using virtual prototypes also has the benefit of cutting away with physical models saving material and making manufacturing more environmentally friendly(Heydaryan et al., 2018).

Finally, one can expect high potential for development of virtual prototyping in the automotive industry in the future. Going forward, as technology advances, there will be added incorporation, application, of more sophisticated tools and methodologies within virtual prototyping. For example, cloud technologies to support the processes of virtual prototyping of products will facilitate cooperation in teams consisting of members located in different countries, which will encourage innovation and internal effectiveness. Further, in the future, newly adopted quantum computing should allow for a practically immeasurable amount of computational power, offering some considerably more complicated methods of simulations/ analyses(Hernández-Chávez et al., 2021).

Therefore, virtual prototyping as a tool can be viewed as revolutionary bringing great solution to futuristic whole car development. As for virtual prototyping, the considerations are as follows: due to advanced technologies allowing to conduct the simulations much faster, with better costs and higher accuracy, the usage of virtual prototyping methods became inevitable in automotive development life cycle. Despite the ongoing implementation issues such as high implementation costs, and data management, the gains that will be received are numerous compared to these challenges(Lanzotti et al., 2020). With technologies like AI, VR and 3D printing advancing as they are, the opportunities afforded by virtual prototyping are still only forging ahead as the technologies grow and evolve. As such, this chapter serves as an introduction to a number of subsequent chapters that explore different aspects of virtual prototyping, including its deployment, uses, and the promising developments likely to define this field.

Objectives

Specifically, the chapter aims to achieve the following objectives: To give clear and holistic conceptual definition of virtual prototyping and its radical application in the automotive field. A key focus on CAD, simulation tools, and digital twins to describe virtual prototyping and raise the question of how that approach fits in vehicle development. This chapter focuses on some of the main uses including crash testing, aerodynamics, and virtual assembly, but also examines newer trends that are quickly developing, including AI, machine learning, and virtual reality. Furthermore, it aims at describing what has hindered the implementation of virtual prototyping and what directions are yet unseen but can open up new horizons for development of virtual prototyping as a tool that can increase the effectiveness of the construction process, reduce costs, and have a positive impact on the environment.

Figure 1. Core technologies in virtual prototyping

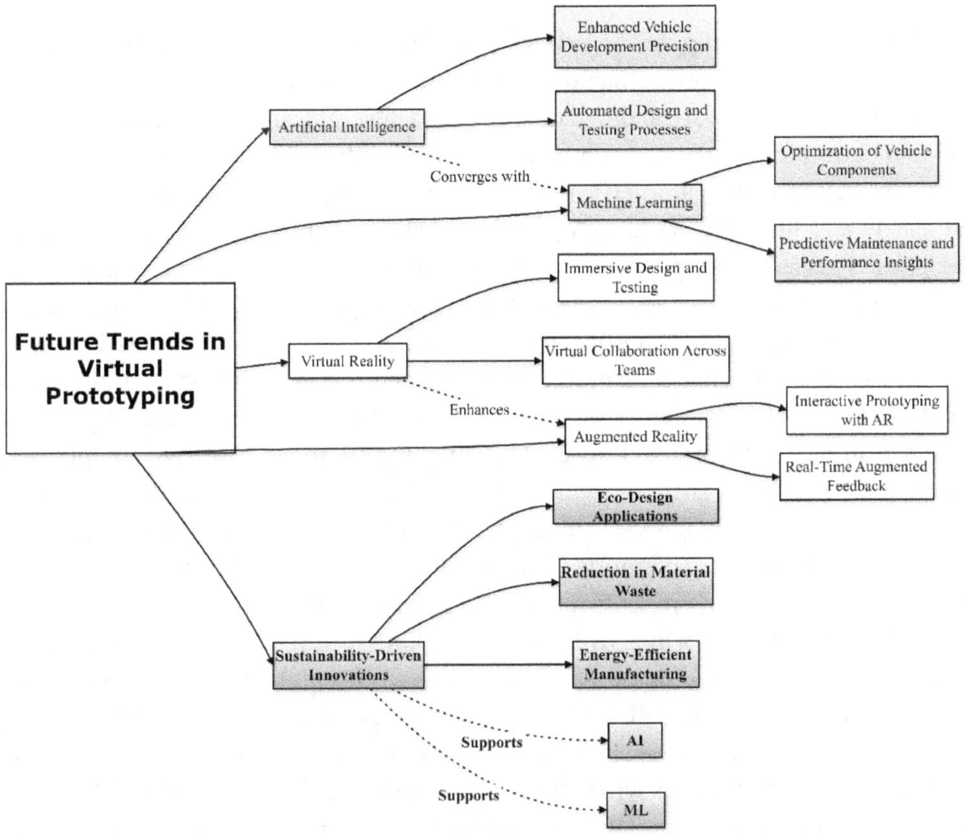

CORE TECHNOLOGIES IN VIRTUAL PROTOTYPING

A virtual prototype is an essential tool in contemporary design for production, notably in automobile manufacturing. That it can boost the design cycles, reel down the cost and improve the innovation quotient is largely attributable to core technologies. Of these technologies, CAD, FEA, CFD, and Digital Twin technologies can be at the most basic level considered as fundamental technologies. All of these technologies play a specific role in the virtual prototyping and to guarantee the performance, safety and sustainability when it comes to complex structures such as automobiles. This section explores these core technologies and explain, how and where they work, as well as the revolution they bring to automotive field(Guo et al., 2020).

CAD stands for Computer Aided Design and it revolves round virtual prototyping. CAD systems allow engineers and designers to draw accurate 2D or 3D representation ofAutomotive part and/or subsystems. These classified digital models are the foundation for nearly all other analyses and simulations. CAD has many strong features, including the parametric modeling where dimension and property changes are easily and instantly achievable; the assembly modeling will guarantee that all the particular parts are appropriately joined and work in the required manner.

CAD enables massive reductions in the time required to develop a design and improves on the accuracy unlike when physical drawings and manually drawn iterations have to be made. also, material parameters can be introduced into CAD models, helping downstream simulation to work with close to real performance. For instance, in auto-mobile designing CAD enables the engineers to design complicated facets like power train, structure and the inside of the car with the best accuracy rates. This allows for the quick review of different designs and optimisations before any physical model is built. Therefore, CAD is a potent instrument of enhancing vehicle performance, decreasing manufacturing cost, and meeting regulatory requirements.

Although FEAs are primarily built-on CAD models, it serves a critical purpose of analyzing automotive parts' structural advisability and fitness for application. Computer aided analysis called FEA involves breaking the system down into tiny components known as finite elements and then utilizing mathematical formulae to estimate how such elements will behave to various forces including stress, strain and temperature influences etc. This technology is most applicable in the automotive industry as safety and reliable auto parts are of essence. Structural analysts leverage FEA for approximate dissection of car crashes, determination of rigidity of utilized materials, and fine-tuning of part shapes to support their functionality load intensities. For instance, the car makers and builders can use FEA to model a frontal impact and study the manner in which the chassis of the car dissipates impact loads hence providing protection to the occupants. Furthermore, FEA plays an important role in reduction strategies that attempt to replace steel with lightweight material such as aluminum or composites. With such conceptual designs being optimized using FEA, manufacturers can redesign their vehicles to be lighter and stronger, therefore enhance fuel economy hence solving the emissions problem. Moreover, FEA creates capability to carry out iterative moves; it is flexible in the sense that engineering designers can easily make changes to their models after doing simulations and thus saving time.

Still, FEA is typically used as a counterpart to Computational Fluid Dynamics (CFD), which handles all the issues related to fluid flow and aerodynamics of the object in design. CFD is the mathematical simulation of the effects of the fluids and gases on surfaces that means it is highly versatile in vehicle design, cooling techniques and lifting of the automobiles and efficient fuel consumption methods. CFD simulations make it easier for engineers to track the flow of air across the car exterior and body, to spot areas of turbulence, and optimize the former in order to reduce the latter. For example, CFD can be used to determine the best aerodynamic design of a car's bodywork in a bid to increase fuel economy or reduce wind induced noise. However, accurately predicting cooling loads, position, and the airflow of a radiator and the general layout of the engine bay are critical, and these are areas where CFD is particularly valuable. It helps keep some of the vital parts at efficient temperature levels while in use in the course of their operation. Furthermore, CFD is also applicable to electric vehicle (EV) where thermal control of battery packs and motor is significant to efficiency and durability. The results of the study indicate that through the use CFD in the virtual prototyping, manufacturers are able to save a lot of costs and deliver high performing products since key faults are solved at the design stage(Guo et al., 2020).

The second disruptive technology in virtual prototyping is the Digital Twin. Digital twin refers to a virtual model that reflects the physical asset, process or system by being updated with data in real time. In automotive, digital twins make it possible for the engineering team to track, model, and improve automobiles at all their phases. Virtual modeling allows for manufacturing productive simulations of a vehicle or its part, in other words, one can test all the outcomes relying on the virtual prototype. For instance, an engine's digital twin can fundamentally reproduce actuation loss throughout the life of an engine and be used to identify time for maintenance or areas likely to fail. Furthermore, through ' digital twins' mechanism, enhanced and more proactive predict and maintain strategies are made through constant data analysis from originated sensors in actual cars. This not only helps to save time but also to increase the durability of essential parts of the machine. In addition to maintenance, digital twins help to innovate as engineers can apply new modifications to the structure and check the results in simulations before using the changes in physical models. For instance, a digital twin is employed by autonomous vehicle developers to test new vehicles and drive scenarios with breakthroughs of ADAS and full autonomy. In addition, applications based on the creation of a digital twin are essential for manufacturing industries since these models can help to determine an effective layout of production lines as well as avoid waste and guarantee the quality of the products manufactured. Because IoT and Industry 4.0 have become the new direction of the automotive industry, the use of digital twins is expected to increase further, which will promote a great leap in performance and development.

These four technologies, namely CAD, FEA, CFD, and Digital Twin, are synchronized to create a form of descriptive model for the revolutionary virtual prototyping within the automotive industry. CAD offers the precise representations of the project required to facilitate all kinds of analyses. FEA provides for sustaining structural integrity for components to withstand safety and durability while CFD analyzes fluid flow and thermal characteristics. Digital twins exist as an overlay between reality and simulation that can be further enhanced during a car's life cycle. Collectively, those technologies allow manufacturers to achieve greater levels of accuracy, productivity, and advancement than ever before possible. Thus, with the help of virtual prototypes, it is possible to adapt faster to market needs, satisfy various standards and regulations, and push automotive development toward sustainability. These technologies are still in their infancy and their synergistic application with principles of artificial intelligence, machine learning and advanced simulation tools look set to transform vehicle design and manufacturing all over the world in the years to come.

APPLICATIONS IN AUTOMOTIVE DEVELOPMENT

Virtual prototyping has emerged as the crucial element in automotive industry innovation to design, test, and manufacture vehicles better. Some of the uses of virtual modeling for automobile manufacturing include virtual car crash, aerodynamics, and heat ferrying, manufacturing, assembly, and the validations of interfaces between car and driver, known as human machine interface or HMI. These applications not only reduce the time and effort in development but also increase safety, efficiency, and the performance of the system(Tubaro & Casilli, 2019).

Figure 2. Applications of virtual prototyping in automotive development

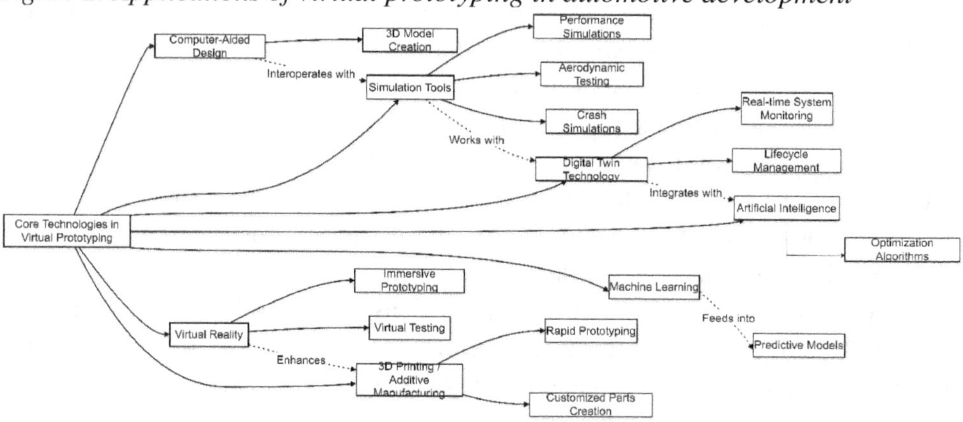

Virtual Crash Testing

Car crash testing has been widely used for over thirty years to ensure the validity of car safety designs. The standard crash tests are nevertheless much longer and definitely more costly, as a physical model has to be constructed and then wrecked. Virtual crash testing occurs in order to solve these challenges by using a virtual model of a car accident. With the assistance of CAE technologies, an engineer is able to simulate the dynamics of a car in the case of different type of accidents including frontal, side and rear ones. Computer generated crash testing enables changes to be made to structures, materials and other design aspects to pass crash safety requirements without the actual production of different models. Another benefit which virtual crash testing has over physical testing is that it encompasses more scenarios than physical testing. For example, the engineers can perform crash analyzes where the certain speed, angle or other conditions of impact will be reproduced in order to find out how exactly the car will respond to various situations in the real life(Lee et al., 2023). Also, virtual simulations give designers a better understanding of interior parts, including the airbag and seat belts and crumple zones as well to guarantee all parts of the automobile add to the safety of the passengers. That is why it is possible to modify those components that contain potential vulnerabilities in the design stage, without constructing realistic prototypes, saving time and money in the process. Also, advanced virtual crash testing also helps the company adhere to the international safety standards. Further to this, there are regional differences in safety standards, with compliance with these standards usually involving a large amount of testing.

Aerodynamic and Thermal Analysis

Vehicle aerodynamics and thermal management system are essential to the performance of a vehicle including its fuel efficiency and overall performance. CAE tools allow engineers to perform aerodynamic and thermal simulations at the preliminary design stage to get an idea about vehicle surface curvature, airflow, and dissipation of heat and heat generated inside the vehicle without going through detailed wind tunnel and thermal chambers respectively. In aerodynamics, CFD of air is simulated in order to evaluate the nature of air flow around an automobile. Through analyzing drag, lift and turbulence engineers are able to adjust the outer car design in order to increase the stability and the fuel efficiency. For example, if drag is reduced, fuel consumption is also minimized, which also corresponds to the international aspect of environmental priorities. It also means that every detail of the car, from spoilers to grilles and underbody panels is tested virtually as well(Rzepka et al., 2018).

Vibration analysis on the other conversely concerns itself with controlling the heat produced by different parts of the vehicle such as the engines, batteries and the braking systems. For BEVs a thermal management system is very important since the battery pack can easily overheat when heavily utilised. Computer models enable engineers to incorporate adverse effects of heat into cooling structures such as heat exchange systems and airflow conduits with less energy use. These analyses also highlight likely high heat zones and thermal issues that should be corrected at the design stage. Thus, the aerodynamic and thermal simulations can be implemented into the virtual prototyping process providing optimal performance, efficiency and reliability. Not only does this integration enable development to proceed at a faster pace than normal but it also decreases the number of physical prototypes that needs to be made cutting costs as well as the harm done to surrounding environment.

Virtual Assembly and Manufacturing Simulations

This stage is one of the most important in automotive development since it occurs on the path from design to manufacturing. This methodology of virtual assembly and manufacturing simulation is critically significant for deliberation of the possibility of efficient manufacture of the vehicles such that they do not have assembly problems or production bottlenecks. These remodel the manufacturing process using virtual models to consider the various possibilities to find the most efficient setup prior to physical production. In virtual assembly simulations, engineers get to find out whether or not given parts will fit together properly and if the various assembly processes will be easy on the hands and wrists. For instance, they can expose some parts as being easily, or not easily reachable to the assembly line production personnel or if the robotic arms can properly execute some certain tasks. This serves to enable manufacturers to design processes of assembly in a way that is both effective and free from defects, which often lead to costly overhauls or time overruns(Rzepka et al., 2018).

Manufacturing simulations also apply to efficient line production. Because the layout of assembly lines can be mapped, manufacturers get insights into issues such as material handling or workload imbalances within the factory. The aspect of work compartmentalization helps in optimizing production process flow and cutting on various operational expenses. Other than that, these scenarios are helpful in providing training to the workers through an assembly line simulation to reduce production time and facilitate a familiarization with the process. The last application of virtual assembly and manufacturing simulations is, however, in quality control. Real-life manufacturing is transformed into a simulation so engineers can find flaws or irregularities in the manufacture of the final product and addresses those problems before full-scale production begins. This makes sure that cars that are manufactured will have acceptable quality to the market while at the same time reducing on scrap or redo.

With the development and gradual implementation of new manufacturing technologies like automation and robotics in the automotive industry, it is expected that engineering virtual simulations, as a tool in designing production lines, will become more relevant. Virtual assembly and manufacturing provide benefits that directly relate to reducing time and friction between design and manufacturing teams.

Human-Machine Interface (HMI) Testing

The way drivers interact with vehicles is quickly changing due to improvements in connectivity, automation and infotainment systems. Human machine interface (HMI) testing is another prominent application of virtual prototyping which allows manufacturers to prototype and evaluate user interfaces that are intuitive, functional, and safe. Meanwhile, virtual HMI testing utilizes simulations and virtual reality environments to evaluate how drivers interact with controls, displays and other interface elements. Using a model of the driver's point of view, engineers can assess the effectiveness of the layout of dashboards, touchscreens, voice commands and haptic feedback systems. This allows for interfaces that are easy to operate and distracting less, thus making driving both safer and a better experience(Chu et al., 2018).

Virtual HMI testing encompasses a varied range of user scenarios, which is one of its key benefits. For example, one might simulate different driving conditions — urban traffic and highway cruising — to measure how effectively the interface supports the driver in each case. Virtual environments can also emulate different user populations, including drivers of varying skill or ability levels, providing a way to make sure interfaces are inclusive and accessible. Human-vehicle interface (HMI) testing is also vital in the development of advanced driver-assistance systems (ADAS) and autonomous vehicles. With these technologies proliferating, the importance of designing interfaces providing information to drivers in a clear and effective way increases. For example, in the context of autonomous driving, Virtual Simulations enable engineers to test driver response to alerts, notifications, and handovers of control, while also ensuring systems behave in an intuitive and reliable manner.

Abstract: Virtual HMI testing also allows for fast iteration and improvement of interface designs. Engineers can easily update layouts, features or interactions from simulation feedback, leading to lower development time and costs. By introducing new features and upgrades through regular over-the-air updates, manufacturers can keep their vehicles competitive and capitalize on changing consumer preferences and technological trends.

That being said, the applications of virtual prototyping across the automotive development vertical are extensive and, indeed, revolutionary. This allows for safer design iteration without the costs and time associated with traditional crash testing. Aerodynamic and thermal analyses, for example, can optimize vehicle performance and energy efficiency, supporting sustainability objectives. Production processes are effortless and frustration-free due to virtual assembly and manufacturing simulations, which reduce errors and delays(Chu et al., 2018). Lastly, HMI testing are used to create user-centric, intuitive interfaces in vehicles, thus facilitating the uptake of technologies like ADAS and autonomous driving. Collectively, these applications highlight the essential contribution of virtual prototyping in defining the future of automotive innovation.

Figure 3. Implementation strategies for virtual prototyping in automotive development

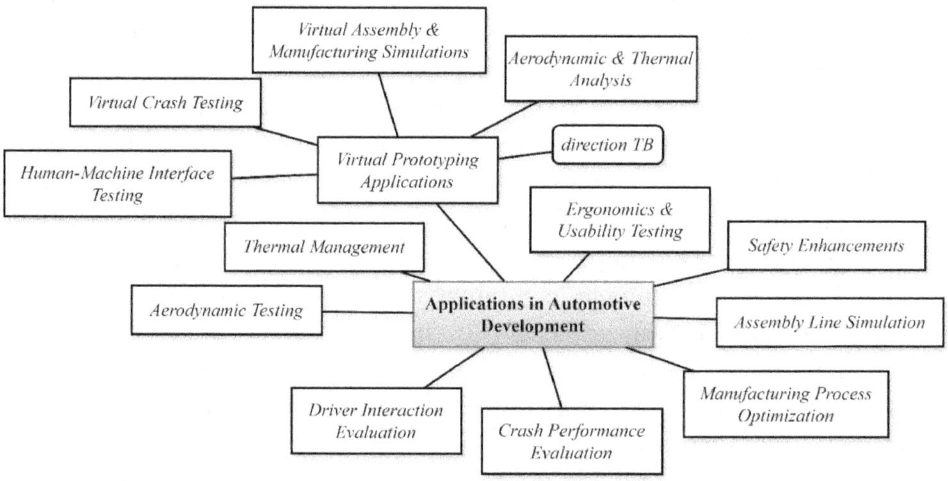

IMPLEMENTATION STRATEGIES

In the automotive sector, this change is due to the fact that virtual prototyping is the most valuable method to enhance the process of design and development, as well as minimize the time and cost associated due to it. It necessitates implementation strategies to fully realize the advantages of virtual prototyping. Examples of these methods include the use of virtual prototyping as part of the design process, collaboration between teams within the virtual sphere, and cost-benefit analysis for executive review and ROI measurement. Each of these strategies ensures that virtual prototyping is able to deliver its full potential in the automotive industry(Halabi, 2020; Stark, 2022; Wolfartsberger, 2019; Wolfartsberger et al., 2020).

Transitional Approach: Integration of Virtual Prototyping in the Design Flow

The process of successful virtual prototyping starts with integration into the automotive design workflow. Traditional design processes were based on the development of several physical prototypes, which was costly and time-consuming. Through digital testing and validation, the designers can do the rapid iterations before they move towards physical iterations. Implementing this integration necessitates the use of cutting-edge software solutions including computer-aided design (CAD), computational fluid dynamics (CFD), and finite element analysis (FEA). With such tools, engineers can create simulations of real-world situations and assess virtual vehicles' performance, safety and durability. Also, it is crucial to use a modular workflow. Modular workflows deconstruct the vehicle development process into smaller, interdependent stages, wherein each module—from chassis design, through engine optimization, to interior ergonomics—can be independently developed and tested in the virtual environment. This not only improves efficiency but also guarantees that any changes made to a functioning module cannot disrupt another.

Another aspect of integration involves linking virtual prototyping tools to legacy PLM systems. PLM systems have been the backbone of car development, providing a single source of truth for all file and documentation. Connecting virtual prototyping tools to PLM allows teams to maintain continuity across the developmental cycle from concept through production. This helps facilitate better decision-making, while also minimizing errors through real-time data syncing. In addition, organizations need to spend on training programs to upskill the workforce on working with these virtual prototyping tools. Engineers and designers need to be adept at using that simulation software, interpreting the results, and applying that to enhance designs. Teams are trained through workshops and certifications to make sure they use virtual prototyping the right way.

Collaboration Across Teams Using Virtual Environments

With agile development and virtual prototyping, collaboration across multidisciplinary team is key, and virtual environments is critical for this collaboration. Many different departments must contribute to the automotive design and development process, including engineering, manufacturing, marketing, and quality assurance(Čujan et al., 2020; Freitas et al., 2020; Wolfartsberger et al., 2020). This is where virtual environments come in; they give a life-like shared digital workspace, where teams can collaborate in real-time, no matter where they might be in the real world; Such is particularly useful for global automotive manufacturers with internationally-distributed teams. The virtual reality (VR) and augmented reality (AR) technologies play a key role in collaboration. AR and VR enable stakeholders to visualize and interact with 3D models by building immersive virtual environments. For instance, designers can walk through virtual vehicle interiors to assess ergonomics and aesthetics, while engineers can view mechanical systems "from all perspectives." Such immersive experiences allow for much quicker identification of design flaws with enhanced collaboration for team members. Moreover, cloud-based platforms are a game-changer for collaboration. These platforms allow teams to store, access, and share virtual prototypes in real time. Cloud-based solutions also provide version control, so all stakeholders are on the same design iterations. These platforms help to minimize communication silos, speed up decision-making, and overall boost project efficiency.

Standardized processes and workflows also forms a part of cross-functional collaboration. By standardizing this process across teams, organizations avoid misunderstandings and ensure a consistent approach within their use of virtual prototyping tools throughout the company. Defining communication protocols and maintaining dedicated points of contact in both teams aids learning through mutual interaction. In addition, creating a culture of open communication and collaboration are keys to success. Such cross-disciplinary conversations and knowledge sharing will allow teams to tackle complex problems more efficiently and effectively. For example, engineers can share material suggestions for manufacturability, and marketing teams can share customer data helping guide well rounded vehicle designs.

Cost-Benefit Analysis & ROI Considerations

In order to implement this technology, the automotive industry need to spend a lot of money on software, hardware, training, etc. So making a thorough cost-benefit analysis is essential to ensure a strong ROI for these investments. A cost-benefit analysis allows organizations to weigh the financial and operational benefits of virtual prototyping against any potential roadblocks. A major advantage derived from virtual prototyping is the decrease in costs related to physical prototyping. Conventional development processes typically rely on numerous physical prototype iterations that can cost millions of dollars. Replacing physical prototypes with virtual counterparts, organizations can reduce these costs substantially. Lastly, virtual prototyping reduces development time, allowing design changes to be made quickly, and enabling problems to be identified sooner. This means quicker turnaround times, product launches, and better market competitiveness(Lee et al., 2023).

A second financial upside is proper resource allocation. With the help of virtual prototyping, teams can detect design problems at an earlier stage in the development process, preventing expensive rework in later stages. Additionally, organizations can also work towards reducing material wastage, optimizing fuel efficiency, and enhancing vehicle performance, all by mimicking real-world conditions — contributing to long-term cost benefits.

But companies must also factor in the upfront cost of setting up virtual prototyping. These costs cover the acquisition of high-end software licenses, the upgrading of hardware to execute resource-demanding simulations, as well as employee education. These costs need to be justified, and they will need to be to quantify the expected benefits including savings, time and quality. During any kind of ROI calculation, you should take into consideration the tangible benefits that can include cost savings and any intangible benefits that include better customer satisfaction and brand reputation. Real-world example: Case studies of successful implementations of virtual prototyping can also be helpful when it comes to ROI considerations. For example, virtual crash test, and aerodynamic simulations have helped automotive companies achieve large cost savings and become more efficient. Such success shops become the role model for other organizations willing to adopt virtual prototyping.

Finally, for the future, the use of virtual prototypes in the automotive sector needs to be approached with consideration of integration into the design process, team work, and cost-benefit analysts. With the help of sophisticated tools, virtual environments, and standardized processes, organizations can leverage the strengths of virtual prototyping and diminish the challenges associated with it. The auto industry is not only constantly evolving but improving, and these implementation strategies will be pivotal in not just driving innovation forward but cutting costs for consumers and ultimately bringing high-quality vehicles to the market.

Figure 4. Emerging trends and technologies (AI, ML, VR, and 3D printing) in virtual prototyping for the automotive industry

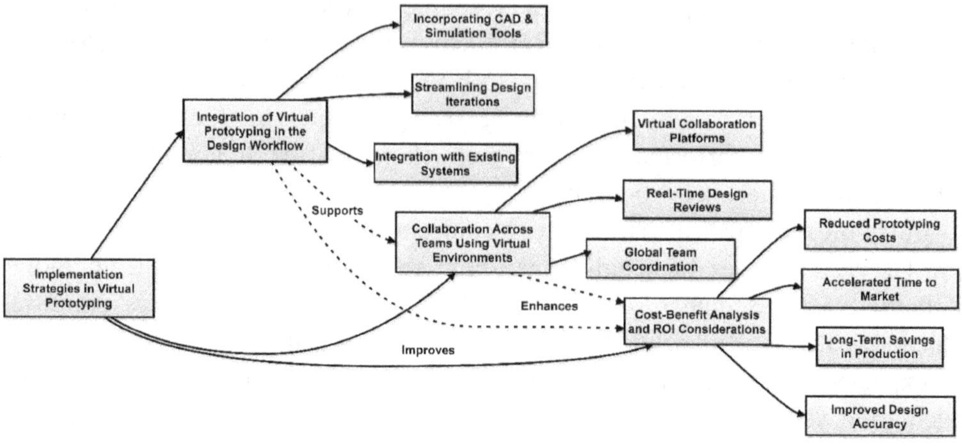

EMERGING TRENDS AND TECHNOLOGIES

AI, ML, VR and 3D printing driving transformation of Virtual Prototyping with emerging technology in automotive The Impact of AI and ML on the Automotive Industry Automobile design and development is undergoing a revolution owing to AI and ML; these technologies not only automate complex tasks but also enhance accuracy and offer predictive insights. These technologies help engineers process massive amounts of data created in the design and testing processes. By understanding how a car might behave under different circumstances, AI-driven algorithms can help optimize safety features, fuel economy, and other aspects of vehicle performance. For example, AI generative design tools are used to explore new structural solutions that would be hard to achieve using traditional techniques. Machine learning models are also reducing reliance on physical simulation and speeding up the design process through pattern and anomaly detection in simulations, increasing accuracy and decreasing time between iterations. Thanks to the automation of processes AI and ML enable; automotive manufacturers are able to develop vehicles with faster time-to-market and lower development costs without sacrificing quality or safety(Freitas et al., 2020; Kim & Oh, 2022; Wolfartsberger, 2019).

At the same time, Virtual Reality (VR) is revolutionizing the world of virtual prototyping. Now, digital content can be brought to life with the help of VR (virtual reality). The potential of this is especially useful in the early stages of the design process, as it allows for a better understanding of a vehicle's ergonomics, aesthetics, and functionality. For instance, via means of VR, designers can test both the interior layout of a car design and the visibility from the driver's standpoint before any physical production ever takes place, helping to catch design flaws that otherwise could have led to issues. Moreover, VR allows collaborative design, as team members located there and then around the globe can interact with the same virtual prototype. This lowers communication barriers and boosts decision-making speed. We are validating VR not only for design but also for virtual training in assembly line processes, where we can expose workers to assemble components of actual vehicles without the risk to life or equipment. By streamlining prototyping and manufacturing using VR, the product development cycle is moving towards a much more flexible, creative, and participative model.

As well as AI and VR, 3D printing and additive manufacturing are part of the game-changers that are redefining prototyping workflows. These technologies serve as a bridge between the digital world and the physical world, complementing the work done with virtual prototyping. This technology is not even a decade old but it allows the producer to print out a three-dimensional physical item directly from a virtual design which creates for them a physical model that they can use to validate their design. This need for additive manufacturing is especially important when creating complex geometries, lightweight components, and customized parts that would be hard or impossible to create with average manufacturing methods. Automotive companies use 3D printing to create functional prototypes that can undergo extensive testing, which minimizes the prodigal lead time that comes with prototyping by other means. The flexibility to create higher-quality components with less waste is also in line with current industry sustainability initiatives, particularly at a time when there is a clear push against traditional manufacturing processes that consume high amounts of energy. For example, manufacturers now can 3D-print individual parts to test their performance, swiftly make improvements based on the findings, and then 3D-print updated designs without the need for a new mold or tooling. The iterative cycle allows for an accelerated development process and also encourages innovation through experimentation.

By integrating these emerging technologies, the automotive sector can significantly improve its prototyping process, leading to faster and more efficient product development. Machine learning and AI, for instance, allow designers to sift through data captured from 3D-printed prototypes, which can be used to improve future iterations, while VR technology can be employed to help teams envision and engage with prototypes that aren't yet in existence. These technologies are also driving the shift to higher levels of autonomy and electrification as more detailed modeling and simulation of complex systems becomes possible. Additionally, with digital transformation taking hold in the automotive domain, these trends are facilitating a step toward more sustainable and efficient vehicle development processes. But, while these technologies do have their advantages, their adoption also raises challenges, including significant upfront costs, the need for specialized skills, and adapting new tools to existing workflows. Overcoming these challenges is essential to harness the full potential of AI, VR, and 3D printing for virtual prototyping. Automotive methods of innovation are poised to be redefined as these technologies mature, leading to a faster, smarter and greener design and development process.

FUTURE TRENDS IN VIRTUAL PROTOTYPING

Virtual prototyping in the automotive industry is deeply integrated with the ever-evolving landscape of artificial intelligence (AI), machine learning (ML), virtual reality (VR), augmented reality (AR), and sustainability-oriented technologies. With automotive makers looking to make their vehicles more efficient, performant, and eco-friendly, these technologies are on the verge of becoming a game-changer(Paul et al., 2024). By leveraging the disparate realms of AI and ML, immersive worlds of AR/VR, and eco-design solutions, vehicle development will become more accurate, faster and attempted in a more sustainable manner.

Figure 5. Future trends in virtual prototyping, emphasizing the role of AI, ML, VR, AR, and sustainability-driven innovations

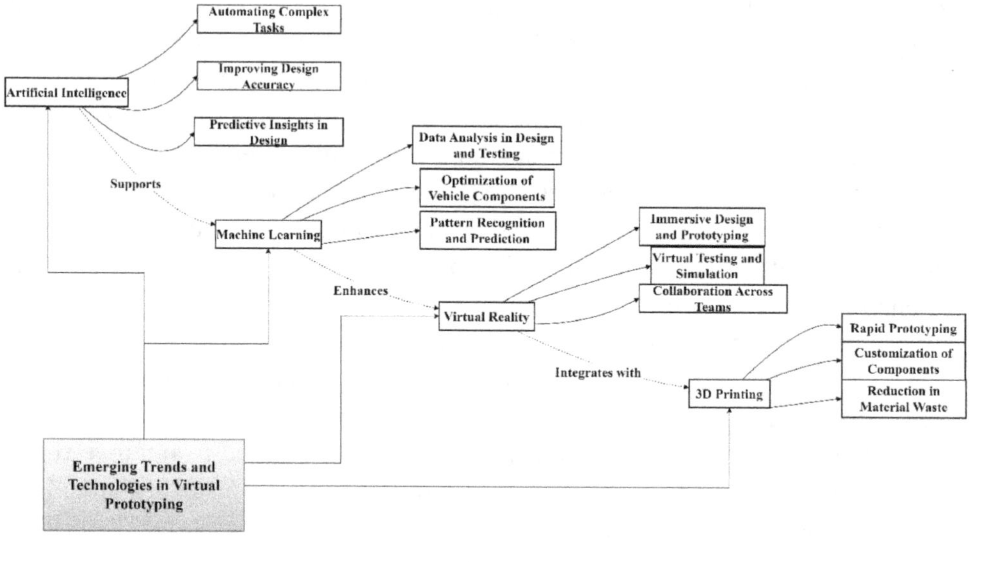

Role of Artificial Intelligence and Machine Learning

Artificial Intelligence as well as Machine Learning are among the best forces which influence the future of virtual prototyping in the automotive. These tools are currently enhancing how we design, manufacture, and test vehicles through more intelligent and adaptive processes. AI and ML will be further embedded into each stage of prototyping, from concept design to production(Glady et al., 2024; Muthukumar et al., 2024; Reddy et al., 2024; Saravanan et al., 2024).

One key domain where AI and ML are influential is predictive modeling and simulation. This AI technology will analyse large datasets through training on data sets from sources, including historical performance of vehicles, and sensor-related information and real-time driving conditions, predicting how designs will perform across multiple scenarios. This will enable engineers to make data-informed decisions earlier in the design stage, minimizing the need for physical prototypes and delivery costs. AI will further improve optimization algorithms to allow automotive manufacturers to discover optimal and cost-conducive designs. The use of AI-enabled generative design methods, for instance, will allow for new vehicle topologies to be generated with minimal weight and maximal strength, which can enhance fuel efficiency and performance.

And, not to be left out, machine learning — a subset of AI — will play a significant role in streamlining and enhancing simulation automation. When ML models are trained on real-world data, they learn to predict vehicle behavior more accurately, whether that's how a vehicle might react in a crash, how well it performs relative to fuel efficiency, or how much energy electric vehicles will use. This will allow for increased testing and iteration speeds, so new designs will be thoroughly validated before physical prototypes are produced. A further application domain in AI and ML as potential enablers is their role in streamlining the manufacturing processes improving design optimization. Leveraging information from sensors implemented in the production line, AI systems can assess the quality of the assembled components and identify defects, and make real-time adjustments dumb beforehand, thereby enhancing efficiency and minimizing waste.

Advancements in Virtual and Augmented Reality (VR/AR)

Virtual and Augmented Reality technologies already have significant potential in terms of virtual prototyping but it will only become more promising in the future. This will allow all engineers, designers, and stakeholders to have immersive experiences where they will be able to visualize and interact with prototypes and processes in real-time before things are built physically. What this should mean for automotive prototyping for VR and AR in the future—harnessing emerging technologies in new ways to better quality new mobility vehicles will pave the way for the next wave of new mobility innovation(Sekhar et al., 2024).

With VR, engineers can create fully immersive environments that closely approximate the real-world scenarios a vehicle will be subjected to. This enables more intuitive evaluation of vehicle design, so designers can effectively "walk around" a prototype, paying attention to details like ergonomics, sight lines and space: This also makes it possible to fine-tune driving dynamics, ensuring that the car provides the best possible driving experience. Along with VR, the technology will also provide even more realistic and detailed simulations with physical feedback in the form of haptics, allowing users to feel the interplay of different vehicle elements in the simulated world.

Augmented reality, conversely, adds digitally rendered elements to the physical world surrounding the engineer and the designer, thereby improving how they consider their physical prototypes. Augmented reality (AR) could further take this a step further, by letting engineers view data and simulations integrated with actual vehicle parts in the real world, making it possible to change and adjust things live. For instance, an engineer could don AR glasses, allowing him to see design specifications and performance metrics while working on a car prototype, effectively merging the digital and physical worlds. AR may also be used as a tool for training, where technicians and assembly line workers are guided through the assembly step by step, to assist them in the assembly job. AR reduces errors and assists technicians in building complex items. Integrating VR and AR will help automotive companies streamline the development process, improve teamwork, and hasten decision-making across design and manufacturing.

Applications in Sustainability & Eco-Design

Sustainability is becoming a key priority for the automotive industry and virtual prototyping will be essential to manufacturers in achieving environmental goals. In a few years, this way of thinking will be oriented towards eco-design, as well as sustainable processes and materials — all of which can be promoted and improved through virtual prototyping technologies. The use of AI combined with VR or AR will allow designers to carry out an early actual environmental impact assessment for vehicles, so that greener options can be employed long before production and in the best possible way(Domakonda et al., 2022; Harikaran et al., 2023; Janardhana et al., 2023).

Machine learning and AI can support sustainability measures by allowing designers to evaluate the environmental impact of various materials and production methods. By modeling the lifecycle of a vehicle from design to production and disposal, AI can help identify solutions that are most energy efficient and have the least emissions impact from cradle to grave. The ongoing developments in machine learning will further enhance and refine the optimization of vehicle designs in terms of weight reduction, energy use and recyclability of materials. With the automotive sector transitioning to electric and hybrid technologies, virtual prototyping can help manufacturers refine the designs of batteries, energy management systems, and charging infrastructure in a virtual environment rather than relying on multiple physical prototypes and consuming resources in the process.

With 3D printing and additive manufacturing technology available, sustainability will improve, as these technologies will allow localized manufacturing and reduced material waste. As prototypes become increasingly representative of real-world properties, 3D-printing technology allows manufacturers to produce global vehicle components and spare parts on demand, reducing the need for mass production and minimizing the overall environmental footprint. Also, virtual prototypes will be used to test and validate sustainable and recyclable materials, enabling automotive manufacturers to adopt eco-friendly solutions without compromising either performance or safety.

With an increasing demand for sustainable vehicles and eco-friendly production methods, the auto sector is pressed to embrace green technologies, and virtual prototyping will be key to fulfilling these needs. Overall, incorporating aspects of AI, VR/AR, and eco-design processes will further catalyze development of environmentally friendly vehicles and propel the industry toward greener innovations.

CHALLENGES AND OPPORTUNITIES

Technological and Organizational Barriers

The benefits of virtual prototyping are there, but it is not that easy to adopt it as there are certain technological and organizational barriers to be broken. On the technology front, one of the significant challenges is integrating new software and tools with the current systems. Numerous automotive manufacturers are heavily dependent on legacy software systems that were not built to integrate with cutting-edge virtual prototyping technologies—such as AI-controlled simulations, VR or 3D printing. It becomes a problem in terms of data compatibility, system integration, and disruption of the workflow. For companies to take full advantage of virtual prototyping, they must modernize their infrastructure, embrace new tools, and ensure that tools integrate smoothly across all stages of the vehicle Lifecyle. Simulations and digital twins also generate vast amounts of data, which require high-performance computing power. As vehicles grow more complex, the need for powerful hardware capable of running in real-time simulations is on the rise. This comes at a cost and poses scalability and resource allocation challenges(Dass james & Boopathi, 2016; Vignesh et al., 2018).

One of the biggest challenges on the organizational level is resistance to change within existing companies. Moving from physical prototyping and testing methods to virtual prototyping has required major cultural and operational shifts for the automotive industry. Executives have to realize that employees, particularly those on the factory floor, will resist incorporating new technologies; whether due to lack of understanding of how it works, or fear of losing their job because of automation. [Virtual Prototyping Challenges & Resistance can be Solved too!; We had learnt a lot from Gen1 generated, Gen1 addressability, Gen1 co-existence with others & with Gen1 era followed with ...] Leadership teams can advocate this transition, ensuring the workforces are onboard with the new targets and technologies. Organizational silos can also be an obstacle for successful virtual prototyping implementation. It is fundamental to integrate departments: design, engineering, manufacturing, etc.; in order to take the complete advantage of all available virtual tools. Companies need to encourage cross-functional collaboration to Map virtual prototypes to real world needs and translate them to replaceable products.

Bridging the Gap Between Virtual and Physical Validation

One of the biggest issues in virtual prototyping is the ability to narrow the gap between virtual and physical validation. However, sometimes those simulations we run in the buildings are no guarantee of a highly accurate model, and the simulation results represent real-world scenarios when we compare to prototypes, especially with respect to the underlying effects. Indeed, physical validation, wherein a prototype is developed and tested in the real world, is still a fundamental step within the automotive design process. In order for virtual prototyping to deliver full value, there needs to be a robust correlation between the digital simulations and physical results. This involves advanced validation techniques, leveraging hybrid testing strategies that integrate both virtual and real-world experiments. An example would be a vehicle crash test which can be simulated by creating virtual prototypes but they need to be compared and validated with real crash tests to ensure the reliability. The same is true for other aspects of a car, like the aerodynamic performance that can be simulated with virtual prototypes, but must be validated with physical wind tunnel testing before being trusted in the real world(KAV et al., 2023; Venkateswaran et al., 2024).

Reducing these gaps takes time, constant work, better simulation, and more detailed inputs for data and modeling. It is anticipated that artificial intelligence and machine learning technologies will contribute to more faithful virtual simulations. Alternatively, the predictive power of the virtual approaches can be enhanced using data produced from physical tests, by training machine learning models on these large datasets. Furthermore, digital twins, real-time digital representations of physical systems, can close this gap because they provide an opportunity for these virtual models to be continuously updated using real-world data. You can use these models to study how components behave over their lifetime, so that virtual testing is in sync with actual performance in the real world. This will reduce the number of costly and time-consuming physical prototypes required, as virtual prototyping will eventually become a more trusted tool for automotive manufacturers as the virtual and physical validation merge.

Industry Adoption and Standardization

Initially, virtual prototyping is being used in some industries, such as the automotive, but there is still individual work being done instead of collaboration and cooperation, another obstacle is also the fact that standardization is not guaranteed and since the automotive sector works mainly with custom-made products, all products contain thousands of non-standard components. Proprietary systems and tools for virtual prototyping, various methods for data exchange and an absence of interoperable tools lead to isolated processes and systems and challenge consistency and data exchange across the supply chain. Lack of common standards for virtual prototype formats, simulation methods, and validation processes may lead to inefficiencies, miscommunications, and error. So to leverage the immense potential of virtual prototyping, the industry must agree on and accept standards that allow industries ease of data sharing and collaboration between stakeholders. Uniformity across the industry will facilitate smoother tool integration, more rapid adoption of innovative technologies, and consistency throughout the design and testing process.

Standardization in virtual prototyping is already making its way in various industry groups and consortia are already working to development common frameworks and guidelines. Standards Organizations – A coalition of organizations, including the Automotive Industry Action Group (AIAG) and the Society of Automotive Engineers (SAE), is driving the development of standards that support greater collaboration and consistency in areas like data exchange, simulation protocols, and validation processes. "This effort is critical to ensuring that virtual prototyping independently becomes the very beating heart of the automotive industry's digital transformation. Moreover, it can set a uniform standard for smaller manufactures that would not have had the budget to create proprietary systems. These shared standards enable smaller manufacturers to utilize the same fees and tools as their bigger counterparts, spurring innovation and competition in the landscape.

That is, beyond technical consistency, we need a business culture that fosters more collaboration and openness." Automotive business is not aloof anymore, it is highly dependent on suppliers and accomplices for parts, programming, and frameworks that coupling into the general vehicle outline. Virtual prototyping also is an excellent way for suppliers and manufacturers to collaborate more closely, giving everyone a common digital platform for design and testing. For this collaboration to take place though, companies need to take a more flexible approach to IP (intellectual property) and data sharing. Most organizations have staked their lives on their proprietary designs and how to make them. In conclusion, while virtual prototyping represents a powerful tool for companies seeking to speed up development and reduce costs, in a highly competitive global marketplace, companies that embrace collaboration and open innovation will be particularly well-positioned to unlock the full value of this exciting technology.

CONCLUSION

The use of virtual prototyping has revolutionized design and development process in the automotive sector offering several benefits such as cost reduction, faster development cycles, and better accuracy. In focusing on core technologies that make virtual prototyping possible, this chapter pointed to the role of defining technologies — AI, machine learning, virtual reality, 3D printing — that are expanding innovation in automotive design. Automotive manufacturers now have access to tools that will allow them to simulate, test, and even improve their designs to a level of exactness we have yet to see, while also creating new potential for personalization, performance, and sustainability.

This widespread adoption of virtual prototyping does come with challenges. The potential for virtual prototyping to become a mainstream approach to product development hinges on overcoming different technological and organizational barriers, bridging the disconnect between virtual and physical validation, and finally achieving industry-wide standardization. The industry needs to come together across the automotive value chain to invest in the infrastructure to make this new world possible as well as disrupt the traditional automotive culture that has no experience with these challenges. But the potential for growth and innovation is enormous. With this data and the assistance of software solutions for prototyping, this new technology in the auto industry is set to revolutionize its design.

The Future of Virtual Prototypes in Automotive Industry As simulation accuracy improves, tools become more integrated, and collaboration across industries increases, virtual prototyping is ever more reliable and powerful. Emerging technologies such as AI, machine learning and VR will play a role in continuing to automate design and manufacturing processes to increase efficiencies. The trend with both sustainability and greener technologies is also a major opportunity for the role of virtual prototyping; as designs are optimized for fuel efficiency and impact on the environment. Consequently, widespread adoption of virtual prototyping across the automotive sector, as seen in the case of Tesla and General Motors, represents a critical step towards higher speed innovation and ultimately safer, more efficient cars with a lighter environmental footprint.

REFERENCES

Besnea, F.-L., Resceanu, I.-C., Cismaru, S.-I., Ganea, I.-E., Pistritu, I.-L., & Bizdoaca, N.-G. (2019). Experiments regarding implementation of a virtual training environment for automotive industry. *2019 20th International Carpathian Control Conference (ICCC)*, 1–6.

Chu, Y., Hatledal, L. I., Zhang, H., Æsøy, V., & Ehlers, S. (2018). Virtual prototyping for maritime crane design and operations. *Journal of Marine Science and Technology*, *23*(4), 754–766. DOI: 10.1007/s00773-017-0509-z

Čujan, Z., Fedorko, G., & Mikušová, N. (2020). Application of virtual and augmented reality in automotive. *Open Engineering*, *10*(1), 113–119. DOI: 10.1515/eng-2020-0022

Dass, A., & Boopathi, S. (2016). Experimental Study of Eco-friendly Wire-Cut Electrical Discharge Machining Processes. *International Journal of Innovative Research in Science, Engineering and Technology*, *5*(1), 666–675.

Domakonda, V. K., Farooq, S., Chinthamreddy, S., Puviarasi, R., Sudhakar, M., & Boopathi, S. (2022). Sustainable Developments of Hybrid Floating Solar Power Plants: Photovoltaic System. In *Human Agro-Energy Optimization for Business and Industry* (pp. 148–167). IGI Global.

Firu, A. C., Tapîrdea, A. I., Feier, A. I., & Drăghici, G. (2021). Virtual reality in the automotive field in industry 4.0. *Materials Today: Proceedings*, *45*, 4177–4182. DOI: 10.1016/j.matpr.2020.12.037

Freitas, F., Oliveira, H., Winkler, I., & Gomes, M. (2020). Virtual reality on product usability testing: A systematic literature review. *2020 22nd Symposium on Virtual and Augmented Reality (SVR)*, 67–73.

Glady, J. B. P., D'Souza, S. M., Priya, A. P., Amuthachenthiru, K., Vikram, G., & Boopathi, S. (2024). A Study on AI-ML-Driven Optimizing Energy Distribution and Sustainable Agriculture for Environmental Conservation. In *Harnessing High-Performance Computing and AI for Environmental Sustainability* (pp. 1–27). IGI Global., DOI: 10.4018/979-8-3693-1794-5.ch001

Gromova, E. A. (2019). Digital economy development with an emphasis on automotive industry in Russia. *Revista Espacios, 40*(06).

Guo, Z., Zhou, D., Zhou, Q., Zhang, X., Geng, J., Zeng, S., Lv, C., & Hao, A. (2020). Applications of virtual reality in maintenance during the industrial product lifecycle: A systematic review. *Journal of Manufacturing Systems, 56*, 525–538. DOI: 10.1016/j.jmsy.2020.07.007

Halabi, O. (2020). Immersive virtual reality to enforce teaching in engineering education. *Multimedia Tools and Applications, 79*(3), 2987–3004. DOI: 10.1007/s11042-019-08214-8

Harikaran, M., Boopathi, S., Gokulakannan, S., & Poonguzhali, M. (2023). Study on the Source of E-Waste Management and Disposal Methods. In *Sustainable Approaches and Strategies for E-Waste Management and Utilization* (pp. 39–60). IGI Global. DOI: 10.4018/978-1-6684-7573-7.ch003

Henriques, A. C., & Winkler, I. (2021). The advancement of virtual reality in automotive market research: Challenges and opportunities. *Applied Sciences (Basel, Switzerland), 11*(24), 11610. DOI: 10.3390/app112411610

Hernández-Chávez, M., Cortés-Caballero, J. M., Pérez-Martínez, Á. A., Hernández-Quintanar, L. F., Roa-Tort, K., Rivera-Fernández, J. D., & Fabila-Bustos, D. A. (2021). Development of virtual reality automotive lab for training in engineering students. *Sustainability (Basel), 13*(17), 9776. DOI: 10.3390/su13179776

Heydaryan, S., Suaza Bedolla, J., & Belingardi, G. (2018). Safety design and development of a human-robot collaboration assembly process in the automotive industry. *Applied Sciences (Basel, Switzerland), 8*(3), 344. DOI: 10.3390/app8030344

Iqbal, D., Abbas, A., Ali, M., Khan, M. U. S., & Nawaz, R. (2020). Requirement validation for embedded systems in automotive industry through modeling. *IEEE Access : Practical Innovations, Open Solutions, 8*, 8697–8719. DOI: 10.1109/ACCESS.2019.2963774

Janardhana, K., Singh, V., Singh, S. N., Babu, T. R., Bano, S., & Boopathi, S. (2023). Utilization Process for Electronic Waste in Eco-Friendly Concrete: Experimental Study. In *Sustainable Approaches and Strategies for E-Waste Management and Utilization* (pp. 204–223). IGI Global.

KAV. R. P., Pandraju, T. K. S., Boopathi, S., Saravanan, P., Rathan, S. K., & Sathish, T. (2023, November). Hybrid Deep Learning Technique for Optimal Wind Mill Speed Estimation. In 2023 7th International Conference on Electronics, Communication and Aerospace Technology (ICECA) (pp. 181-186). IEEE.

Kim, J. Y., & Oh, J. M. (2022). Opportunities and challenges of metaverse for automotive and mobility industries. *2022 13th International Conference on Information and Communication Technology Convergence (ICTC)*, 113–117.

Lanzotti, A., Vanacore, A., Tarallo, A., Nathan-Roberts, D., Coccorese, D., Minopoli, V., Carbone, F., d'Angelo, R., Grasso, C., Di Gironimo, G., & Papa, S. (2020). Interactive tools for safety 4.0: Virtual ergonomics and serious games in real working contexts. *Ergonomics*, *63*(3), 324–333. DOI: 10.1080/00140139.2019.1683603 PMID: 31648616

Lee, J., Chua, P. C., Chen, L., Ng, P. H. N., Kim, Y., Wu, Q., Jeon, S., Jung, J., Chang, S., & Moon, S. K. (2023). Key enabling technologies for smart factory in automotive industry: Status and applications. *International Journal of Precision Engineering and Manufacturing-Smart Technology*, *1*(1), 93–105. DOI: 10.57062/ijpem-st.2022.0017

Muthukumar, R., Pratheep, V. G. G., Kunduru, K. R., Kumar, P., & Boopathi, S. (2024). Leveraging Fuel Cell Technology With AI and ML Integration for Next-Generation Vehicles: Empowering Electric Mobility. In A Sustainable Future with E-Mobility: Concepts, Challenges, and Implementations (pp. 312-337). IGI Global.

Paul, A., & Thilagham, K. T., KG, J. C., Reddy, P. R., Ravishankar, S., & Boopathi, S. (2024). Multi-criteria optimization on friction stir welding of aluminum composite (AA5052-H32/B4C) using titanium nitride coated tool. *Engineering Research Express*, *6*(1), 015402.

Reddy, P. Y., Madan, B. S., Anandaram, H., Rathod, P., Vasanthaseelan, S., & Boopathi, S. (2024). Analyzing Fuel Cell Vehicles Through Intelligent Battery Management Systems (BMS): AI and ML Technologies for E-Mobility. In *Advances in Web Technologies and Engineering* (pp. 335–355). IGI Global. DOI: 10.4018/979-8-3693-1487-6.ch016

Rzepka, S., Otto, A., Vogel, D., & Dudek, R. (2018). Application-Driven Reliability Research of Next Generation for Automotive Electronics: Challenges and Approaches. *Journal of Electronic Packaging*, *140*(1), 010903. DOI: 10.1115/1.4039333

Saravanan, S., Khare, R., Umamaheswari, K., Khare, S., Krishne Gowda, B. S., & Boopathi, S. (2024). AI and ML Adaptive Smart-Grid Energy Management Systems: Exploring Advanced Innovations. In *Principles and Applications in Speed Sensing and Energy Harvesting for Smart Roads* (pp. 166–196). IGI Global. DOI: 10.4018/978-1-6684-9214-7.ch006

Sekhar, K. Ch., Ingle, R. B., Banu, E. A., Rinawa, M. L., Prasad, M. M., & Boopathi, S. (2024). Integrating VR and AR for Enhanced Production Systems: Immersive Technologies in Smart Manufacturing. In *Advances in Computational Intelligence and Robotics* (pp. 90–112). IGI Global. DOI: 10.4018/979-8-3693-6806-0.ch005

Stark, R. (2022). *Virtual product creation in industry*. Springer. DOI: 10.1007/978-3-662-64301-3

Tubaro, P., & Casilli, A. A. (2019). Micro-work, artificial intelligence and the automotive industry. *Economia e Politica Industriale*, *46*(3), 333–345. DOI: 10.1007/s40812-019-00121-1

Venkateswaran, N., Kiran Kumar, K., Maheswari, K., Kumar Reddy, R. V., & Boopathi, S. (2024). Optimizing IoT Data Aggregation: Hybrid Firefly-Artificial Bee Colony Algorithm for Enhanced Efficiency in Agriculture. *AGRIS On-Line Papers in Economics and Informatics*, *16*(1), 117–130. DOI: 10.7160/aol.2024.160110

Vignesh, S., Arulshri, K., SyedSajith, S., Kathiresan, S., Boopathi, S., & Dinesh Babu, P. (2018). Design and development of ornithopter and experimental analysis of flapping rate under various operating conditions. *Materials Today: Proceedings*, *5*(11), 25185–25194. DOI: 10.1016/j.matpr.2018.10.320

Wolfartsberger, J. (2019). Analyzing the potential of Virtual Reality for engineering design review. *Automation in Construction*, *104*, 27–37. DOI: 10.1016/j.autcon.2019.03.018

Wolfartsberger, J., Zenisek, J., & Wild, N. (2020). Supporting teamwork in industrial virtual reality applications. *Procedia Manufacturing*, *42*, 2–7. DOI: 10.1016/j.promfg.2020.02.016

Yildiz, E., Møller, C., Bilberg, A., & Rask, J. K. (2021). Virtual prototyping: Evaluating the digital twin based virtual factory for new product introduction. *Complex Systems Informatics and Modeling Quarterly*, *29*(29), 1–16. DOI: 10.7250/csimq.2021-29.01

Chapter 3
Smart Sensors-Based Autonomous Vehicle Safety With Auto Alert Using Wireless Network

K. Avinash Babu
Department of Electronics and Communication Engineering, UCET, Mahatma Gandhi University, Nalgonda, India

G. Eswara Rao
Department of Electronics and Communication Engineering, VITAM, Berhampur, India

G. Aparna
Electronics and Communication Engineering, Pentagon Rugged Systems India Limited, Hyderabad, India

N. Sowmya
https://orcid.org/0009-0002-2415-2685
Electronics and Communication Engineering, Pentagon Rugged Systems India Limited, Hyderabad, India

ABSTRACT

A major contributor to accidents is alcohol, and shift workers, business drivers, and truck drivers are the most susceptible to dozing off while operating a motor vehicle. Along with putting lives in danger, these problems put a heavy financial strain on business owners, who are held responsible for these kinds of accidents. In this research, a mobile application-integrated adaptive driver and business owner alert system for tracking driving behaviors is presented. The system uses advanced sensors

DOI: 10.4018/979-8-3693-7287-6.ch003

to detect drunkenness, tiredness, and seat belt usage, triggering a buzzer when risky behaviors are detected. The system uses GPS and GSM modules to communicate the car's location to the designated individual, deactivating the motor if critical thresholds are crossed.This system, controlled by a Cortex M3 microprocessor, proactively addresses major crash causes, reducing accidents, injuries, fatalities, and property damage. The virtual and augmented reality-related future trends in vehicle safety with Auto Alert have also been explained.

INTRODUCTION

One of the biggest causes of accidents is drowsy driving. Road accidents account for about 50% of accidents. Increasingly, traffic accidents are being caused by drowsy drivers. To keep the driver from falling asleep at the wheel, some techniques must be created. The creation of a system for the prevention of this problem has become a significant challenge. Visual analysis of eye condition and head position (HP) was utilized to continuously evaluate driver attentiveness (Mbouna et al., 2013). Using the Raspberry Pi camera and the Raspberry Pi 3 module, the amount of driver sleepiness was measured. The use of visual data and artificial intelligence was implemented in a module for the Advanced Driver Assistance System (ADAS) to decrease the number of accidents caused by driver weariness. The following methods have been used by researchers to assess driver drowsiness: (1) vehicle-based methods; (2) behavioral methods; and (3) physiological methods. The development of a driver drowsiness detection system prototype is the goal of this paper.

The main goals of this device are to keep track of the driver's blink rate and body temperature. Additionally, the driver's heart rate is monitored. The necessary sensors are used to measure these variables. The microprocessor compares the sensor readings to the supplied reference readings. If these values depart from the reference value range, an alarm is given to the driver. The GSM module also notifies the concerned parties about the driver via message(Khunpisuth et al., 2016).

The project is developed for one of the greatest road safety problems, which is due to negligence caused by or impact on the driver. Each year thousands of lives are lost in crashes caused by drowsy driving, drinking, and other basic safe practices like seat belts. These happenings not only bring about human suffering; they're also injuries to individuals, corporations, and states. The list of the most vulnerable includes shift workers, long-distance truck drivers, and business drivers, which is mainly due to their long hours on the road and fatigue. This study demonstrates that smart sensors and wireless networks can be effectively utilized in preventing roadside dangers preemptively, thus obtaining a safe driving environment(Alshaqaqi et al., 2013).

Advancements in technology have enabled systems to track human behavior and monitor environmental conditions in real time. This research explored a driver alert system that would use various cutting-edge sensors to detect hazards and respond similarly. Generally, the entire system gets structured over the most looked-after vital data points, such as the alcohol consumption, fatigue levels, seat belts, etc. These sensors are integrated with the wireless communication technologies of GPS and GSM, through which the system can provide real-time updates in case of an emergency to the relevant personnel (the business owner or emergency contacts) so that quick actions can be taken at the time of the emergency. This approach represents a shift from reactive to proactive safety practices, wherein risks are identified and mitigated early, prior to them becoming life-threatening problems(Sahayadhas et al., 2012).

And consumption of alcoholic beverages is still a top assassin of road smashes across the universe. Despite extensive rules and public awareness campaigns, driving under the influence of drugs and alcohol is a deadly act. Adding on to that is an alcohol detection system that continuously checks the driver's breath for signs of alcohol. If the concentration exceeds a threshold, the alert is escalated. Simultaneously, the vehicle's motor is disabled, stopping any further movement and stopping the threat at the source. This practice ensures that lives are not put at risk—and it also reduces liability for business owners, as they are routinely prosecuted for accidents caused while an employee is engaged in work-related duties. The system can stop the driving of vehicles in case of drunken driving and can have a further impact on the road safety and liability management(Boyraz et al., 2008).

Another important function provided by this system is the detection of fatigue. Driving long hours due to a lack of, or inadequate, rest is a significant factor leading to drowsy driving, which is a leading cause of crashes involving commercial drivers. Sensors monitor its drivers—head position, rapid blinking of the eyes, and steering inputs—to detect early signs of fatigue. Once drowsiness is identified, the system can sound a buzzer to alert the driver and bystanders. By making their condition known to the driver, the driver has the opportunity to stop or seek assistance. Also, the online notice through GSM, which is received in real time by the business owner(s) or person(s) responsible, is an extra protection measure, as well as having to take some action(Saini & Saini, 2014).

Seatbelt usage is a common driver safety protocol that is often overlooked. Musicians have been shown again and again to significantly improve their chances of getting out of serious, or even fatal, accidents by simply seat belting. The system also monitors whether the driver and passengers are buckled up. In case the passenger is not buckled up for safety, the system gives a warning. If the driver does not comply, the alerting can be joined by communications classifies when inquiring wireless to promote the alert to the established contact. It also ensures compliance with the most basic but effective safety practices while developing a culture of responsible driving(Singh & Kaur, 2013).

The addition of GPS and GSM modules enables the system to retrieve the needed information, thus providing tracking and real-time communication. Knowing the car's whereabouts on the map is beneficial—particularly in emergencies. Where one is driving and exhibiting unsafe behavior, or the system detects a crash, location coordinates are automatically sent to the nominated person or organization. This kind of communication is used to send emergency service vehicles, notify business owners, or assist law enforcement with locating the vehicle. As GSM technology is utilized for this purpose, alerts will reach the desired recipients regardless of where the vehicle is located, making the system robust and functional for the most diverse set of use cases(Lee & Chung, 2012).

The system is directed in real time by the Cortex M3 microprocessor, the key controller to which the various elements of the set are integrated. The microprocessor processes the data from the sensors in real time and makes rapid decisions to set off alerts, turn off the motor, or send notifications. This low power drain, along with high processing capability, makes it an ideal candidate for the application since the system needs to be available always, without tapping the vehicle resources. This also shows how hardware selection is a critical factor in defining the system characteristics—performance versus reliability versus cost—that are available for an application, also demonstrated by a Cortex M3 microprocessor (Lenskiy & Lee, 2012).

The system not only addresses the technical complexity of recognizing risky driving behaviors and how to respond to them, it also may bring up concerns on broader scales of how application road safety & accident prevention can be handled. Instead of using computing power to study every little aspect of a crash scene, it zooms in on the biggest contributors to accidents; alcohol use is almost always the most common cause, along with fatigue and seatbelt use. Such a system with many layers might help reduce injury, death, and the economic costs of road injuries and deaths. Top reasons businesses will want to implement the technology are reduced liability, increased employee safety, and reduced operational costs related to fewer company vehicle incidents(Churiwala et al., 2012).

Hence, this study proposes a novel system that can enhance vehicle safety by integrating smart sensors and wireless technology (in the form of USW technology)—the latter having recently been shown in the laboratory (and proposed as a viable solution to this dangerous issue). Specifically, this system is touted as a preventative measure for some of the most common reasons for collisions, with features like alcohol detection, fatigue monitoring, seatbelt enforcement, and live tracking all rolled into one. The integration of GPS, GSM, and Cortex M3 microprocessors makes the system efficient as well as reliable(Dasgupta et al., 2013; Ueno et al., 1994). The new driver alert system is a huge breakthrough in the prevention of accidents on the stretch of road, as it can not only save lives and avoid injuries but also avoid financial losses. All of these technologies will advance us into an era of safer roads for everyone.

RESEARCH BACKGROUND

Road safety devices play an important role in saving human lives; thus, they need to be implemented in an environment such as cars, trucks, and various vehicles running on the road. One of the main reasons for deaths in the world still continues to be road accidents, which are mostly caused by human errors such as driving under the influence of alcohol, fatigue, and negligence towards wearing safety measures like seatbelts. The issue is particularly acute among shift workers, long-haul drivers, and business vehicle operators, who are frequently put on strenuous schedules, deprived of downtime, and subjected to high-stress pressures. All of these factors together lead to higher chances of dangerous driving habits, which not only put lives at risk, but also lead to various financial as well as legal problems for both individuals and companies(Alsayaydeh et al., 2023).

In recent years, smarter technologies have been developed that provide the next generation of driver monitoring systems that are capable of tracking driver attentiveness in real time and taking action to prevent dangerous conditions. Integrating alcohol detection, fatigue monitoring, and seatbelt monitoring sensors with GPS and GSM wireless communication technology as an approach to lessen road hazards. These technologies enable the identification of unsafe practices in real time, the generation of alerts, and the tracking of the physical space to facilitate preventive measures in case of an accident. Such systems ensure complete coverage, resulting in solutions that improve road safety by identifying the cause of crashes, thereby preventing the emotional and financial trauma associated with inculpating others in traffic collisions.

Car makers and safety researchers have spent significant effort on implementing artificial intelligence, IoT-based sensors, and microcontrollers to improve safety standards in vehicles and help with safe riding. Based on a Cortex M3 microprocessor, the system indicates a robust controller, enabling the processing of several sensors' data accurately and promptly. It enables the system to detect the signs of impairment or drowsiness or seatbelt non-compliance and react by sounding a buzzer, alerting concerned persons, or switching off the motor of the vehicle. By integrating with GPS and GSM modules for wireless communication, essential information can be communicated to appropriate users immediately, forming an effective action method that extends beyond geographic limitations(Lin et al., 2015).

In this chapter you are trained on the understanding of "smart sensors, real-time data processing, and adaptive driver alert systems using wireless communication." The research aims not to be purely theoretical and practical in understanding these technologies but to contribute to the greater good of developing safer roadways and mitigation for poor decision-making by drivers resulting in accidents. The driver status monitoring system is depicted in Figure 1.

Figure 1. Driver status monitoring system

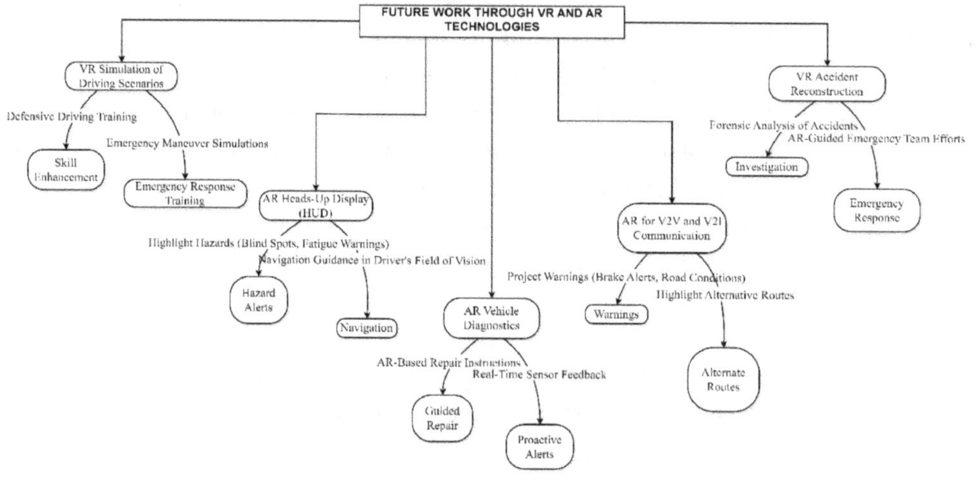

SYSTEM ANLAYSIS

The system that was proposed combines multiple state-of-the-art technologies to combat the leading causes of accidents on the road, which include alcohol, fatigue, and non-use of seatbelts. This study is a systematic review to assess its functioning, effectiveness, and road safety implications. In this section, we will explain the details of the components, operations, and performance of the system to fulfill the desired goals of the system, which are accident prevention and improvement of the driver's behavior(Biswas et al., 2006).

The control unit is a Cortex M3 microprocessor that communicates with the rest of the system components. This microcontroller then processes the real-time data from various sensors like the alcohol detection sensor, fatigue detection module, and seatbelt compliance sensors. The alcohol detection sensor checks the breath of the driver for signs of alcohol, the fatigue module checks for fatigue by looking for eye blinking and head nodding, while seatbelt sensors check for the seatbelt.

When any unsafe condition is detected, a microprocessor quickly analyzes the data and can trigger appropriate warnings such as a buzzer or motor disabling. It is interfaced by GSM and GPS modules to enrich the features of the system. The GPS module is used to find the location of the vehicle, and the GSM module is used to communicate to the concerned person (business owner or any emergency contact). If the driver exhibits behavior that suggests they are driving unsafely—by getting behind the wheel under the influence of drugs or alcohol, failing to secure a seatbelt, or getting drowsy—the system sends an alert and shares the location of the vehicle. This guarantees relevant influencers must act to avoid potential disasters in a timely manner.

Most likely, its greatest strength is that it gets ahead of the curve. The system isn't waiting for accidents to happen before it reacts; it simply monitors how the driver is performing on the road and tries to prevent risk from getting to that point. For instance, the motor is disabled, and the vehicle cannot move if the alcohol detection sensor detects harmful levels of alcohol. Fatigue detection works just like this, giving timely alerts and advising the driver to take a break. The technology is scalable and can even be adapted for personal and commercial vehicles. Through real-time processing of said data, facilitating good wireless communication as well as enabling automated responses, it proves to be an effective and powerful method to reduce accidents. This system is leveraging all major risk-contributing factors and their corresponding system approaches to facilitate a broad-based but effective movement that would be towards making roads safe and adjusting to responsible driving behavior to achieve the goal of enhanced driving safety and road safety.

Limitation of Existing System

The support vector machine (SVM) splits a series of video clips into driving events that are alert or not alert. It is used in the current systems to identify tiredness. The other approaches depend on artificial intelligence and visual intelligence(Singh & Kaur, 2013). The frequency of head tilting, face recognition, and eye blinking employing image processing are other factors that affect it.

Problem Definition

The driver's heart rate, body temperature, and the rate at which his or her eyes blink are all measured with the right sensors, such as a heart rate sensor, a body temperature sensor, and an eye blink sensor. Using things like a heartbeat sensor and a temperature monitor, the driver's health state is being monitored. In an emergency, the location of the driver and vehicle can be determined using a GPS module(Churiwala et al., 2012).

Proposed System Features

Here, any car can quickly and easily integrate the suggested system. The driver's eye blink sensor has been installed using eyewear. The eye twitch sensor can detect eye movement. The sensor output is connected to a microcontroller. This prototype's propulsion is a DC motor. The microprocessor regulates the motor directly. If the sensor detects that the ocular is not moving, it will send a signal to the microcontroller. The microcontroller sends out a warning signal and uses an LCD to display the cause. If the eyeball does not move repeatedly, the microcontroller quickly shuts off the engine. The system also makes use of temperature and heartbeat sensors(-Li et al., 2020). Since the sensor's outputs are analog, we must utilize an ADC to transform them into digital signals that a microcontroller can process. Initially, The microcontroller stores all sensor values and phone numbers. The microcontroller sends GSM location data to the saved number when any of these three parameters falls beyond the specified range (Dasgupta et al., 2013).

Figure 2. Block diagram for the proposed methodology

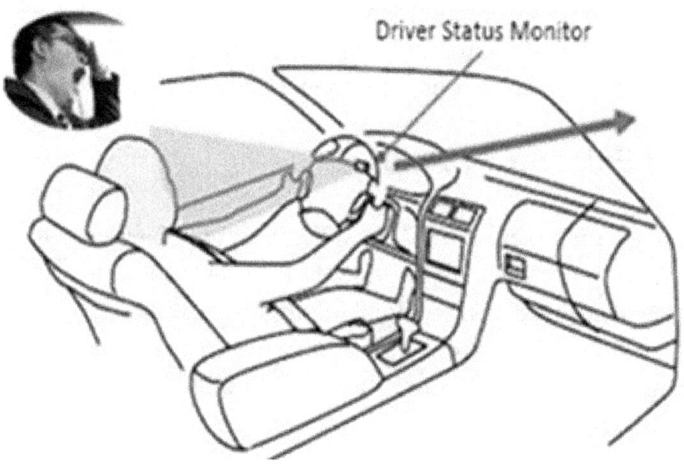

METHODOLOGY

There are countless truck accidents every year, which result in serious injuries and fatalities, costly insurance claims, and protracted traffic backups as the wreckage is cleaned. The driver is frequently in the news after a serious collision involving a commercial truck. Due to the size and weight of trucks, the driver must be highly skilled and laser-focused in order to manage these gigantic vehicles (Lee & Chung, 2012). Commercial truck accidents usually result in substantial claim amounts because of the severity of the property damage and injuries. Blind areas, also referred to as "no zones" in the trucking industry, are one of the most frequent mistakes made by truck drivers when turning or changing lanes. They are on each of the four sides of the semi, and a lot of collisions occur when the truck driver maneuvers into the spaces without first thoroughly examining the space for clearance(Churiwala et al., 2012). Driving the rig defensively is a component of the truckers' "expanded duty" to safeguard us because they are operating a hazardous vehicle. Other common mistakes made by truck drivers include failing to pay attention to their surroundings, speeding, not knowing their way around, becoming exhausted, and operating a vehicle while impaired by drugs or alcohol. This idea comprises preventing accidents, preserving the lives of the drivers, and solving the problem of the owner having to make repairs each time, even when the driver is at fault. Figure 2 depicts the chain of events leading to truck accidents, their consequences, causes, and the strategies used to address these issues.

Every year, thousands of truck accidents cause devastating injuries, fatalities, property damage, and financial loss. The wrecks result in expensive insurance claims and extended traffic delays while debris is cleared. The truck driver is tentatively blamed, and of course, if it is a commercial truck, this is usually public. These are vehicles so large and so heavy that piloting them safely and effectively takes a tremendous amount of skill, attention, and awareness. But many accidents stem from avoidable mistakes, which makes a robust system for monitoring and intervention all the more essential(Kumar et al., 2019).

Blind spots are one of the top causes of accidents involving commercial trucks, and they are referred to as "no-zones" by the trucking industry. Semis have blind spots on all four sides (front, rear, right, left) that impede a driver from seeing smaller vehicles or objects in close proximity. Several of these accidents occur when the truck driver takes another vehicle by surprise because the driver failed to confirm that there was sufficient space for the truck. This problem is compounded by a lack of situational awareness and poor defensive driving practices. When you are behind the wheel of such a dangerous vehicle, truck drivers have a much higher responsibility to the people on the road around them, and you will learn to drive defensively. Failure to HONOR this responsibility causes severe accidents that lead to large claims due to either severe injury and/or damage.

The trucking mistakes mentioned are responsible for many accidents. These are too fast, not paying enough attention to what is around you, getting lost without learning the lay of the land, driving while you are asleep, and drugs or alcohol. It is particularly challenging for long-haul drivers who spend long hours on the job without sufficient rest. In addition, a constant road peril is driving while creamy whatever is ticketed to impairing in liquor or drugs. However, to combat this challenge, there is a need for a unified system that can detect and respond to risky activities while they occur to minimize accidents and enhance road safety for all. This framework uses advanced sensors, wireless communication, and real-time processing that all act in concert to stop accidents before they happen and keep drivers and other users safe. Smart Sensors to Check Drivers: The system deploys alcohol detection sensors to screen the drivers for their impairment, fatigue detection mechanisms to identify signs of tiredness on the face of the driver, and seatbelt sensors to ensure that the driver takes the basic safety measures. Solving these major issues, the system aims to prevent accidents rather than react to them after they have occurred.

To improve the efficiency of the system, GPS and GSM modules are implemented for tracking as well as returning the current status of the system. Not only does the system set off local alarms, e.g., a buzzer, if any dangerous actions are detected, but it also sends up alarms to designated personnel, e.g., fleet managers or emergency contacts. This attribute permits the timely intervention, which makes getting things to be corrected Conclusion In extraordinary situations, like if the driver was drunk or had neglected the safety precautions, the system can even shut off the motor of the vehicle itself. By bringing these technologies together within a single methodology, the truck accident prevention framework addresses not only accident prevention but also finance and operations problems faced by truck owners and fleet managers. The system reduces repairs and liability claims as they ensure that drivers comply with safety standards and defensive driving. So, based on the visit of this exclusive solution, the number of commercial truck accidents will reduce, thus helping to save lives and reducing damage as well as making highways and paths safer.

SYSTEM DESIGN

A 230V, 50Hz AC signal is supplied to the step-down transformer from the main supply board. The transformer is chosen based on the output voltage range of 10 to 12 volts. The primary purpose of this power supply is to produce +5V voltage (Dasgupta et al., 2013). The proposed block diagram of the system design is depicted in Figure 2.

The transformer receives an input voltage of 230 volts, which it uses to scale down the input voltage. In the beginning, a capacitor filter is utilized to convert alternating current voltage to direct current voltage (Abdulwaheed & MASHOOD, 2019; Alvi et al., 2020). Most of the time, this DC voltage has some wave or AC voltage changes. A regulator circuit is employed to reduce this ripple voltage, resulting in a controlled DC output. To generate voltage regulation, the appropriate number of voltage regulation IC units should be employed.

Architecture of a smart sensor vehicle safety The proposed wireless vehicle safety system architecture is because the various components are robust and efficient, and all the components should be running simultaneously. This necessitates a constant power supply to power the sensors, processing units, and communication modules that identify unsafe driving behaviors and respond accordingly. Since the power supply circuit is one of the critical paths of a system, this chapter describes how the power supply circuit for the project works.

The main supply voltage is a 230V, 50Hz AC voltage, which is at the input of the circuit. This AC signal is then filtered down using a step-down transformer mode, dependent on the jump up between the voltages across the transformer. The transformer is selected to provide an output of 10 to 12V, which must be appropriate to the next stage of rectification and regulation. The power transferred to the power grid will be the one handled by the step-down transformer, which is necessary to lower the voltage and also to avoid losses through energy transfer so that it can be used in future processing. The proposed block diagram of the system design is depicted in Figure 3.

That means even after it has stepped down the voltage, it'll be outputting an AC shape from the transformer, which is not convenient for normal powering of the electronic components of the model. To correct this, we use a correction process. Usually, a bridge rectifier is used to convert the AC voltage to pulsating DC voltage. But this throbbing DC has peaks, or ripples, that can make working with delicate components like microprocessors and sensors challenging. Therefore, DC output needs to be processed using filtering and regulation stages to maintain constant output.

Step 1: The first step in the process is to convert the rectified DC voltage to the smooth DC voltage, which is done with the help of a capacitor filter. This filter stores electrical energy, returning it slowly to smooth the ripples in the pulsating DC voltage. This is where our capacitor filter will come in; it will filter out the AC component from the voltage to make the output DC as smooth as possible. However, filtering can still leave behind some residual ripples. To eliminate these ripples and provide a very stable and regulated DC voltage, a voltage regulator circuit is incorporated into the design. The Application of the Voltage regulators are one of the core components in electronics power supply circuits that enable getting a stable output voltage in alternate input voltage or load scenarios. In this system, a voltage regulator IC (like the 7805 IC) provides the regulated +5V DC output. The 7805 IC is a 3-term regulator that regulates voltage to a constant 5V, useful in driving the Cortex M3 microprocessor, sensors, and other modules. This is an IC; this helps in stabilizing the voltage, as well as protecting the components from the voltage spikes.

Transducers convert one form of energy into another. This is how the DC circuit produces a stable +5V DC, which is shared in the system. Because of this regulated voltage, it ensures the sensor operates correctly, the microprocessor can process the data efficiently, and the communication modules can transmit the signal without any glitches. The system design shows that the power supply stability is the main condition for the reliability and effective implementation of the smart sensor-based vehicle safety system. Thus, by way of such a power supply circuit and design solution, the system will be able to deliver sufficient power and adjustability to reduce accident rates and enhance the complete road performance along with a complex recognition and action mechanism.

Figure 3. Proposed block diagram of the system design

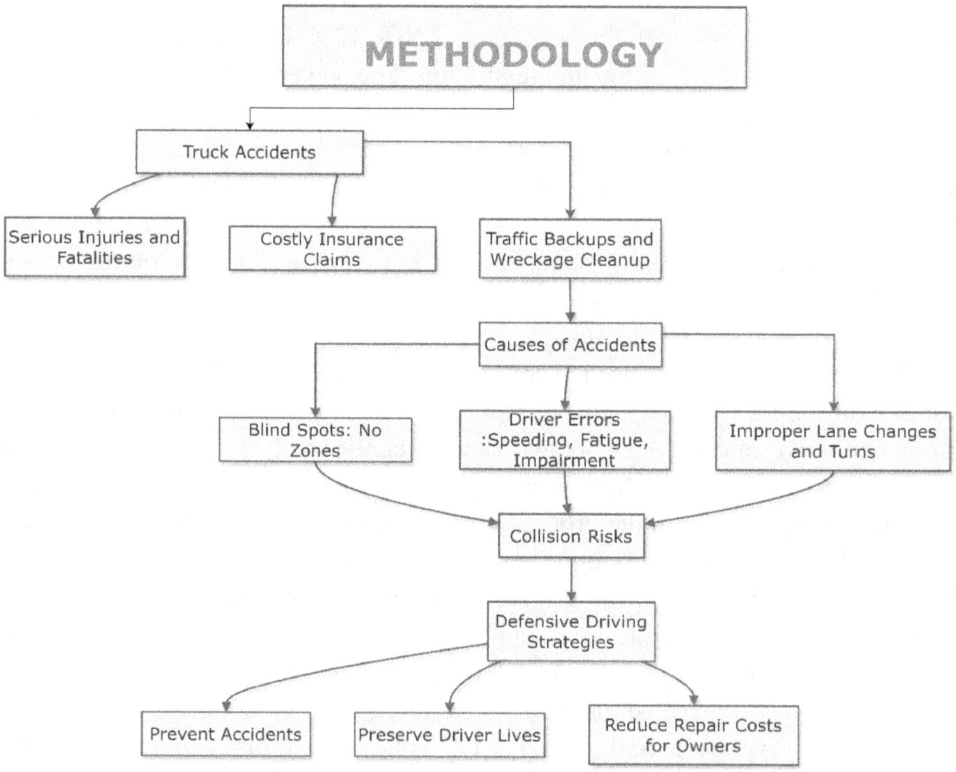

Cortex M3

Microcontrollers, car body systems, industrial control systems, wireless networking, and sensors are some of the many gadgets available that may be built on platforms with high performance and low cost thanks to the Cortex-M3 processor, which was developed expressly to support this (Smith, 2008). The quickest, easiest, and risk-free path to bespoke silicon success is offered by Arm Design Start.

- Create the best System-On-Chip possible, complete with system interfaces, integrated debug and trace components, and a CPU that perfectly balances space, performance, and power.
- Create solutions using a full-featured Armv7-M instruction set that has been proven in a variety of embedded applications for a variety of markets.
- Capture a global developer community with considerable experience to hasten the adoption of new Cortex-M3-powered devices and to save support costs.
- Due to robust clock gating, integrated software-controlled sleep modes, and customizable state retention, you may get outstanding 32-bit performance with minimal dynamic power.

Powerful debug and non-intrusive real-time trace: Developer productivity is significantly increased with comprehensive debug and trace features. Embedded software development is quite effective when done properly.

Memory Protection Unit (MPU): When each module is only given access to the exact parts of memory that it needs to function, Software dependability increases. This safeguard stops unauthorized access that might replace important data.

Integrated nested vectored interrupt controller (NVIC): An independent external interrupt controller is not required. The NVIC handles interrupt handling, eliminating the burden of manually managing interrupts via the processor.

Thumb-2 code density: When compared to 8bit and 16bit architectures, a mix of 16bit and 32bit instructions results in a higher code density (Khunpisuth et al., 2016). With fewer memory requirements and better use of priceless on-chip Flash memory, this provides many benefits.

GPS

Civilians can use the space-based radio navigation system GPS for free worldwide positioning, navigation, and timing. GPS receivers offer location and time. In any weather, day or night, and for an infinite number of users, GPS offers precise location and timing information (Santamaria et al., 2014; Sreevishakh & Dhanure, 2015).

Three components make up the GPS:

- Satellites orbiting the Earth
- Control and monitoring stations on Earth
- The GPS receivers owned by users.

GPS receivers pick up and recognize the signals that GPS satellites broadcast from space. Each GPS receiver sends time, latitude, longitude, and altitude.

GSM

Voice and data services for mobile phones are provided by the open, digital GSM network. GSM lets you make talk calls, send SMS messages, and send data at up to 9.6 kbit/s. In Europe, 900MHz and 1.8GHz are used for GSM, while in the US, 1.9GHz and 850MHz are used. Australia, Canada, and many South American countries use 850MHz for GSM and 3G. Harmonization of a lot of the world's spectrum has made it possible for GSM's foreign roaming feature to let customers use the same services when they travel as when they are at home (Fogue et al., 2012; Santamaria et al., 2014). Customers can now connect smoothly and on the same number in over 218 countries.

GSM is mainly built on 3 building blocks as shown in below figure

- GSM Radio Network – The system's signaling is addressed in this. There are handovers in the radio network. An assortment of frequency channels is allotted to each BTS.
- GSM Mobile switching Network - The storage of data necessary for routing and service provision is a concern for this network.
- GSM Operation and Maintenance – Its duties also include managing security, network configuration, operation, and performance, as well as administrative and commercial operations.

Buzzer

An electrical, electromechanical, or mechanical bell or beeper is a sound-flagging device. Ringers and beepers are often used as alarms, clocks, or to confirm user input like a mouse click or keystroke. When opposition and arrangement are combined (as shown in diagrams (an) and (b)), abnormal trembling may occur when adjusting the sound volume. Put a capacitor in parallel with the voltage wavering board in this case, as shown in depiction (c). So, by establishing one side, unexpected wavering can be stopped. However, the voltage VB added to the voltage wavering board must be within the maximum information voltage range, and a capacitance of at least 3.3 F must be connected. A swinging electrical circuit or another source of sound signals could be used to drive a piezoelectric component. Ticks, rings, or signals are frequently employed to indicate that a catch has been squeezed. Today, there are many uses for electronic signals.

Alcohol Sensor

In pure air, SnO_2, the MQ-3 gas sensor's sensitive component, exhibits minimal conductivity. The conductivity of the sensor goes up as the goal alcohol gas concentration goes up. Please utilize a simple electrical circuit to signal gas concentration-dependent conductivity changes. The MQ-3 petrol sensor detects alcohol well and resists fuel, smoke, and vapor. It can detect different alcohol concentrations and is cheap and versatile.

Eye Blink Sensor

The sensor used to determine whether an eye is open or closed is called an eye blink sensor. A drowsy driver can be identified in a variety of ways. An eye blink sensor is being used in this case. An IR-based eye blink sensor is what it is. It operates on the same principles as an IR sensor. The sensor's output alters if an obstruction is found. The output also fluctuates depending on eye blinks. To determine if the driver is aware or sleepy, we are employing an eye blink sensor in this research. When one eye blinks repeatedly for a predetermined amount of time, it is assumed that the driver is tired. This sleepiness could result from a lack of Eye Gerber eye blink sensors that were utilized in this research. It has three pins: a signal pin, a ground pin, and a Vcc (supply) pin. This eye blink sensor produces a digital output.

RESULT ANALYSIS

It is created a platform based on the Cortex M3 and Keil programming languages to interact with the sensor tags. Despite being modest, it managed to collect the majority of the data that we needed. By adding essential segments and enhancing the functionality and usefulness of the C language, the ability to embed Keil code is possible. A worldwide community is still developing programming languages, by the way. As a result, perhaps one day we won't need to include any C code. Our objective at this point was to construct a prototype sensor-using gadget, comprehend how sensors operate and interact with one another, and gather concepts for the project's future development. Figure 4 displays the output monitoring on a web page.

Figure 4. Output monitoring on web page

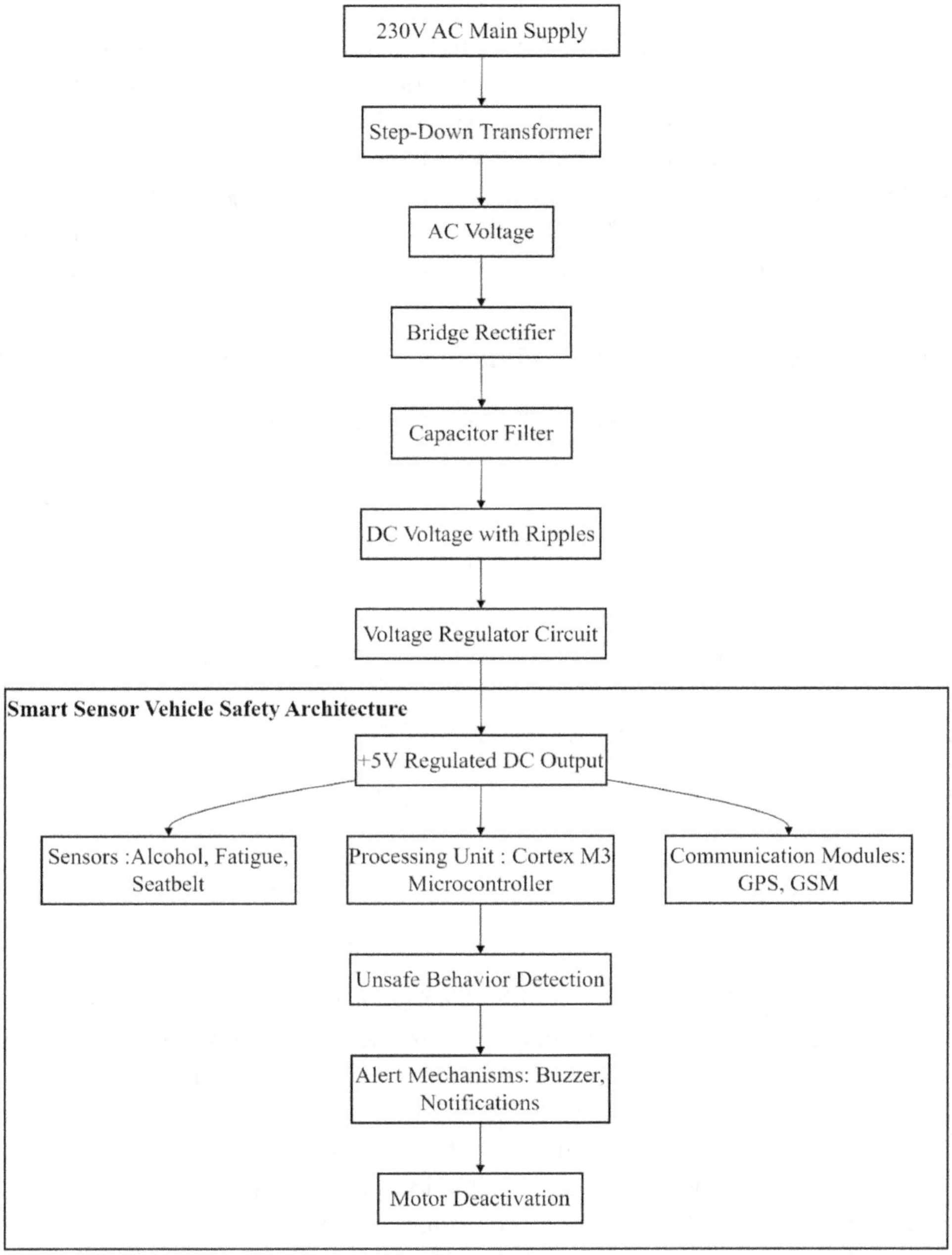

FUTURE WORK THROUGH VR AND AR TECHNOLOGIES

state of the art in VR and AR tech is rapidly evolving, and it opens new doors to advance vehicle safety systems and driver-environment interaction. Utilizing VR and AR in future versions of the smart sensor-based vehicle safety system opens a whole new realm of safety, awareness, training, and more. These immersive technologies are able to tackle existing system drawbacks and take their performance to an entirely next level(Alsayaydeh et al., 2023). Figure 5 illustrates potential future collaborations with AR and VR in the Smart Sensors-Based Vehicle Safety System context.

Figure 5. Future work with AR & VR in the context of the smart sensors-based vehicle safety system

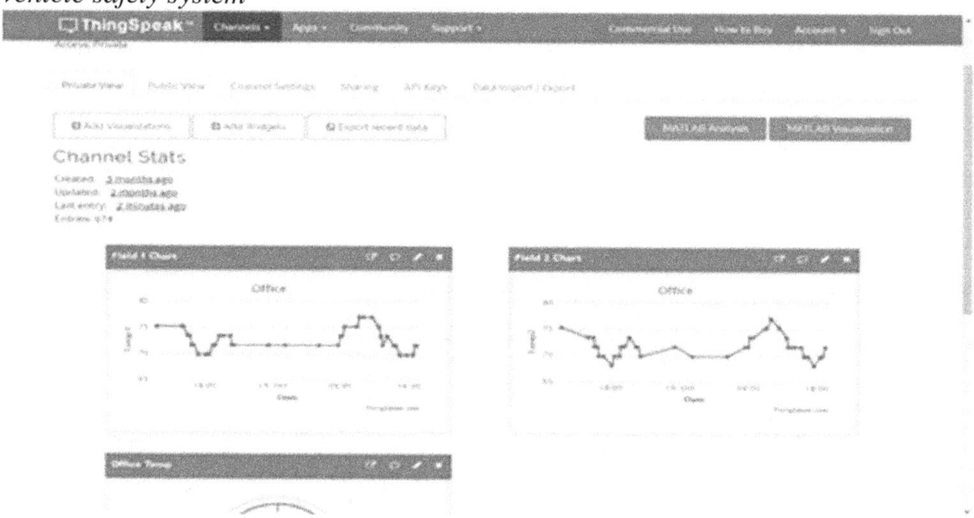

A potential application of VR within this field is the development of advanced driver training simulators. Traditional training methods tend not to have drivers ready for the most extreme situations and for unusual trail conditions. In a safely engineered environment lie replicas of real driving moments—fantastically enshrined in a VR simulator. Drivers are trained in defensive driving, emergency maneuvers, and responses to potentially dangerous situations, like unexpected vehicle failures and extreme weather. This technology could also be used to develop immersive simulations that let them skip straight to the good part and tailor the gear to information issues for truck drivers, such as blind spots and fatigue on long hauls. This could significantly lower the risk of accidents, as, with realistic feedback and performance metrics, driver interests and awareness can be enhanced using VR-based training (Sekhar et al., 2024).

while AR technologies may contribute to setting in place driving data in order to raise real-time situational awareness whilst in motion. By overlaying critical details right in the driver's line of sight, AR encourages reducing distractions and assists with timely decisions. For instance, you could have navigation directions, speed restrictions, and warnings about nearby cars or pedestrians overlaid directly on the glass with an AR-enabled head-up display (HUD). This allows drivers to pay attention to the road ahead while still being aware of their environment. AR systems can also be used to alert the driver to hazards, such as vehicles in the driver's blind spots, or warn the driver about when they are feeling drowsy, whether they are wearing their seatbelt, or when alcohol is detected. Now since augmented reality can provide the users with intuitive and contextually relevant information, this can avoid user behavior that can be risky and can have a positive impact on road safety.

AR inspecting and diagnosing cars. Such as AR glasses or mobile apps that provide drivers and technicians with overlaid step-by-step instructions for performing repairs on real-world elements of a vehicle. Fleet owners can leverage this technology. The sensor system is so useful for eliminating mechanic check-ups by providing real-time diagnosis through augmented reality (AR), a better diagnosis of pump temperature (engine temperature), and real-time diagnosis of wheel pressure (EX) bird, which are the basis for eliminating breakdowns. Moreover, advanced vehicle-to-vehicle (V2V) and vehicle-to-infrastructure (V2I) communication can be integrated into augmented reality (AR) and virtual reality (VR) to make accident prevention easier. This data, sent from other vehicles or infrastructure to AR systems, allows GNCs to visualize warnings in advance of the road conditions, traffic congestion, or oncoming emergency vehicles. If a car suddenly brakes in front of the driver, for example, the AR system can issue a warning for the driver to see on their windshield, giving the driver time to respond. Thus, in soil closures/traffic snarls, AR can suggest alternative routes with minimal lag in time and suboptimal use of transport networks.

Future developments in the space of VR and AR can also change the way people respond to an accident. AR also could assist first responders who must collide by giving them real-time data about the vehicle's occupants, which can help them find out where they could be in the car, their medical history, or how severely they may have been injured. Rescue teams with AR could use this information to deprioritize or expedite their rescue efforts, possibly saving lives. Additionally, VR can be used to recreate accident scenes for forensic study, allowing for the investigation of causation and the identification of factors involved.

Beyond single vehicles themselves, AR and VR can also have an impact on how urban design and infrastructure are considered. And by visualizing the impact of new road layouts, intersections, or signage on driver behavior, an AR simulation can be used to determine the impact of proposed routes on safety and efficiency for city planners. Through simulating traffic flows and maintaining the interaction between pedestrians and vehicles, VR aids urban planners in refining and testing safety measures before they come into force, leading to the making of urban environments tailored around safety and efficiency. From data collection to hardware design, these technologies will face hurdles to overcome as they move into the vehicle safety space. In spite of these challenges, the future of VR and AR in automotive is bright. These innovations could supplement existing smart sensor systems with deeply immersive and interactive tools that make your drive safer, more intuitive, and most cost effective. Long-term, this could lead to broader use cases for VR (Virtual Reality) and AR (Augmented Reality) in terms of reducing road safety issues.

CONCLUSION

This work analyzes and designs a system for alerting and detecting driver drowsiness. The suggested approach is utilized to prevent serious accidents that are brought on by driver weariness and sleepy driving. The device has an eye blink sensor that assesses whether the eyes are open or closed and a heart rate sensor that measures the heart rate every minute. The buzzer is activated to warn the driver when the parameter value exceeds the threshold value. As a result, utilizing such a system can help minimize accidents brought on by tiredness.

Instead of employing an alert, we can use an autonomous braking system to slow the car down. By using the automatic braking system, the vehicle's speed will be first decreased and the parking lights will be turned on at the same time. If a driver is drowsy, the automated braking system can be configured to use a pressure sensor. If the car is stopped in between, it can be transported from one location to another via the CAN protocol. The virtual and augmented reality-related future trends in vehicle safety with Auto Alert were also explored.

REFERENCES

Abdulwaheed, M. U. S. A., & Mashood, S. (2019). Vehicle tracking and accident alert system using gps and gsm modules.

Alsayaydeh, J. A. J., Yusof, M. F., Abdillah, M. A. A., Al-Gburi, A. J. A., Herawan, S. G., & Oliinyk, A. (2023). Enhancing Vehicle Safety: A Comprehensive Accident Detection and Alert System. *International Journal of Advanced Computer Science and Applications, 14*(11). Advance online publication. DOI: 10.14569/IJACSA.2023.0141104

Alshaqaqi, B., Baquhaizel, A. S., Ouis, M. E. A., Boumehed, M., Ouamri, A., & Keche, M. (2013). Driver drowsiness detection system. *2013 8th International Workshop on Systems, Signal Processing and Their Applications (WoSSPA)*, 151–155.

Alvi, U., Khattak, M. A. K., Shabir, B., Malik, A. W., & Muhammad, S. R. (2020). A comprehensive study on IoT based accident detection systems for smart vehicles. *IEEE Access : Practical Innovations, Open Solutions, 8*, 122480–122497. DOI: 10.1109/ACCESS.2020.3006887

Biswas, S., Tatchikou, R., & Dion, F. (2006). Vehicle-to-vehicle wireless communication protocols for enhancing highway traffic safety. *IEEE Communications Magazine, 44*(1), 74–82. DOI: 10.1109/MCOM.2006.1580935

Boyraz, P., Acar, M., & Kerr, D. (2008). Multi-sensor driver drowsiness monitoring. *Proceedings of the Institution of Mechanical Engineers. Part D, Journal of Automobile Engineering, 222*(11), 2041–2062. DOI: 10.1243/09544070JAUTO513

Churiwala, K., Lopes, R., Shah, A., & Shah, N. (2012). Drowsiness detection based on eye movement, yawn detection and head rotation. *International Journal of Applied Information Systems, 2*(6), 45–49.

Dasgupta, A., George, A., Happy, S., & Routray, A. (2013). A vision-based system for monitoring the loss of attention in automotive drivers. *IEEE Transactions on Intelligent Transportation Systems, 14*(4), 1825–1838. DOI: 10.1109/TITS.2013.2271052

Fogue, M., Garrido, P., Martinez, F. J., Cano, J.-C., Calafate, C. T., & Manzoni, P. (2012). Automatic accident detection: Assistance through communication technologies and vehicles. *IEEE Vehicular Technology Magazine, 7*(3), 90–100. DOI: 10.1109/MVT.2012.2203877

Khunpisuth, O., Chotchinasri, T., Koschakosai, V., & Hnoohom, N. (2016). Driver drowsiness detection using eye-closeness detection. *2016 12th International Conference on Signal-Image Technology & Internet-Based Systems (SITIS)*, 661–668.

Kumar, S. V., Vadivel, M., Penchalaiah, U., Ganesan, P., & Somassoundaram, T. (2019). Real Time Embedded System for Automobile Automation. *2019 IEEE International Conference on System, Computation, Automation and Networking (ICSCAN)*, 1–6. DOI: 10.1109/ICSCAN.2019.8878820

Lee, B.-G., & Chung, W.-Y. (2012). Driver alertness monitoring using fusion of facial features and bio-signals. *IEEE Sensors Journal, 12*(7), 2416–2422. DOI: 10.1109/JSEN.2012.2190505

Lenskiy, A. A., & Lee, J.-S. (2012). Driver's eye blinking detection using novel color and texture segmentation algorithms. *International Journal of Control, Automation, and Systems, 10*(2), 317–327. DOI: 10.1007/s12555-012-0212-0

Li, H., Zhao, G., Qin, L., Aizeke, H., Zhao, X., & Yang, Y. (2020). A survey of safety warnings under connected vehicle environments. *IEEE Transactions on Intelligent Transportation Systems, 22*(5), 2572–2588. DOI: 10.1109/TITS.2020.3026309

Lin, J.-R., Talty, T., & Tonguz, O. K. (2015). A blind zone alert system based on intra-vehicular wireless sensor networks. *IEEE Transactions on Industrial Informatics, 11*(2), 476–484. DOI: 10.1109/TII.2015.2404800

Mbouna, R. O., Kong, S. G., & Chun, M.-G. (2013). Visual analysis of eye state and head pose for driver alertness monitoring. *IEEE Transactions on Intelligent Transportation Systems, 14*(3), 1462–1469. DOI: 10.1109/TITS.2013.2262098

Sahayadhas, A., Sundaraj, K., & Murugappan, M. (2012). Detecting driver drowsiness based on sensors: A review. *Sensors (Basel), 12*(12), 16937–16953. DOI: 10.3390/s121216937 PMID: 23223151

Saini, V., & Saini, R. (2014). Driver drowsiness detection system and techniques: A review. *International Journal of Computer Science and Information Technologies, 5*(3), 4245–4249.

Santamaria, A. F., Sottile, C., De Rango, F., & Voznak, M. (2014). Road safety alerting system with radar and GPS cooperation in a VANET environment. *Wireless Sensing, Localization, and Processing IX, 9103*, 120–133.

Sekhar, K. Ch., Ingle, R. B., Banu, E. A., Rinawa, M. L., Prasad, M. M., & Boopathi, S. (2024). Integrating VR and AR for Enhanced Production Systems: Immersive Technologies in Smart Manufacturing. In *Advances in Computational Intelligence and Robotics* (pp. 90–112). IGI Global. DOI: 10.4018/979-8-3693-6806-0.ch005

Singh, K., & Kaur, R. (2013). Physical and physiological drowsiness detection methods. *International Journal of IT. Engineering and Applied Sciences Research, 2*, 34–43.

Smith, A. E. (2008). *Emergency Vehicle Wireless Routing and Notification System*. SAE Technical Paper.

Sreevishakh, K., & Dhanure, S. (2015). A Review Paper on Automotive Crash Prediction and Notiication Technologies. *2015 International Conference on Computing Communication Control and Automation*, 999–1002. DOI: 10.1109/ICCUBEA.2015.197

Ueno, H., Kaneda, M., & Tsukino, M. (1994). Development of drowsiness detection system. *Proceedings of VNIS'94-1994 Vehicle Navigation and Information Systems Conference*, 15–20. DOI: 10.1109/VNIS.1994.396873

Chapter 4
Virtual Reality Training for Technicians and Factory Workers for Skill Development in Automotive Sectors

S. Pushpa
Department of Computer Science and Engineering, St. Peter's Institute of Higher Education and Research, Chennai, India

R. M. Sathiyamoorthy
Department of Mechanical Engineering, Easwari Engineering College, Ramapuram, India

S. Anita
Department of Science and Humanities (General Engineering), R.M.K. Engineering College, Kavaraipettai, India

R. Sathishkannan
Department of Computer Science and Engineering, Bannari Amman Institute of Technology, Erode, India

M. Renuga
Department of Electronics and Communication and Engineering, Prince Shri Venkateshwara Padmavathy Engineering College, India

S. Boopathi
https://orcid.org/0009-0000-3045-3834
Muthayammal Engineering College, Namakkal, India

ABSTRACT

This chapter explores an overview of using virtual reality (VR) to train technicians and factory personnel in the automotive industry. It also shows how immersive virtual reality technologies can simulate real environments, providing a safe, controlled area for the skills to be developed and practiced on. VR enables workers to prac-

DOI: 10.4018/979-8-3693-7287-6.ch004

tice how to operate complex machinery, perform repair procedures, and work on assembly lines without risking injury to themselves or damage to equipment—often at a fraction of the cost of training the old-fashioned way. This can allow industrial corporations within automotive design and restoration to boost their performance and security. The chapter continues explaining how useful it is to implement VR to engage employees with the application, where it may even enhance learning, reduce training time, and increase the general output quality of the workers.

INTRODUCTION

The best jobs in the entire industry, driven, technically competent technicians and assembly workers have never been in more demand in the rapidly-evolving automotive industry, reliant on their expertise to create an efficient and dependable range of vehicles. Automotive technologies are evolving, and hence, the workforce in the automotive industry has to adapt to the ever-expanding complexity in automotive systems, machinery, and assembly processes. For any years, the traditional training, which has on-the-job learning and instructor-led sessions, used to be a good approach. However, they present significant challenges, including risk, inefficiencies, and a lack of real-world interaction. Such demands are driving the high demand towards effective, safe, and scalable training solutions to augment the skill of the automotive technicians and factory workers. Training comes with its own pain points that cannot be overlooked by organizations that want to be successful. One solution that has been trending for the past few years is immersive training with virtual reality (VR)(Garcia et al., 2019).

Virtual reality is a modern technology that represents actual surroundings in a completely interactive virtual environment. It has a reality feature through which users are able to interact with 3D models and environments that provide experiences that are real and hyper-realistic. This immersive quality is one of the key reasons that VR is a revolutionary training method in industries as diverse as automotive manufacturing and maintenance, among others. In the auto industry, VR allows technicians and factory floor workers to train in inline environments that can mirror on-the-road situations, such as malfunctions and assembly of complex auto parts. Such depth of immersion may allow for significant hands-on lessons without the built-in dangers (or costs) of traditional methods—tangible prototypes, equipment, or materials(Naranjo et al., 2020).

VR technology has enormously exciting capabilities for skills development, and the automotive sector is on the council already. Factory workers and automotive technicians may need to learn. managing accurately functioning machinery and systems, maintaining cleanliness classes, and following layered systems while meeting stringent safety organizational requirements. Virtual reality technology allows trainees to practice a task multiple times without concern, dramatically increasing their competence and confidence. Furthermore, VR training can also be customized to each worker, as it is based on a particular set of skills of individual workers and the type of work they perform. For example, a technician might specialize in engine diagnostics, and a factory worker might specialize in factory production line techniques. This way the training method can be tailored in such a way that each entity gets the required training, which would be more relevant to them, leading to increased productivity at all levels of a team(Shamsuzzoha et al., 2021).

A further big advantage of training in virtual reality is the potential to recreate environments too difficult, expensive, or hazardous to replicate in real life. In automotive, this can be hazardous materials, machinery, breaking, or practice emergencies. Historically, these circumstances demanded specialized hardware, extra safety measures, or controlled environments. On the other hand, with VR, these experiences can be replicated in the virtual world, and employees can be introduced to any hazardous scenario without any danger. This not only increased safety but also instilled problem-solving skills and critical thinking as workers were placed in a high-pressure, realistic in-training scenario in which fast thinking and action were required(Smith & Salmon, 2017).

One more key aspect of VR training is a faster learning curve. The worker demands much of her time in a classic learning environment and often takes much observation, practice, and errors before a task can be mastered. But with VR, that process has been streamlined. Trainees can practice as many times as necessary without incurring the cost of real-world resources and can also receive instant feedback on their performance. And this allows them to accelerate their learning process and reduce the time you need to spend to master your skills. VR fitness can be customized too, so it starts simple and gradually gets more difficult, keeping those workers on their toes and allowing them to gauge their skills. Cleaving ideas into palatable pieces allows for less stress and more focus during the learning process(Gavish et al., 2015).

Along with being a great way to improve your skill, VR training can lead to substantial financial savings for an organization. Traditional training programs often require a heavy investment in materials, equipment, and infrastructure. VR training, on the other hand, requires few physical resources and can be executed remotely, which helps to avoid expensive training venues or costs associated with travel. As VR technology can be used simultaneously by more than one user, even in the case of distributed teams, it has been found that companies can hold multiple training sessions in a week at a lesser cost than a single session in the case of tumble-down training, so this form of training has proven to be an extremely scalable solution for all companies having a large workforce. Moreover, using VR training, it can also train staff remotely, in multiple locations, and at the same time, without the need for staff to travel to the training center(Watanuki, 2008).

The benefits of VR training are evident, but there are several considerations that must be made for a VR program to be successfully brought to life in the automotive space. They need to be incorporated with existing training systems and processes where they augment (as opposed to entirely replacing) traditional methods. VR is an adjunct to hands-on experience, not a substitute for it. Moreover, content creation should take place, ensuring adherence and relevance to the industry requirements. Collaborate with subject matter experts to ensure the VR training scenarios are as realistic and effective as possible. And finally, investing in the hardware and software necessary for VR training and the expertise to develop and sustain these training programs. While these initial costs can be quite substantial, the long-term benefits—better skills development, shorter training time, and improved safety—typically outweigh the initial investment(Watanuki, 2008).

Especially, last but not least, we generalize such technology not only as training but some sort of skills improvement as VR. It has already started in some organizations only as a high-level skill improvement, but it will definitely bring the whole industry near to an extent of checked and competent factory people/technicians. Leveraging the unique ability of VR to create realistic, immersive environments, training services are often faster, cheaper, and safer than traditional training methods, providing custom experiences to cater to the individual learning curve of the user(Norris et al., 2019). This will include VR simulations and tutorials on familiarizing workers with the tools they need to work on electric components, like safely engaging with a battery or electronics and putting together components, which is one of the most rudimentary things to learn.

Objectives

This chapter aims to examine how time is affected by Virtual Reality (VR) as a way of training technicians and factory workers in the automotive industry. The aim is to show how VR can craft skills-building experiences that are more engaging and retainable than rote instruction, teaching people how to operate everything from a machine to how to do a complex repair. The chapter aims to demonstrate the benefits of VR, like increased productivity, reduced training costs, and improved safety. This in whole will also include training in automobiles and VR and how VR can aid and assist students not just in the context of knowledge retention but to bridge the gaps and build up on proficiencies, which will lead to the students being more employable and being able to deliver in their jobs.

BACKGROUND: VIRTUAL REALITY (VR) TECHNOLOGY

VR Virtual reality is a 3D technology that creates virtual surroundings a user can experience and engage with as if they were real. VR is built on the premise of placing people inside compelling, immersive virtual worlds through content and devices. Users explore the virtual world through devices—such as VR headsets, motion trackers, and gloves—that enable them to see, hear, and even touch in real time virtual objects. The functioning of VR systems includes generating a 3D environment and monitoring the user position/movement to provide the illusion of presence in that environment. That technology relies on detailed follow-and-rendering and visual, auditory, and, in some cases, haptic responses that all merge into a life-like simulation(Chiang et al., 2022; Webel et al., 2013).

Looking at these types, there are many types of VR systems available. In summary, these systems can be grouped into non-immersive, semi-immersive, and fully immersive VR systems.

1. **Non-Immersive VR**: This is more passive VR, however, accessing an environment rendered on a screen (like a desktop or monitor) and using interaction in that space. The user remains unimmured inside the environment, interacting via standard input devices like a keyboard or mouse. An example of non-immersive VR would be screen-based technology such as video game consoles or computers, where users can visualize and navigate a digital environment and is usually seen in use for things such as training simulations or virtual tours.
2. **Semi-Immersive VR**: HoloLens offers a high-interaction, moderate immersion, semi-immersive experience. A VR system may include a monitor, projector, or 3D display to show the simulated environment and motion-tracking devices (like controllers) to interact with the environment. Compared to non-immersive VR, users can react to the virtual environment more dynamically, but the user does not have a completely enclosed virtual experience. This is a VR suitable mostly for sim practice, be it through games or education where the balance between sensation and utility can be achieved.
3. **Fully Immersive VR**: This, then, is the highest level of VR and the richest, most immersive, and interactive type of VR that replaces users in a virtual reality. A fully immersive system has a headset, motion sensors, gloves, or other devices that track the user's movements as he or she moves about a lifelike virtual world. Essentially, these systems holistically deliver 360-degree visuals, 3D audio, and haptics factors that mimic actuality in unison. Fully immersive VR is used in high-end training programs, gaming, design visualization, and other applications that require total engagement.

Each type of VR system has its benefits and is chosen according to the needs of the application. As VR technology matures, these categories blur, creating new interactive and immersive experiences.

Advantages of VR for Training

Virtual reality (VR) is one of the most advanced and fastest-growing technologies in the world, offering immersive experiences to its users. This advantage makes VR better than traditional ways finally, when it comes to industries like automotive maintenance or manufacturing. The first and foremost advantage of VR-based training comes from the level of immersion that enables a highly interactive learning experience. A virtual environment allows learners to practice skills and commit solutions to memory in situations that are similar to those they will encounter in the real world(Norris et al., 2019).

VR training significantly reduces many of the risks and dangers that accompany physical training in potentially dangerous settings. For example, in the automotive industry, employees could practice complex repairs or machine operations without risk of damaging expensive property or making costly mistakes. The immersive nature of a virtual environment allows for complete individualization and enables trainees to practice the same task time and time again until they could perform it without error, which is practically impossible in the real world due to resource constraints and time commitments.

The flexibility of accessibility and scalability is another major benefit of VR. 24/7 Availability: VR training programs can be accessed by workers from any location, allowing for remote training and less reliance on being physically present for training. It's particularly strong for large companies with many locations, where even a disseminated training push can be difficult to reach all sites. Furthermore, you can tailor VR training according to a worker's capacity as well as learning style; they don't have to risk lagging behind when training at their own pace compared to in-person classes.

VR training therefore is a game changer in today's high-paced technological society by leading to sky-high skill acquisition, reduced training costs, and protection from injuries through the rich integration of VR systems in automotive teaching.

Figure 1. VR in automotive sector training: VR for vehicle assembly and maintenance

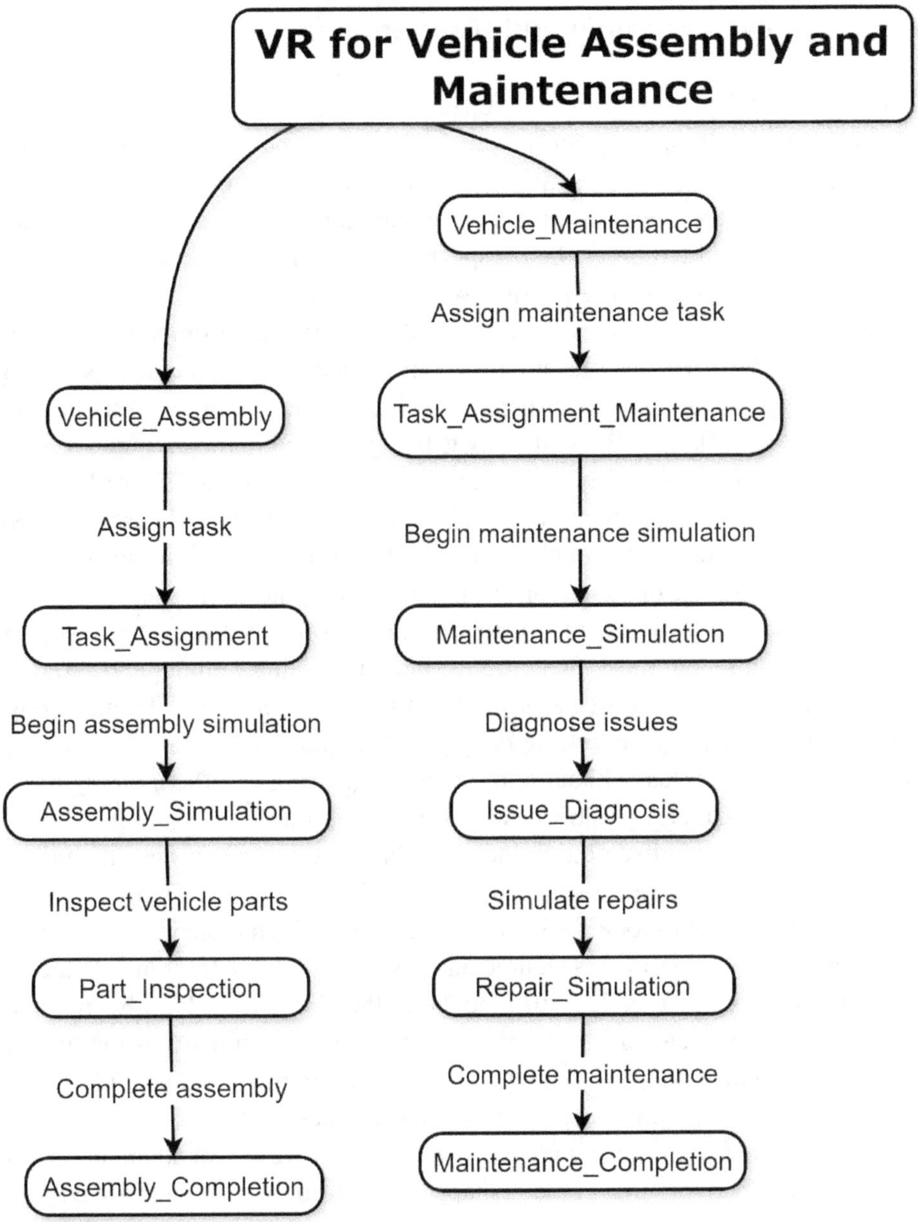

APPLICATION OF VR IN AUTOMOTIVE SECTOR TRAINING

VR for Vehicle Assembly and Maintenance

Most specifically in the automotive field, utilizing of VR has advanced into a profoundly crucial thing in order to train technicians and engineers to tegrow or service vehicles. As modern vehicles tend to be tightly integrated and complex systems, hands-on training is an expensive and logistically rigorous exercise. Luckily, with Virtual Reality comes the prospective solution in the shape of a realistic, state-of-the-art simulation of a prospective vehicle assembling line, the reparing, and maintenance works. Allows trainees to enter a virtual environment interactively, so they can learn all the skills and procedures needed to operate the technology without using physical vehicles or expensive equipment, saving its users time and capital(Firu et al., 2021). VR technology is being utilized in the automotive sector for training in vehicle assembly and maintenance, as depicted in Figure 1.

One of the common applications of VR would be in the vehicle assembly process, aiding the workers in getting familiar with the complex and intricate processes such as the installation of the components along with the wiring and integration of the different automotive components within the vehicle. Like a real assembly line, workers can repeat those processes in the virtual environment and refine every step until it's complete, regardless of physical environments. Trainees might practice installing an engine, setting body panels in place, or connecting electrical systems—all essential tasks in automotive manufacturing—without worrying that they would damage a part or slow down a production line.

Vibration This forced version brands Samdani how to tell stories. Technicians can log into a virtual environment and practice diagnosing and fixing those vehicles, discovering what to look for in areas like wiring, engine operation, and brake system components. This allows technicians to see intricate systems in 360 degrees, which isn't always practicable with models on the shop floor. It's also relatively indifference to the practical limitation of a physical environment, as the growing use of VR-based maintenance training, which can cover a wide range of scenarios, creates opportunity for common problems to more exotic failures to be experienced, as we would always prefer the employee to be ready before any incident in the field.

This makes VR a tool that could reshape how automotive workers are educated at assembly facilities, at repair shops, and such, as it enhances access to hands-on practice while lowering the expected no-shows of priceless employees.

Simulating Complex Repair Tasks

Automotive industries are leveraging on virtual reality (VR) technologies to simulate intricate repairs for their trainees, who learn and practice without something going wrong. Use of automotive training simulations enables workers to exercise complex repair procedures, such as engine diagnostics or electrical system troubleshooting, with no fear of making a costly mistake or damaging expensive equipment. VR builds ultra-detailed, interactive models of the vehicles being serviced so that trainees can inspect every orifice, from engine compartments to the most delicate areas of the car's electrical systems(Langley et al., 2016).

VR simulations allow trainees to disassemble and reassemble components virtually, perform diagnostics and repairs, all under the same conditions in which it would happen in the real world. However, it is the authentic experience every technician must experience repeatedly to become accustomed to advanced automotive repairs whilst familiar with all tools, processes, and procedures, further enhancing their problem-solving skills and technical competency. Additionally, VR enables technicians to try out rare or complex repairs, allowing them to experience scenarios that they otherwise might not encounter often in their jobs. Most VR simulations do the worst job at most basic to complex repair. VR can also involve feedback systems to receive immediate adjustments on their technique and guide trainees through best practices and on to handling fewer errors when they are in charge of repairs in the real world. Such a form of training is bound to reduce the chances of making any mistake and enhance job efficiency, improving the way you deliver quick turnaround time and quality automotive repair services.

In a nutshell, automotive technicians are able to skillfully practice intricate automotive repairs with VR technology, being unhindered by time, safety, and physical restraints that would have previously affected the developing technician, meaning capable workers are prepared to make an impact sooner in their careers.

VR in Factory and Assembly Line Training

Decades of Heartbreak, Waiting, and Relief: Automakers and the Training VR Technology has Sparred the Training of Factory and Assembly Line Workers. Traditional training methods typically necessitate practice on real production lines, exposing employees to risk and costing potential mistakes—and long learning curves. With AR, trainees can see in real time what they're supposed to do; with VR, they are put inside the simulated factory and assembly line and can practice their work in an environment where they won't be risking out valuable products or equipment by making mistakes. Employees can train on advanced equipment and assembly lines and rehearse safety protocols without stopping the production line or risking damage to costly equipment(Hernández-Chávez et al., 2021; Lawson et al., 2016).

From managing parts and assembling their components to navigating the layout of the factory floor, VR training programs can mirror any aspect of an assembly line. Install components, use tools, quality assurance, etc. in a highly immersive environment. This approach not only gets workers familiar with the job but it also conditions them to understand where hazards may be found and problem areas. With VR-based training, workers can interact with 3D models of vehicles, machines, and assembly stations, enabling them to practice the precise and efficient assembly of parts.

VR allows for the simulation of production line conditions that would be hard to simulate in traditional training environments. Certain scenarios can be rare or complicated, like equipment failures or safety issues, which can be practiced in the VR environment to help workers get accustomed to an approach and practice for it. In addition, because VR training can be tailored to individual needs, workers—whether new hires or seasoned technicians looking to brush up on their abilities—can have the specific level of training that best suits their needs. If factory and assembly line training is done using VR, it makes it both safer and faster.

BENEFITS OF VR TRAINING FOR TECHNICIANS AND WORKERS

The rise of virtual reality (VR) training and upskilling has completely changed how technicians and factory floor workers gain skills and has provided a number of advantages over traditional training techniques. These advantages, such as improved learning experience with cost and time efficiency, are the reasons why they are being implemented into the automotive industry(Carnazzo et al., 2024; Quevedo et al., 2017). Figure 2 illustrates the advantages of VR training for technicians and workers.

Improved learning outcomes are one of the major benefits of VR training. Training in virtual reality (VR) allows the participants to perform real-world functions in a virtual environment that can be simulative in nature and interactive. This "gamification" accelerates the processing of knowledge and skills by workers compared with passive learning via video or lectures. It provides a platform where procedural interaction with machines, assembly lines, and repair processes can be practiced until competence and confidence emerge through the dynamic integration of skills. Flexible training with the possibilities of virtual reproduction of less predictable components of the profession VR training allows users and subjects to learn in a controlled environment, simulating high-risk scenarios (e.g., mock equipment malfunctions, emergency protocols). This not only guarantees that workers are fully prepared to intelligently deal in this type of situation but also develops crucial problem-solving and decision-making skills. In this setting, VR creates an immersive experience that removes distractions in the work environment, giving workers a 100% immersive learning experience, thus improving retention of material learned and, subsequently, performance.

Figure 2. Benefits of VR training for technicians and workers

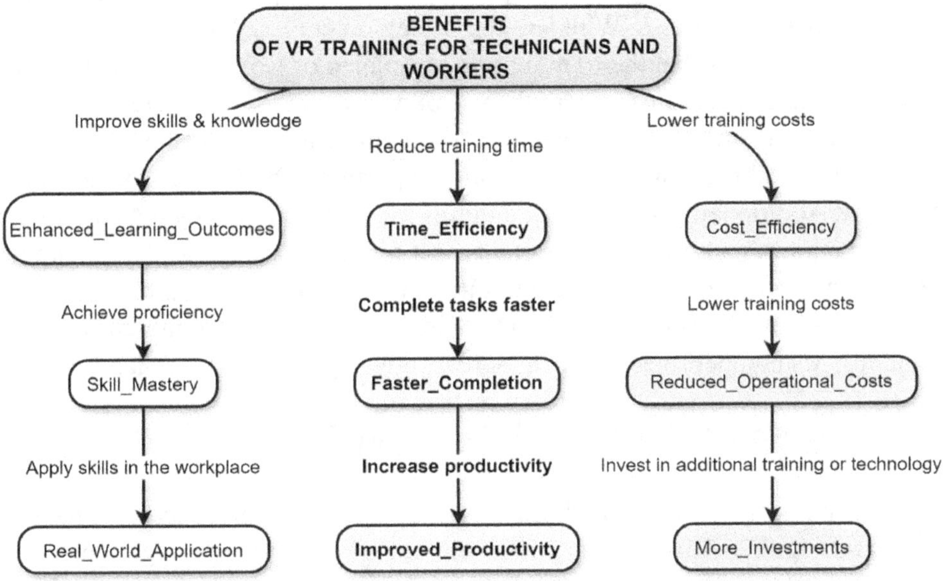

It's economical and efficient. One of the significant advantages of VR training The traditional training plans consume a great deal of cash for materials, hardware, and arrangements, and the relating inactivity for down-to-earth exercises. On the flip side, there are no physical resources needed; hence, the costs for VR training are decreased. Workers are able to use virtual tools and machinery as opposed to the physical machines, preventing wear and tear on the expensive equipment. One of the other things about how VR helps is that it helps with such training in a controlled environment where they did not have any impact on the production schedule. The flexibility to train at any place and any time helps companies deploy training across different sites with minimum travel or additional infrastructure costs, while also ensuring consistency.

VR training also minimizes the learning curve of employees with custom-tailored, self-paced modules. Technicians and factory workers can, on what they need to work on, develop competency quicker than a one-size-fits-all model. It also rewards training time, which companies can reduce, saving time for new hires and increasing productivity across the board." Also, VR can record very detailed statistics of performance in training, which can help trainers determine where skills may be lacking and provide programs to those who fall short in specific areas. The training process becomes more efficient, encouraging workers to reach the competency levels necessary to perform their tasks, all while relying upon data used to provide education in a more individualized fashion. Over time, these advantages create a very competent workforce that can improve the flow of operations and the quality of its products(Henriques & Winkler, 2021; Lawson et al., 2016).

The bottom line is that the training of a technician or worker in the automotive sector is almost revolutionized with the use of VR training. VR creates immersive and interactive environments that improve retention and transfer of knowledge, subsequently enabling employees to build important job skills faster and more effectively. Additionally, VR training requires less time and money than traditional training methods, making it an important asset for organizations as they train their workers while ensuring business continuity and minimizing operating costs. As a result, VR is now viewed as a game-changing tool for growing and ensuring the auto industry's competitiveness in an increasingly sophisticated tech landscape.

Improving Worker Safety and Reducing Errors

VR training, which is revolutionizing the skill acquisition process, enhancing safety, and significantly reducing job-field mistakes, is delivering tremendous benefits to technicians and factory workers. Demonstrating Simulating high-risk environments is one of the significant benefits of VR training, so workers can practice tasks and procedures without subjecting them to real-world hazards. For certain types of equipment, attaching VR technology may reduce the need for such recurring operational trials, allowing technicians to practice in a virtual environment. Automotive manufacturing and maintenance technicians need to be trained to operate heavy machinery, powertrain electrical systems, or hazardous materials if they're very specifically detailed in virtual reality, and for many industries those standards must be kept to ensure safety for a person being trained. This reduces the chance of workplace accidents in every step, including training or on-the-job work. This means that workers receive a much greater understanding of safety protocols and what to do in an emergency with VR training. In this way, trainees get to experience situations similar to real-life situations, like equipment failures or fire hazards, and practice the right responses, all without the risk. This methodology goes beyond theory, aiming to build the muscle memory and confidence required to tackle real-life emergencies efficiently(Henriques & Winkler, 2021; Lawson et al., 2016).

Moreover, VR can help decrease human error, which is a common error due to a lack of experience and lack of exposure to different equipment. Condensing the time it takes for workers to get through multiple iterations of a task to an optimal performance, countdown, and improving performance as many times as needed before a live performance using VR. This is particularly critical in precision-related roles, such as performing tasks like assembling intricate automotive parts or assessing mechanical malfunctions. Most crucially, VR is an interactive experience—trainees receive immediate and immersive feedback about their actions, helping them to experience common mistakes early on in the learning curve.

VR and AR training avoid mistakes and improve safety awareness, therefore increasing organizations productivity while lowering operational costs. It ensures that every worker is adequately prepared, confident, and capable of performing the task more accurately and efficiently, creating a sustainable and reliable workplace in the automotive industry.

Figure 3. Designing effective VR training programs

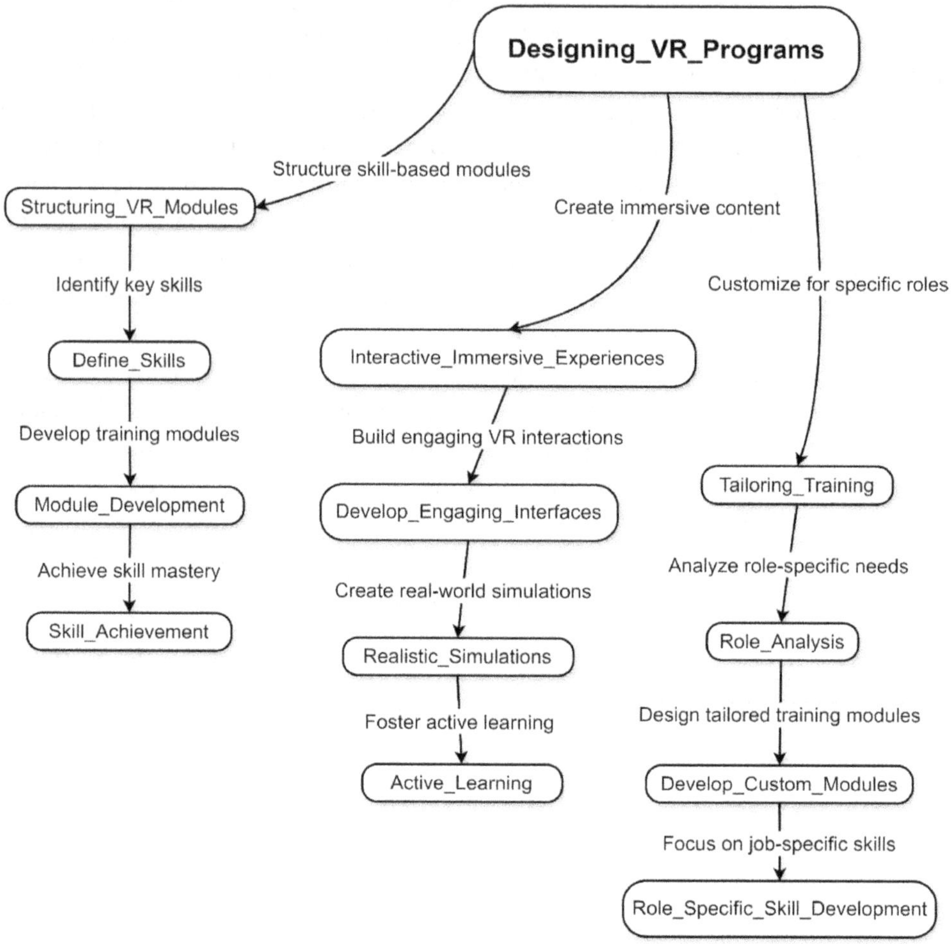

DESIGNING EFFECTIVE VR TRAINING PROGRAMS

Structuring VR Modules for Skill Development

VR training modules need to be realistic and should mirror real-world situations where trainees are trained and perform tasks in an extremely interactive and immersive setting. The content of these modules should align with instructional design principles, beginning with introductory material and progressing to advanced skills in a way that promotes gradual learning and builds confidence(Powell, 2008). Figure 3 illustrates the process of designing effective VR training programs.

Even a basic module should include a phase known as the orientation phase, at the beginning of the session, where trainees are oriented with the virtual environment, tools, and destination of the session. In this phase, the workers get accustomed to the VR interface and seamlessly navigate the environment. Then, train on task-specific instructions and how to break down complex processes into individual tasks, where possible. In the automotive industry, for instance, modules can include recognizing tools needed for a job, assembling vehicle components, or diagnosing mechanical problems. In world practice, extra interactivity aids in cementing these skills. Including real-time feedback is an important component of designing VR training modules. The system should give instant feedback to trainees about their performance, including their correctness, speed, and protocol compliance, as they perform tasks. It allows workers to learn from mistakes and hone their techniques before deploying their skills in the real world.

This engages participation and excitement, mapping your progress, skills, and milestones with gamification of VR modules. It can also include emergencies or other technical scenarios that require multiple iterations around infrequent problems to sharpen and pressure test solutions under pressure, for example. A thorough VR training course will accelerate the acquisition of the necessary skills while ensuring the workers involved are in the right mindset, ready to pursue the demands of the role. VR training modules and solutions can lend themselves to nurturing a highly skilled workforce with a consistent confidence level in their ability.

Interactive and Immersive Learning Experiences

VR training programs are ideally interactive and very immersive, so creating good examples is challenging. With its interactive possibilities, VR offers training capability that goes beyond the mere mental practice of traditional methods, enabling trainees to immerse themselves in an interactive, virtual environment that can be practiced in a more hands-on way. Immersion brings facets of validity to education in that it really makes employees feel as if they are doing tasks that will soon be manifested in the real world, which adds to how well they retained the material and further bolsters their confidence(Roberson et al., 2001).

Secret Sauce of Successful VR Training: Interaction Elements In this controlled environment, workers can access virtual tools, drive machinery, and move through assembly lines. In this interactive environment, trainees became not only an audience but also a part of their own learning process. Or in a simulation of an automotive assembly line, the workers can rehearse installing vehicle components, checking parts for defects, or troubleshooting machine malfunctions. These programs give employees the opportunity to experience skills in a safe environment without incidents, providing them with a unique hands-on experience.

Immersive learning experiences also enable personalized training, adjusting to the skill levels and paces at which each individual worker learns. In real-time, this means that trainees can get immediate feedback on their performance by marking mistakes and providing correct guidance. This enables a self-reinforcing feedback loop, promoting best practices while identifying pitfalls and providing a holistic learning experience. Moreover, VR programs can recreate a variety of experiences, from the mundane to high-stakes or emergency events, which means preparing for all work-related possibilities. Virtual reality (VR) builds an effective and engaging training environment through realistic visuals, sound effects, and interactive components. In the end, VR courses prioritize creating interactive and immersive training programs that can truly arm workers with the skills and confidence they need in order to thrive in their jobs, all the while improving safety and efficiency on the job.

Tailoring Training to Specific Roles

Aligned Sphere with Role-Based & Personalized Learning is One of the biggest advantages of virtual reality (VR) training is its ability to deliver role-based, personalized learning experiences. This is especially the case in the automotive sector, where engineers and factory workers or assembly line workers have very different roles and responsibilities, and with VR training programs tailored to specific roles, you can provide the most effective and relevant training for employees. Not only does this mean employees are trained on the service-specific skill sets for their roles, but it also cuts down on irrelevant training material(Carliner, 2015; Roberson et al., 2001). Also, VR training is immersive and offers technicians a 3-D view of such jobs as deciding and disputing complex automotive systems, such as automotive engines, electrical wiring, and braking systems. A program might start by teaching some troubleshooting techniques, showing how the right tools are used, and then walk you through a repair step by step in a safe virtual environment for realistic training. Simulation covering role-specific scenarios will help technicians build confidence on the job and an enhanced level of competence to ensure they are job-ready.

For example, factory employees may require training related to assembly line processes, quality control standards, or operating equipment. VR programs can be tailored to the equipment and workflows students will encounter, helping to train them for tasks like assembling vehicle components, inspecting individual parts for defects, or operating heavy machinery safely. Standard operating procedures are shared, and reinforced production techniques are delivered with real-time feedback through interactive modules. Moreover, the VR training can be tailored for skill because it can provide different modules for beginner, intermediate, or advanced experience based on the worker's experience. Novices can focus on fundamentals while experienced employees use VR to adopt new technologies, improve productivity, or train for other positions.

IMPLEMENTATION OF VR IN AUTOMOTIVE TRAINING

Virtual reality (VR) is a game changer for the automotive industry in terms of collaboration and training. VR—Virtual Reality—allows businesses to employ immersion and interaction to enhance their training environments, turning them dynamic and lifelike. Thus, VR training must offer a practical environment that would enhance the productivity, safety, and performance of their processes(Carnazzo et al., 2024; Henriques & Winkler, 2021). The integration of Virtual Reality (VR) into automotive training is depicted in Figure 4.

Figure 4. Integration of Virtual Reality (VR) into automotive training

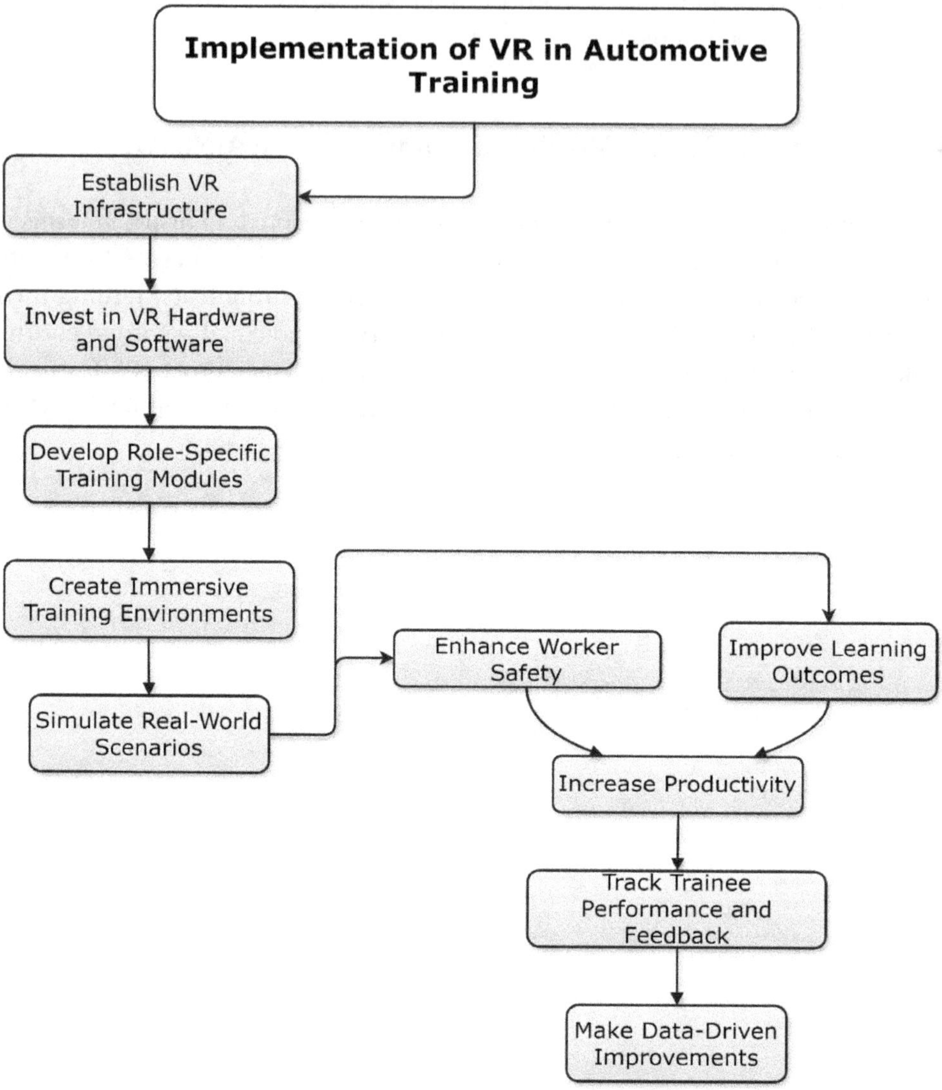

Setting Up VR Infrastructure

VR-equipped automotive training all begins with proper infrastructure. This includes both the hardware of VR—headsets, controllers, and motion-tracking systems—and the software platforms that offer tailored training exercises. Companies might have to collaborate with VR developers to customize training programs for those types of jobs, such as assembly line work, machinery maintenance, or vehicle diagnosis. In addition, the adoption and expansion of VR technology for immersive events relies on its compatibility with existing systems while maintaining a stable network to provide the most effective immersive experience.

Role-Specific Training Modules

The VR implementation involves developing training modules tailored to different roles within the same company, as technicians and factory workers have different training needs. Example: VR simulations, practice the test of failure to diagnose the engine, electrical systems testing, pulse advanced troubleshooting, automotive industry scenario architecture framework in the automotive domain scenarios are requested to use. On the other hand, factory personnel can be trained on assembly line processes, quality inspections, etc., as well as the operation of the equipment. The custom-fit modules provide workers with experiential, hands-on, etc., real-life experience within their own roles, giving them the edge to feel confident and competent enough to do their jobs.

Enhancing Worker Safety

One of the most significant benefits of VR training in the automotive field is better worker safety. VR filters out high-risk tasks into a compressed, far less threatening virtual space, allowing trainees a chance to carry them out without making expensive, real-world mistakes. For instance, workers can learn how to operate heavy machinery or react to emergencies like equipment failures or hazardous material spills without the risk of real-world harm. Thus, encouraging safety training in advance leads to fewer workplace accidents and an improved culture of safety.

Scalability and Accessibility

With ever-evolving and more sophisticated VR capabilities, this stream of AR & VR provides an innovatively scalable option for organizations of different sizes. Once these programs are developed, they can be installed and used in multiple locations, offering some consistency as far as the training is concerned across all employees. This, combined with the fact that VR makes for excellent remote training, which is excellent when companies have teams based across many different locations, as employees can work through modules from anywhere they have access to the internet. It also means workers have consistent, high-quality training, despite their destination on the globe.

Tracking Progress and Providing Feedback

While all of these features make for a streamlined and efficient training experience, one more key aspect of VR training is the capability to track trainee progress with the ability to give real-time feedback. VR systems can monitor performance-based metrics such as time to task completion, accuracy, and adherence to safety protocols. Trainers can use this data to gain insight on where the trainer wanted to improve as well as customize their upcoming training sessions. It offers an omnipresence that leads to continuous improvement, to progressive learning gains, inherently relying on data for decision-making, and the ever-expanding skill set needed to make it happen.

Overcoming Challenges

Although there are boons to VR training, there are challenges to implementing it efficiently. These factors include the initial cost of implementing VR solutions, a learning curve associated with implementing new systems, and the necessity for regular updates to training content due to the technology employed in contemporary vehicles. Organizations can leverage VR training solutions to drive real value, but overcoming these challenges requires strategic planning and collaboration. All in all, VR can streamline training programs, avoid on-the-job injuries, and even save time and money throughout the automotive industry. This will enable the companies to cultivate workforce preparedness with automotive industry needs while creating a culture of innovation and excellence by creating immersive and role-based experiences to learn.

CHALLENGES AND LIMITATIONS OF VR IN TRAINING

While Virtual Reality (VR) has changed the way training is done across a range of industries (automotive included), it also brings its own challenges and potential limitations. A better understanding of these obstacles could help organizations draw up more realistic plans for integrating VR training into their curriculums. Key challenges include technological and infrastructure barriers, resistance to the adoption of new training methods, and a learning curve for new technology adoption(Chiang et al., 2022; Hernández-Chávez et al., 2021; Shamsuzzoha et al., 2021). Figure 5 provides a comprehensive analysis of the potential and limitations of utilizing virtual reality (VR) in training.

Figure 5. Challenges and limitations of VR in training

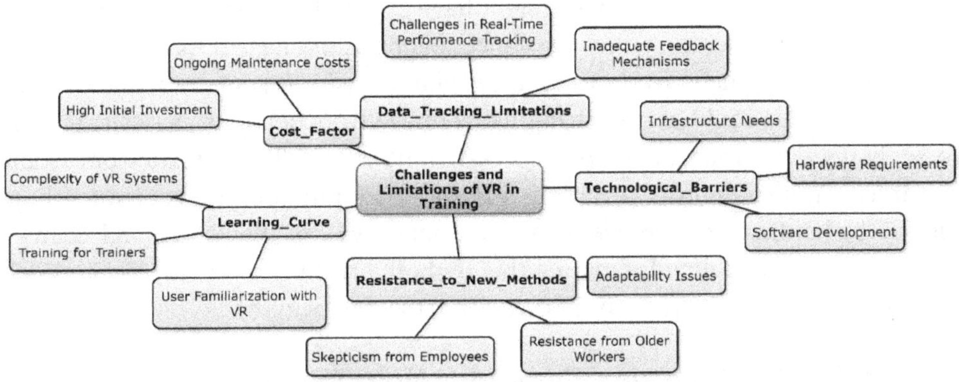

Technological and Infrastructure Barriers

One of the most readily apparent challenges of using VR in training is the technology and infrastructure needed to support it. Training in VR does demand a good up-front investment in keeping the top-tier, high-end/reality-class VR hardware (headsets, motion sensors, and computing systems with power resources to run heavy-duty simulations) in operation. These costs can be prohibitive for smaller companies or those with limited training budgets. Furthermore, the fact that creating software systems for customized VR modules by business and profession is a time-consuming and costly process that was normally done with specialized creators.

A much larger issue is the infrastructure needed to facilitate VR training. Smooth, immersive VR training experiences require high-speed internet and advanced networking capabilities, especially for organizations with multiple training sites. Deficient infrastructure can lead to lag, glitches or other technological challenges that disrupt the training. Moreover, there is the need for constant upkeep, as well as occasional replacements of hardware and software, for the system to be useful, as well as in touch with the technological evolution, burdening the operations with a long-term cost and business.

Overcoming Resistance to New Training Methods

It is human nature to help trainers design VR-based training programs keeping their targeted audience in mind. Some other employees may not consider VR to be credible, viewing it instead as a nifty experience rather than a meaningful training opportunity. Some may be challenged by the immersive nature of the tech, and others may fear they won't know how to use it. Older workers, or workers who have not been using digital tools in the past, might have especially strong resistance.

So, in order to overcome this resistance, people don't need a lot more training; what organizations need to do is educate them on why it actually works to train stuff in VR." This could help in changing its image from being just a flashy tech or, worse, being dangerous and help in creating a trust in it. To alleviate worries and encourage an open mindset regarding VR adoption, pilot programs or workshops can also be held where employees can experience VR for themselves. Also, management should model and reinforce this by both creating and going through VR training to ensure employee buy-in on a fundamental level.

Addressing the Learning Curve for New Technologies

Incorporating VR into training programs is a learning curve for both trainers and trainees. Initially, users must familiarize themselves with VR, and many workers new to this technology may get a little frustrated to begin with, and it might slow down the learning process. They need specialized behavior-learning skills to use VR systems and to develop training appropriately and so need time and resources themselves to be trained to perform competence in VR. Fortunately, there is an approach to tackle this learning curve systematically. Sessions to acclimate employees to new VR hardware and software can ease the transition. Hope this helps for onboarding users; their first experiences will be comfortable and confident with step-by-step guidance and support in the initial steps of training together with what is on advantage for trainers. They can invest and spend on scholarships for these specialized courses, which will give them a technical edge to their expertise, which can give them a better VR in teaching.

Moreover, a virtual reality device must be easy to use so that an average person can use one without needing to acquire too many technical skills. Offer clear user mappings, guided walk-throughs, and interactive how-tos, which can greatly reduce the studying curve and accelerate the adoption switch. It is equally important for me also to collect the regular user feedback to enhance the system towards the challenges faced by the user as a training process. In conclusion, while VR has game-changing benefits in training, it is important to overcome the challenges and limitations for an effective deployment. By tackling the technological and infrastructure issue, reducing skepticism, and solving the adaptation challenges for users, organizations can utilize the potential of VR training and lead its development toward the sustainable implementation of effective learning methods within the automotive sector and beyond.

FUTURE TRENDS IN VR FOR AUTOMOTIVE TRAINING

Automotive training through virtual reality (VR) has already cemented its importance, yet there is plenty of room for growth. Emerging technologies, integration with augmented reality (AR) and artificial intelligence (AI), and advancements in immersive learning techniques are likely to shape the future of VR in this sector. This will accelerate training, decentralize learning, and increase workforce response to continuously changing industry patterns(Carnazzo et al., 2024; Chiang et al., 2022; Mendes Monteiro et al., 2023). Figure 6 depicts future trends in VR for automotive training.

Figure 6. Future trends in VR for automotive training

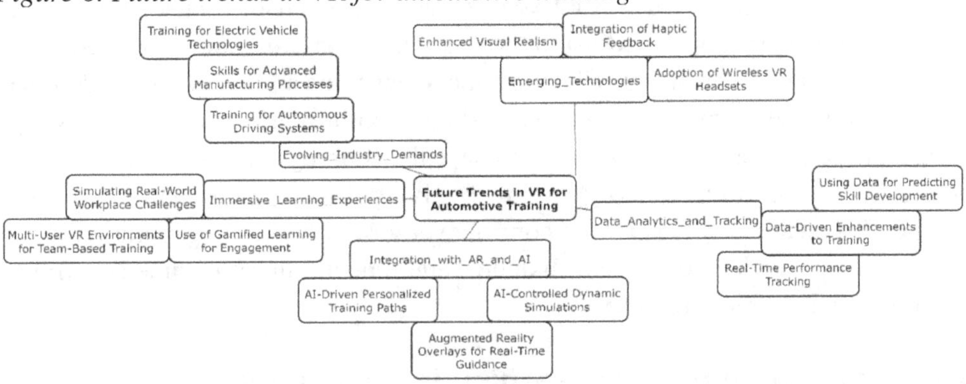

Emerging VR Technologies

As VR technology continues to advance, these new applications are creating increasingly more realistic, accessible, and gamified training programs. One of these advancements is haptic feedback, or the sensation of touch—such as vibrations or pressure—when interacting with virtual objects. For automotive tinkering, that translates to an incredible level of realism when using tools, operating machinery, or assembling parts.

The other trend is to use cutting-edge graphics and rendering techniques to create hyper-realistic virtual worlds. These improvements create a sense of realism and excitement in training scenarios by enhancing the immersion experience. Finally, we are seeing the rise of wireless VR headsets and portable systems, enabling more mobility and flexibility in the training environment. As hardware continues to improve and become much more affordable, VR solutions should become available to a broader spectrum of organizations; think smaller businesses with lower budgets.

Integration with Augmented Reality (AR) and AI

The integration of VR and AR with AI is the second biggest contributor to the future of the automotive industry. Augmented reality (AR), which superimposes digital content onto the real world, can be used to enrich and augment VR by providing an immersive mixed-reality experience for education. AR, for example, would allow technicians to view 3D representations of the parts they are working on, marrying theoretical education to practical application. It enables smooth transitions between theory and practice, and vice versa, and also greatly adds value to the experience.

Unlike VR training, which is a fixed experience, AI-enabled customer support allows intelligent systems to adjust to the learner. VR environments powered by AI can evaluate trainee performance in real-time, identifying strengths and weaknesses to help customize training programs accordingly. For instance, if a worker is struggling to do a particular task, AI could adapt the training module, focusing more on the area that a worker is struggling with, giving them more guidance and more practice." AI will also help provide control over virtual worlds to generate realistic simulations with added challenges like new equipment failures or time constraints to put you in a problem mindset.

The Future of Immersive Learning in the Industry

Emerging VR Technologies Collaborative VR/VR-AR Tech: The future of immersive learning in the automotive industry will rely on the greater utilization of collaborative virtual environments. Multiplayer VR platforms will train teams of employees in "shared spaces," putting to use communication skills and creating environments of collaboration. This can be particularly useful in assembly line-style businesses, as many assembly line-style type businesses have workers who leverage each other to maintain efficiency and quality.

Moreover, VR analytics may assist institutions in more accurately assessing their training program's effectiveness. Analytics about the performance, participation, and learning outcomes of the trainees can be used to gradually improve training content and strategies. Businesses could also recreate workplace scenarios and analyze them to identify areas for improvement, drive higher productivity, or find possible bottlenecks. As the automotive industry embraces digital transformation, VR training is set to play a vital role in preparing the workforce for transformational ecosystem trends like electric vehicles, autonomous driving systems, and advanced manufacturing processes. This guides the industry on a track that doubles up on the promising new frontier of VR, AR, and AI tech to make training programs far more efficient, effective, and enjoyable than ever before while being future-tough.

In a nutshell, VR is an exciting, evolving technology in automotive training because it can change and evolve just as new technology does, creating immersive, adaptable, and collaborative learning opportunities. These developments will give technicians and factory workers the skills they need to thrive in an increasingly sophisticated and increasingly fast-moving industry.

CONCLUSION

Using virtual reality (VR) in automotive training is one great leap forward for the skill enhancement of the technicians and the workers on the factory floor. One will face the shortcomings of traditional training methods in virtual reality through immersive and interactive experiences that aim to address the challenges posed by limited accessibility, safety hazards, and ongoing learning outcomes. So far we have looked at how VR can be used to create real-world scenarios for training purposes so that workers can rehearse procedures in environments where the potential cost of error is greatly reduced and risk is minimized.

The automotive industry is one field where the use of VR can be seen in many different aspects, from ensuring factory and assembly line mechanics to the comprehensive vehicle diagnostics and repair. It uses a personalized, role-based model of learning that ensures employees are learning only what skills they need to do their jobs. Furthermore, VR training allows the industry to rise to the occasion and embrace its constantly changing demands with the scalability, flexibility, and integration of the latest technologies, such as AR and AI, that VR training solutions provide. But, in order to enjoy all the advantages offered by VR, these challenges should be met; namely, technological barriers, reluctance to adapt, and the learning curve associated with new technologies. However, these challenges can be overcome with careful planning, investment in infrastructure, and a user-centered design approach.

This work will be a huge element in the field of automotive training. By integrating VR as a crucial statement of training, establishments can inspire a more talented, assured, and future-proof personnel—one that's paramount to novelty and excellence in earnest withinside the automotive landscape.

REFERENCES

Carliner, S. (2015). *Training design basics*. Association for Talent Development.

Carnazzo, C., Spada, S., Lamacchia, S., Manuri, F., Sanna, A., & Cavatorta, M. P. (2024). Virtual reality in ergonomics by wearable devices: Experiences from the automotive sector. *Journal of Workplace Learning*, *36*(7), 621–635. DOI: 10.1108/JWL-03-2024-0064

Chiang, F.-K., Shang, X., & Qiao, L. (2022). Augmented reality in vocational training: A systematic review of research and applications. *Computers in Human Behavior*, *129*, 107125. DOI: 10.1016/j.chb.2021.107125

Firu, A. C., Tapîrdea, A. I., Feier, A. I., & Drăghici, G. (2021). Virtual reality in the automotive field in industry 4.0. *Materials Today: Proceedings*, *45*, 4177–4182. DOI: 10.1016/j.matpr.2020.12.037

Garcia, C. A., Naranjo, J. E., Ortiz, A., & Garcia, M. V. (2019). An approach of virtual reality environment for technicians training in upstream sector. *IFAC-PapersOnLine*, *52*(9), 285–291. DOI: 10.1016/j.ifacol.2019.08.222

Gavish, N., Gutiérrez, T., Webel, S., Rodríguez, J., Peveri, M., Bockholt, U., & Tecchia, F. (2015). Evaluating virtual reality and augmented reality training for industrial maintenance and assembly tasks. *Interactive Learning Environments*, *23*(6), 778–798. DOI: 10.1080/10494820.2013.815221

Henriques, A. C., & Winkler, I. (2021). The advancement of virtual reality in automotive market research: Challenges and opportunities. *Applied Sciences (Basel, Switzerland)*, *11*(24), 11610. DOI: 10.3390/app112411610

Hernández-Chávez, M., Cortés-Caballero, J. M., Pérez-Martínez, Á. A., Hernández-Quintanar, L. F., Roa-Tort, K., Rivera-Fernández, J. D., & Fabila-Bustos, D. A. (2021). Development of virtual reality automotive lab for training in engineering students. *Sustainability (Basel)*, *13*(17), 9776. DOI: 10.3390/su13179776

Langley, A., Lawson, G., Hermawati, S., D'cruz, M., Apold, J., Arlt, F., & Mura, K. (2016). Establishing the usability of a virtual training system for assembly operations within the automotive industry. *Human Factors and Ergonomics in Manufacturing*, *26*(6), 667–679. DOI: 10.1002/hfm.20406

Lawson, G., Salanitri, D., & Waterfield, B. (2016). Future directions for the development of virtual reality within an automotive manufacturer. *Applied Ergonomics*, *53*, 323–330. DOI: 10.1016/j.apergo.2015.06.024 PMID: 26164106

Mendes Monteiro, J., Torres, D., Ramos, A. L., & Pimentel, C. (2023). The use of virtual reality as e-training tool for dies' changeover in stamping presses: A case study on automotive industry. *International Journal of Lean Six Sigma.*

Naranjo, J. E., Sanchez, D. G., Robalino-Lopez, A., Robalino-Lopez, P., Alarcon-Ortiz, A., & Garcia, M. V. (2020). A scoping review on virtual reality-based industrial training. *Applied Sciences (Basel, Switzerland)*, *10*(22), 8224. DOI: 10.3390/app10228224

Norris, M. W., Spicer, K., & Byrd, T. (2019). Virtual reality: The new pathway for effective safety training. *Professional Safety*, *64*(06), 36–39.

Powell, M. B. (2008). Designing effective training programs for investigative interviewers of children. *Current Issues in Criminal Justice*, *20*(2), 189–208. DOI: 10.1080/10345329.2008.12035804

Quevedo, W. X., Sánchez, J. S., Arteaga, O., Álvarez, V. M., Zambrano, V. D., Sánchez, C. R., & Andaluz, V. H. (2017). Virtual reality system for training in automotive mechanics. *Augmented Reality, Virtual Reality, and Computer Graphics: 4th International Conference, AVR 2017, Ugento, Italy, June 12-15, 2017. Proceedings*, *4*(Part I), 185–198.

Roberson, L., Kulik, C. T., & Pepper, M. B. (2001). Designing effective diversity training: Influence of group composition and trainee experience. *Journal of Organizational Behavior: The International Journal of Industrial. Journal of Organizational Behavior*, *22*(8), 871–885. DOI: 10.1002/job.117

Shamsuzzoha, A., Toshev, R., Vu Tuan, V., Kankaanpaa, T., & Helo, P. (2021). Digital factory–virtual reality environments for industrial training and maintenance. *Interactive Learning Environments*, *29*(8), 1339–1362. DOI: 10.1080/10494820.2019.1628072

Smith, J. W., & Salmon, J. L. (2017). Development and analysis of virtual reality technician-training platform and methods. *Proceedings of the Interservice/Industry Training, Simulation, and Education Conference (I/ITSEC)*, *6*, 1–12.

Watanuki, K. (2008). Virtual reality based job training and human resource development for foundry skilled workers. *International Journal of Cast Metals Research*, *21*(1–4), 275–280. DOI: 10.1179/136404608X362098

Webel, S., Bockholt, U., Engelke, T., Gavish, N., Olbrich, M., & Preusche, C. (2013). An augmented reality training platform for assembly and maintenance skills. *Robotics and Autonomous Systems*, *61*(4), 398–403. DOI: 10.1016/j.robot.2012.09.013

Chapter 5
VR and AR Technologies Integrated Intelligent Transport System

Sarika Ghanshyam Jadhav
https://orcid.org/0000-0002-3685-5129
School of Computer Science, Engineering, and Applications, D.Y. Patil International University, Pune, India

N. Zahira Jahan
https://orcid.org/0009-0000-2119-2142
Department of Computer Applications, Nandha Engineering College, Erode, India

S. Karthiga
https://orcid.org/0009-0004-6076-566X
Department of Computer Applications, Nandha Engineering College, Erode, India

A. Punitha
https://orcid.org/0000-0001-5095-0694
Department of Electronics and Communication Engineering, K. Ramakrishnan College of Technology, Trichy, India

K. Neeraja
Department of Physics, Narasaraopeta Engineering College, Palnadu, India

Ravi Kannan
https://orcid.org/0009-0003-1456-2709
KKR Institute of Engineering and Technology, India

ABSTRACT

Virtual Reality (VR) and Augmented Reality (AR) Technologies in Intelligent Transport System.Presently, VR and AR technologies play a significant role in enhancing the transport system. Their applications include traffic simulation, route planning, and driver training. Furthermore, AR-enhanced mappings and navigation systems also boosted the efficiency of commuters through trusty Return to Base. Applications of virtual reality allow engineers to create planning for the smart infrastructure

DOI: 10.4018/979-8-3693-7287-6.ch005

by embedding future transport projects within the built environment. Out of these technologies, semiconductor and wireless technologies form the foundation for vehicle-to-everything (V2X) communication, the cornerstone of intelligent traffic management systems through seamless data exchange. The fusion of VR with superbusiness forms of AR and ITS then leads into adaptive and interactive and smart transportation systems which were also sustainable transport solutions to urban mobility challenges.

INTRODUCTION

The inception of modern vehicles, the transportation industry has been revolutionized by technological advancements in their dynamic. Among them, Virtual Reality (VR) and Augmented Reality (AR) have emerged as some radical tools that are embedded into Intelligent Transport Systems (ITS), making the mobility ecosystem smarter and more adaptive. Intelligent transportation systems (ITS) were developed based on communication, computing, and control technologies that have greatly contributed to reducing issues stemming from urbanization, such as traffic jams and environmental pollution. The current capabilities of ITS have grown with the incorporation of VR and AR technologies, which allow for real-time, immersive, and user-centric solutions that transform the very nature of transportation systems(Ejichukwu et al., 2024).

There are VR and AR devices specifically that leverage these interaction techniques to improve the user experience. The technology simulates a real environment and places the user into it, allowing for fully interactive virtual reality experiences that bestow a real-world feel. In contrast, augmented reality enhances real-world environments with digital information, adding layers of interaction to one's perception. In the field of transportation, these technologies offer incredible opportunities for innovation, from traffic simulation through route optimization and smart infrastructure planning to driver training. The propulsion into the role of ITS from these tools has created it into a dynamic ecosystem of information sharing for safe travel solutions for both commuters and the involved authorities(Sadeghzadeh et al., 2024).

Traffic management and simulation are other crucial elements of ITS that VR and AR technologies can make a big difference in. Many traditional traffic management systems are based on static data and cannot respond to high dynamic changes. VR ligate you simulation process traffic situations in the utz network hardware in a simulation mode, allowing planners and police to assess the impact of different ingredients without using them in the real world. AR, however, has been employed to display real-time traffic updates and navigation directives in a user's line of vision, greatly augmenting situational awareness. The impact of these new human-face technologies on driver distraction, a concern and an area of research in their own right, can best be illustrated with an example: augmented reality (AR) systems embed navigation routes, traffic signals, and other data inside the drivers view on the windshield and/or smartphone screen, therefore enabling the driver to stay focused on the road and eliminating the need to look down or away from the road at another display. These applications not only increase effectiveness but also reduce traffic demand and accidents(Rzepka et al., 2018).

Traffic management isn't the only place in which VR and AR technology have transformed; there has also been a shift in how we teach and train drivers. Traditional modes of driver education often employ blueprinted texts or glorified driving, which do not actually equip people for experiences that happen on the road. For VR-based driving simulators, that distance is closed by building virtual, immersive, and interactive environments that allow students to act from highly progressive scenarios in a safe environment, from bad weather to high speeds to emergency maneuvers. These platforms are extremely flexible—allowing instructors to tailor scenarios to specific learning objectives and to scale them. Similarly, AR has been used to extend the car-like guidance that illustrates where there are hazard areas, reminders of the speed limits, and what exactly to drive to improve the state of awareness of the driver and therefore improve safety(Besnea et al., 2019).

Virtual technologies applied in smart infrastructure development planVR and AR technologies have been proven beneficial in this crucial aspect of smart infrastructural planning. New urban transportation infrastructure such as bridges, roads, and public transit systems takes a long time to create and requires painstaking planning and visualization to make sure cities can develop sustainably for decades(Besnea et al., 2019). With the aid of VR tools, planners and engineers can create and manipulate highly detailed 3D models of infrastructure projects and simulate the long-term impact of the proposed design on traffic flow, environmental impact, and the likely aesthetics of urban space. A templated virtual one of these gives all stakeholders a comprehensive overview of what the projects deliver and helps with decision-making further down the line while minimizing the risk of costly mistakes in implementing them. Using AR, technicians could now view utilities buried underground, identify structural flaws, and receive instructions on repairs in real time, allowing for even better maintenance of infrastructure. Incorporating AR and VR in infrastructure management continually empowered the life cycle management of transportation assets(Tubaro & Casilli, 2019).

VR and AR technologies are also contributing to facilitating the improvement of vehicle-to-everything (V2X) communication, an underlying technology of Intelligent Transportation System (ITS). V2X generally refers to vehicle-to-everything communication, which is generally a class of networked technologies referred to as intelligent transportation systems that allow networked vehicles to communicate with moving parts of infrastructure. In a few older studies, virtual reality (VR) applications were created in order to simulate vehicular interactions in a virtual environment where developers could tune their algorithms and test their communications protocols. AR, by comparison, has focused use cases around providing real-time feedback to drivers—everything from alerts about nearby vehicles or pedestrians to potential hazards in the road ahead. These technologies have made roads safer, reduced the possibility of accidents, and set the stage for the rise of self-driving vehicles(Boboc et al., 2020a).

Despite this, their adoption for transport systems, and ITS in particular, has been hampered by many challenges. Despite the enormous excitement around AI technology—it has established itself as one of the hottest topics of our time—we are not living in a world where AI is a mainstream product. Implementation costs, hardware limitations, and risks to data privacy remain fundamental barriers to AI, which appear impervious to breaks in established business models. We need to develop low-cost, lightweight devices with reasonable data security protocols to bypass these barriers. Additionally, the application of VR and AR in ITS can only be successful by relying on high-speed communication networks (especially 5G) to achieve fast processing and transmission of large amounts of data. Only through continued enhancement and experimentation can we address these constraints and understand the entire range of options immersive technologies have in helping the transportation industry(Wolfartsberger et al., 2020).

The potential of cutting-edge technologies such as virtual and augmented reality is phenomenal in the future of ITS as the cities keep growing along with public transport needs. Such tools have already shown the capacity to improve the efficiency, safety, and sustainability of transportation systems. This integration will not only lead to smarter, safer transportation systems but also an inclusive transportation network by facilitating collaboration between technology developers, urban planners, and policymakers. Here is a human-like style sentence: With the ongoing development of these technologies, transportation systems are likely to be completely transformed as they seek innovative solutions for major issues in urban mobility(de Freitas et al., 2022).

Objectives

The heart of this integration of VR and AR technologies for ITS vessels will aim at enhancing transportation systems efficiency, safety, and sustainability in modern transportation lanes. We will also discuss how real-time visualization along with the use of data until October 2023 has helped to enhance traffic management with the design of context-aware AR guidance systems for either advertisement and navigation systems or the use of immersive training that acts as driver training solutions for safer roads. Additionally, VR and AR also attempt to allow smart infrastructure planning due to their ability to get visualizations and maintenance of transport assets. This ensures seamless V2X communication and solves urban mobility challenges through enhancing intelligent, demand-responsive, user-focused transportation systems for a sustainable future.

BACKGROUND: VR AND AR TECHNOLOGIES

VR and AR are disruptive technologies that revolutionize the way humans engage with virtual spaces and deliver immersive and interactive experiences in a variety of contexts. While each is implemented and applied differently, they aim to converge the physical and digital worlds. They are used in a wide variety of sectors, including healthcare, education, entertainment, manufacturing, and transportation, making them versatile and innovative. Their definition and applications will be explored as a basis for the integration of these technologies into Intelligent Transport Systems (ITS designs)(Lee et al., 2023).

Virtual Reality: Definition and Applications

Virtual reality is an immersive technology that makes a real-world environment in which a person can interact in that environment with the maximum level of experience, as in that AR it cannot be limited. So, VR is the individual who is an exact replica of the real one and its role with the reality contact world, Using devices like VR headsets, motion controllers, haptic gloves, etc., in this as HRV, it is completely disconnecting the image or In a copy or in the world of the same copy in a micro-oocentric way, Marther personalizes experience, so this type of communication is seen in the language and language, highlighting or virtualizing real experiences or recreating experiences from the aesthetic perspective of nearby location. What defines VR as VR is that sense of presence—the sense that you are in the made-up world. VR can be found in a wide array of fields, such as training, education, entertainment, manufacturing, design, therapy, and transportation. Whether tomorrow testing for skills through real-life simulations, immersive experiences for gaming or virtual tourism, reducing time and cost for manufacturing and design, therapy and rehabilitation, it works for a variety of reasons. This enables training procedures that would be effective and practical without jeopardizing other drivers and facilities with the training procedures(Bale et al., 2024).

Augmented Reality: Definition and Applications

Augmented Reality (AR) refers to the imposing of digital information, such as images, text, etc., on the surrounding real-world environment, thus enhancing the perception of the real world. This differs from VR because instead of immersing users in a different world entirely, AR allows a mix of both the real and digital worlds and gives users a room that can be occupied by both. Most AR applications attach digital content to the real world through a device such as a smartphone or an augmented reality type of smart glasses or heads-up displays (HUDs).

VR (Virtual Reality): a complete immersive technology that integrates a simulated element for users to experience a simulated, 3D-created environment. By creating a barrier between users and their local environment, it provides a sense of presence. VR is used in training and education, entertainment, manufacturing, design, therapy, and transportation. It helps with skill development by simulating the real-world scenario, provides a thrilling gaming experience, reduces design and manufacturing costs, and is best suited for therapy & rehabilitation. VR is also used for driver training, traffic simulation, and even infrastructure planning. However, there is one dimension to VR and AR that renders these instruments acts of innovation: VR builds simulated experiences, while AR enhances real-life experiences with digital elements. This has led to the evolution of intelligent transport systems; the convergence of their potentials has opened new possibilities for smart, cashless, and efficient mobility solutions.

Figure 1. Integration of VR and AR in Intelligent Transport Systems (ITS)

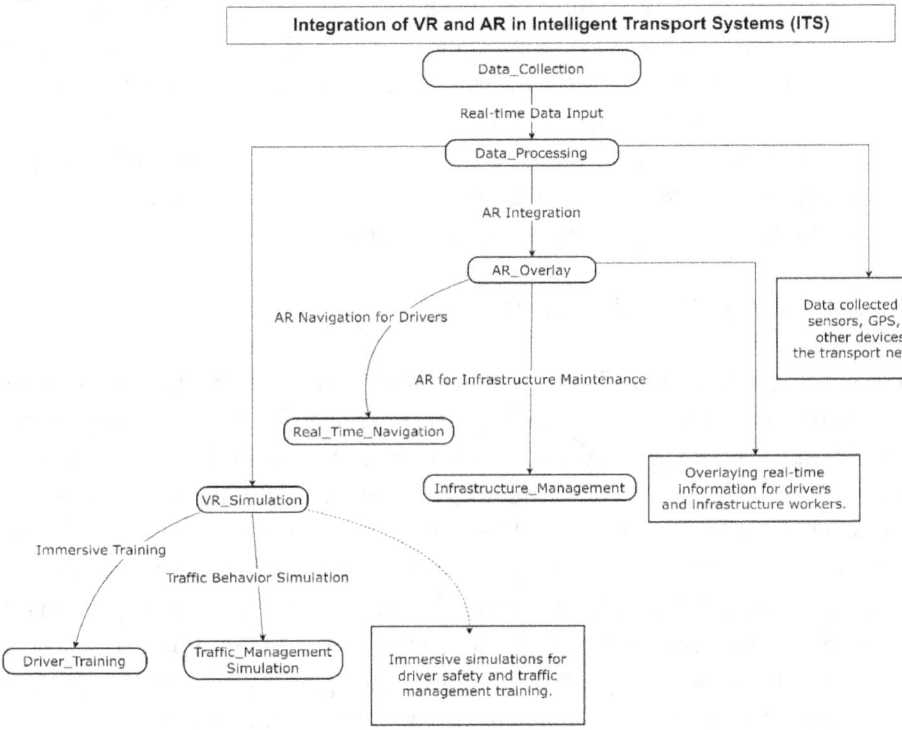

INTEGRATION OF VR AND AR IN INTELLIGENT TRANSPORT SYSTEMS

VR and AR for Traffic Simulation and Management

Intro: VR and AR offer the flexibility of low-cost intelligent transport system (ITS) analysis as part of new traffic simulation and management that solves problems arising from urbanization, road safety, and congestion(Gabbard et al., 2019; Xiong et al., 2018). The integration of VR and AR in Intelligent Transport Systems (ITS) is depicted in Figure 1.

Planners can test the result of changes to their design on realistic VR simulations of traffic on the road and auspicious in a safe, controlled environment. It creates 3D images of roadway grids, interchanges, and district designs, allowing stakeholders to assess the impacts of changes without needing them to work on the roads. VR simulations could help researchers go about determining the behavior of drivers and how to design policies and traffic networks for safety and efficiency. AR technology has elevated real-time traffic management sales via context-aware overlays within users' fields of point of view. This enhances situational awareness and reduces navigation errors. In addition, AR will prove to be a powerful tool in helping the road authorities visualize the traffic congestion patterns, identify trouble spots, and create a window for them to intervene in time. AR is also being used widely in public transportation systems, which are providing real-time schedule updates and assisting in directions for enhanced commuter experience.

Real-Time Navigation Using AR

One emerging field where ITS has gained new momentum is through the infusion of new technologies such as virtual and augmented reality. For high accuracy, safety, and convenience in navigation, AR is the best choice. This digital information overlays the user's field of vision, thus reducing distractions and improving situational awareness, which is a valuable tool in ITS(Boboc et al., 2020b; Fang & An, 2020).

Real-time navigation is provided by AR technology through electronic devices such as smartphones, tablets, smart glasses, or HUDs. It leverages sophisticated sensors, GPS, and computer vision to position digital content within the real world. AR guides drivers by projecting arrows or highlighted pathways, provides real-time updates on traffic conditions, and enhances safety by alerting drivers to potential hazards. This preemptive action prevents accidents. Over the years, AR navigation has truly transformed public transportation systems as well as logistics. For users, it provides real-time schedule changes, directions, and interactive maps while helping fleet operators optimize delivery routes and effectively manage cargo. This AR technology has sprung up an entirely new spectrum of commuting experience, right from safety, comfort, and interactive experience while commuting with ITS. It allows for smarter, safer, more efficient transportation systems for modern cities.

Enhancing Driver Training with VR Simulations

Use of virtual Reality(VR) for driver training in intelligent transport systems (ITS) has revolutionized the entire concept of driver training by providing virtual, realistic, and interactive simulation. In fact, most of today's driver training techniques use static material or do not include much real-world practice. Virtual reality (VR) goes beyond these limitations by simulating controlled and customizable virtual environments in which drivers can experience a wide variety of scenarios, practice crucial skills, and gain confidence with no risk to themselves or other road users. There are VR simulations for driving as well, which could create a safe training environment for trainees and give them a trial with poor weather conditions and crash scenarios that cannot be replicated so easily without VR training. These immersive environments allow learners to make mistakes in the safest environments, so they will be better prepared when dealing with challenges in real life while providing control and safety when they encounter or are confronted with adverse weather conditions(Manimuthu et al., 2022; Vashisht & Rakshit, 2021).

Using advanced analytics, virtual reality driver training systems track performance, delivering rich feedback and personalized recommendations. All of these are data-driven systems, so the training is individual and focused. Throw in all of the above and combine that with specialized vehicles (buses, trucks, emergency vehicles, etc.). optimally trainable, the immersion of the VR simulation saves real-world time and can erode physical muscle memory. Through the intersection of the academic and industrial sides, the permeation of VR on ITS leads to improvements in the quality and efficiency of driver training programs; this, in turn, leads to an increase in the behavior of road users in a way compatible with road safety. It has translated into decreasing road traffic accidents, better driver proficiency, and leading to safer mobile respective systems.

IMPLEMENTATION PROCESSES IN ITS

VR and AR integration in Intelligent Transport Systems (ITS) is a combination of multiple domains, mainly hardware infrastructure, software development, data process management, and user experience design. Their solutions for traffic management, driver training, route optimization, and V2X communication are revolutionary. Their successful deployment hinges on overcoming some technological, logistical, and regulatory rigors and ensuring that the systems generated are scalable, efficient, and user-friendly. In this part, we first present the stages of VR and AR technologies involved in the intelligent transportation system from ideation to operation(Kaladgi et al., 2021; Yıldız et al., 2020). Figure 2 depicts the implementation processes of VR and AR in Intelligent Transport Systems (ITS).

Figure 2. Implementation processes of VR and AR in Intelligent Transport Systems (ITS)

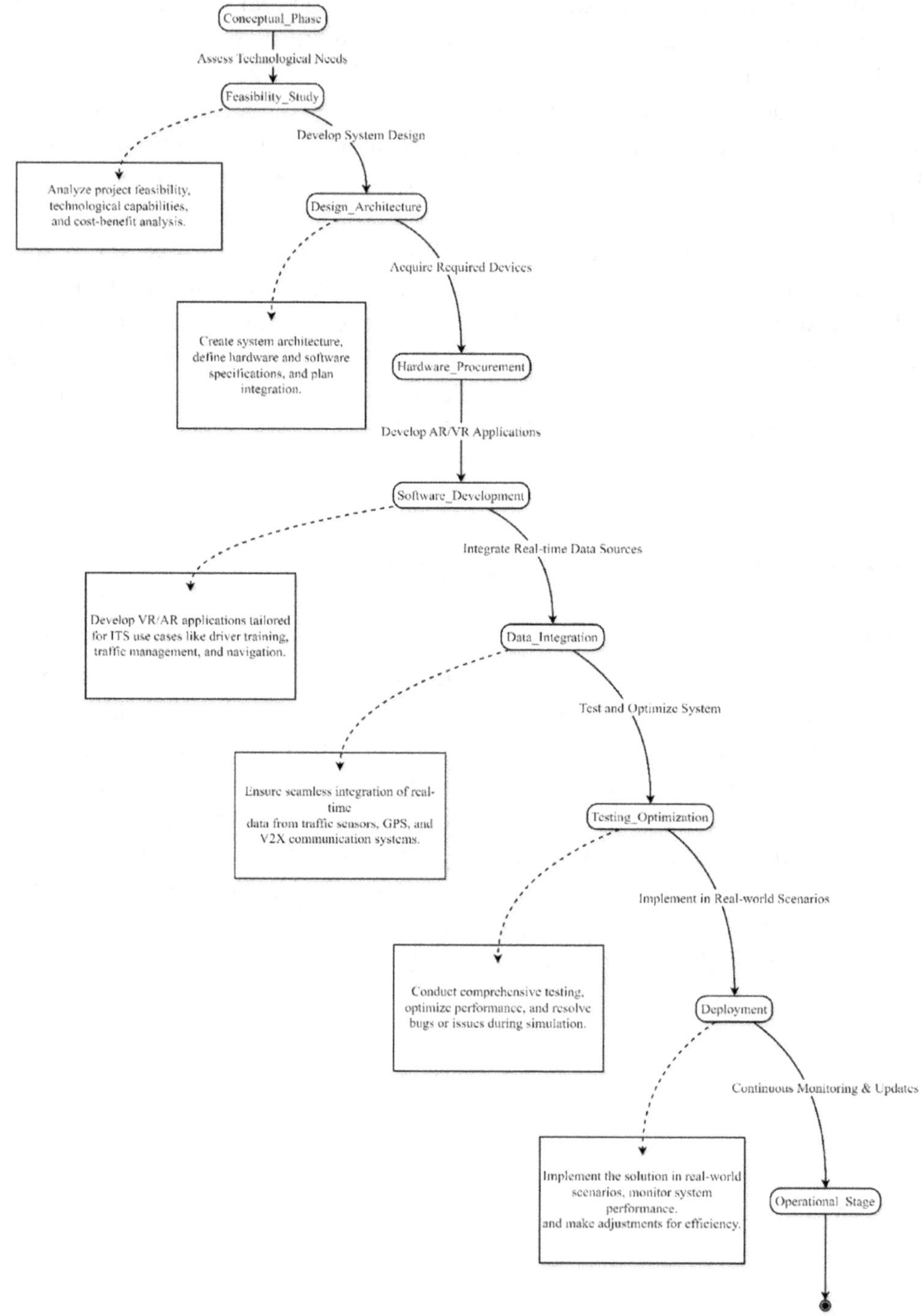

1. Identifying Use Cases and Objectives

The first step in applying these technologies in ITS is the identification of the particular uses and dividends that VR and AR should deliver. These objectives may be determined by the specific goals of an ITS system, like improving traffic flow, reducing accidents or injuries to drivers, improving real-time navigation, or bettering transportation infrastructure. Specifying what exactly the goals are leads stakeholders to prioritize what needs to be solved first. Different fields will have different goals and, therefore, different ways to use simulation in training—for example, with driver training, the main goal is to prevent accidents, so immersing the driver in potentially dangerous road environments via simulation is a very important part of the process. Here is a very similar idea in building a dynamic visualization tool for traffic management: a real-time tool for city planners to observe and manage the traffic(Dalle Mura & Dini, 2021; Kaladgi et al., 2021).

2. Selecting the Appropriate Technology and Tools

They are based on the use cases when we choose VR and AR technologies. Augmented reality (AR) is generally experienced using a pair of smart glasses, a smartphone, or a heads-up display (HUD) to overlay digital information over reality, while virtual reality (VR) is experienced wearing light, comfortable headgear to enjoy high-quality graphics and spatial accuracy. Devices available: The choice of devices will depend on preference, cost, compatibility, and the type of interaction needed by the application. Software platforms allow for the integration of VR and AR into ITS by providing the infrastructure for complex virtual environments, real-time data processing, and multi-user experiences across multiple devices. A VR traffic simulation platform, for instance, needs to connect with city traffic management systems to visualize the real-time info available. There are multiple reasons why software can't just sit still.

3. Data Integration and Real-Time Analytics

In the context of Intelligent Transport System (ITS) applications, the integration of real-time data plays a significant role in the success of VR/AR applications. In applications such as traffic management or navigation, this means connecting the VR or AR systems with various external data sources such as traffic sensors, GPS data, weather forecasts, and vehicle-to-everything (V2X) communication systems. Real-world data, such as closures, accidents, or detours, must be updated in real time and presented to the user, for example, in an AR-based navigation solution. VR simulations used for driver training simulated reality, with changing traffic patterns, weather, and vehicle behavior, but needed to be as realistic as possible. This is where a robust data management system comes into play as it helps manage, process, and store data in real-time for decisions and user interactions. Advanced data analytics and AI algorithms are used to provide predictive traffic graphs, hazard identification, and path optimization. ML may become another well of VR and AR systems that are trained for new instances or user experiences derived from interactions.

4. Designing User-Centric Interfaces and Experiences

User interfaces (UIs) that are intuitive and easy to navigate are critical to the successful integration of virtual reality (VR) and augmented reality (AR) in intelligent transportation system (ITS) applications. The goal is to create a continuous virtual space for interaction. Interfaces should be responsive to users so users are able to complete tasks, such as 'steering a vehicle' or 'casting a vote.'. They use haptic feedback and motion tracking to create a detailed experience. AR must impart distinctly clear digital information in bite-sized form without inundating users with an overload of data. Naturally, they also have applications in navigation systems, supplying route directions, speed limits, and hazard warnings without obstructing the driver's line of sight. In urban areas, AR interfaces could call out points of interest, construction sites, or even available parking spots. Hence, the user-centered design principles are a very, very important part of building a fulfilling system, which needs to be fulfilled to cover up the needs of all stakeholders by this simple approach, like usability testing.

5. System Integration and Testing

The step two implementation deals with the integration of the VR and AR systems into already existing ITS infrastructure after the design and development phase. This might include linking up to traffic management systems, GPS networks, public transportation systems, and other transportation-related tech. VR and AR data must move freely with these kinds of systems, and VR and AR must enhance existing capabilities like traffic control signals, route optimization algorithms, and vehicle sensors. Testing is an integral part of implementing, as it allows us to pinpoint if any problems occur, such as latency, system errors, data discrepancies, etc. The systems that have been developed will need to go through a rigorous testing phase in a simulated environment before they can be implemented in the real world. This encompasses validating the accuracy of data overlays in average systems and the quality of virtual reality training simulations. Additionally, testing should involve security and privacy considerations to protect sensitive data such as personal information and vehicle telemetry as well.

6. Deployment and Continuous Monitoring

Thus they may be slowly deployed in stages, either in pilot programs or in larger areas, after development and integration and testing. After deployment, however, continuous monitoring becomes key, delving into user feedback, system diagnostics, and real-time data to understand areas that need improvement. It requires regular updates to accommodate new data sources, learn new traffic patterns, and add new features as per user demand and technological trends. The infusion of VR and AR technologies in ITS is a multi-layered process that encompasses identification of objectives, selection of technology, integration of data, user interface evaluation, and iterative enhancement of integrated solutions. Technologies that could enhance traffic control, driver instruction, and prevent road disasters.

APPLICATIONS IN SMART INFRASTRUCTURE AND URBAN MOBILITY

Virtual Reality (VR) and Augmented Reality (AR) technologies integrated into smart infrastructure and urban mobility These technologies are changing the way cities map transportation networks, making them more sustainable, effective, and better able to guarantee security. VR and AR are now essential tools used by urban planners, engineers, and transportation authorities to visualize projects, conduct building maintenance, and develop long-term strategies for urban mobility. This part features transport projects visualization, infrastructure maintenance and planning, and sustainability in urban mobility, which use VR and AR(Kaladgi et al., 2021; Manimuthu et al., 2022; Vashisht & Rakshit, 2021). Utilization of VR and AR in smart infrastructure and urban mobility is illustrated in Figure 3.

Figure 3. Applications of VR and AR in smart infrastructure and urban mobility

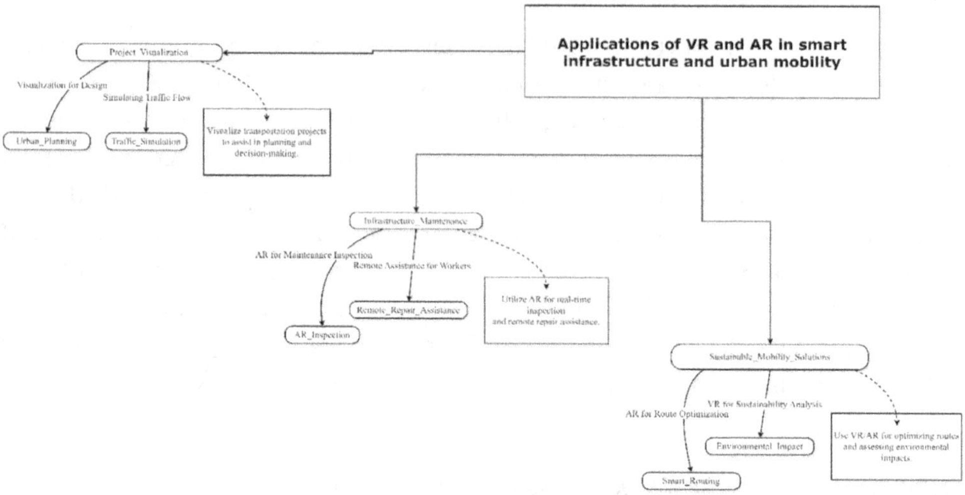

1. Visualization of Transport Projects with VR

The use of Virtual Reality (VR) in the visualization and simulation of transport projects, especially for early design stages, is becoming more prevalent. Engineers and urban planners can now map out things like traffic flow, environmental impact, and public safety, and stakeholders can walk through 3D-yestimations of infrastructure projects before they are realized. By providing them with the visual representation that they need, VR simulations can allow urban planners and decision-makers to also visualize and have a deeper understanding of transportation systems. They enable stakeholders to simulate virtual stations, trains, and interactions with surrounding areas and spot potential issues early on, like congestion hotspots, accessibility challenges, and integration with other modes of transport(Zhang et al., 2019; Zhou et al., 2019).

VR can help to facilitate stakeholder buy-in as local residents, policymakers, and investors can see the potential effect of transport projects on their neighborhoods or municipalities. More informed public consultations, town hall meetings, and often achievable consensus help create more successful projects. Project visualization is heavily used in the transport sector, and with VR, planners can visualize and model future scenarios on an interactive basis, adapting the simulation based on changes such as traffic volumes, seasonal variations, and environmental factors, which helps provide a comprehensive, adaptable view of infrastructure performance. This aids stakeholders in understanding the long-term surface and sustainability of transport systems.

2. AR for Infrastructure Maintenance and Planning

Augmented reality (AR) is a fabulous way to asset manage and field plan in all transport systems. It is essentially a tool that lays a digital layer over the real world to facilitate real-time maintenance and management of critical infrastructure. AR makes this much easier, allowing real-time overlayers to be put over physical world objects from which engineers, maintenance workers, and planners can troubleshoot services and inventory to plan repairs(Eswaran & Bahubalendruni, 2022; Geng et al., 2022).

Provided workers wear AR glasses or contact lenses painted with digital overlays of information (e.g., age, maintenance history, structural integrity), AR technology can better deliver real-time information (at the point of inspection) on existing infrastructure. Such context also enables complex repairs that minimize faults and impact in a rapidly shrinking urban built environment and allow for minimal disruption to transportation networks because internal processes in transportation are predicated on immediate response. Augmented reality apps can assist with the upkeep of subterranean infrastructure, overlaying information about pipes, cables, and conduits onto the visible field of underground workers. This reduces the risk of damaging when excavating or working on, bottom line, and worker safety and integrity.

AR technology can really be of help in urban planning, as it visually allows the interaction of systems impacting each other. It allows planners to test spatial constraints, environmental impact, and traffic patterns before making any physical changes. AR also provides insight on how work will impact space while under construction or renovations. It also enhances collaboration among stakeholders such as engineers, urban planners, and municipal authorities by allowing users to enter shared AR environments to visualize, modify, or annotate virtual models of their infrastructures, thereby enhancing the communication and the decision-making process and efficiency in urban mobility governance.

3. Sustainability and Urban Mobility Solutions

The predominant nature of most cities worldwide is transitioning. Encouraging greener transport: The use of VR and AR technologies can help streamline infrastructure, reduce emissions, and promote greener forms of transport. The deployment of these technologies modernizes public transport systems, electric mobility, traffic management, and environmental issues, promoting overall urban quality of life. For public transport and sustainability, VR and AR can offer solutions. Good urban planners can use these simulations to optimize their networks for access for the greatest number of people while minimizing environmental sustainability. They can model traffic congestion, emissions, and energy use—and help identify ways to reduce a carbon footprint or improve the efficiency of a service. The buses are electric; bike lanes are part of the street plan; sidewalks are pedestrian zones(Geng et al., 2022; Wang et al., 2021).

It aids the infrastructure to be more effective, which boosts sustainability. It steers clear of congested routes, reducing fuel consumption and emissions. Real-time information about traffic conditions, parking availability, and eco-friendly routes helps to prevent the usage of cars, improving the air and reducing the urban heat island effect. By providing visibility into a range of ride types, AR enables carpooling and similar shared mobility solutions. Improved urban mobility has been possible through services powered by the capabilities of AI in processing data from virtual and augmented reality (VR & AR)-based technologies that connect the data sources for an integration of AR & VR. VR is used to design smart intersections with dynamic traffic lights to reduce congestion and optimize the flow of traffic. Post the augmentation of the EV charging stations, real-time location and optimization suggestions can be adapted to harness cognitive technology to make EV adoption more intelligent and accessible.

These virtual and augmented realities are essential to building smarter and more sustainable cities to enhance the planning and maintenance of the different infrastructures and to improve the important utilization of infrastructures while saving precious resources and leading to a lower environmental footprint and the needed resilient long-term mobility where all systems talk to each other to help minimize attention on the availability of further resources. The urban mobility will change on the basis of efficiency, safety, and sustainability in the transportation networks in the light of VR and AR. They help with visualizing projects, maintaining infrastructure, and promoting sustainability. Taken together, rapid growth, congestion, and environmental challenges confront the frontiers of a sustainable urban future—one in which these technologies will undoubtedly become an increasingly important part of the urban mobility mix. As a result, more intelligent, more sustainable, and more connected cities emerged, improving the quality of life of their citizens.

TECHNOLOGICAL CHALLENGES AND LIMITATIONS

While the integration of Virtual Reality (VR) and Augmented Reality (AR) into Intelligent Transport Systems (ITS) promises significant advancements in urban mobility, infrastructure management, and driver training, there are several technological challenges and limitations that need to be addressed for successful deployment. These challenges range from integration and compatibility issues to hardware and cost constraints and data privacy and security concerns. As transportation systems continue to evolve, overcoming these hurdles is essential to realizing the full potential of VR and AR technologies in ITS(Aftab et al., 2020; Aslan & Yıldız, 2020; Boboc et al., 2020b). Figure 4 illustrates the challenges and limitations of integrating VR and AR with ITS.

Figure 4. Technological challenges and limitations of integrating VR and AR with ITS

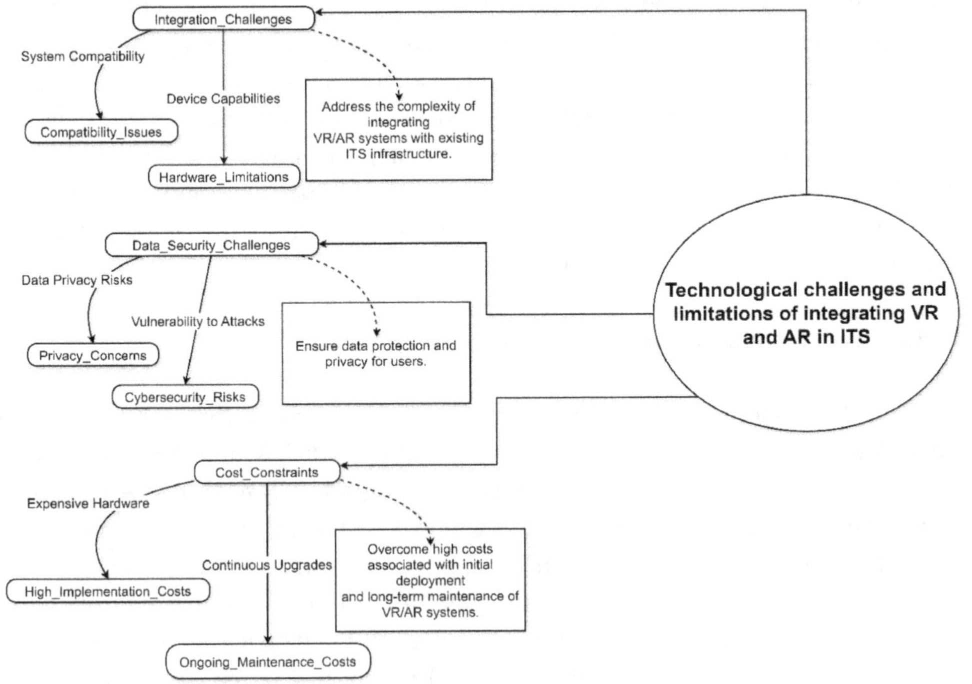

1. Integration and Compatibility Issues

VR and AR integrated into Intelligent Transport Systems (ITS) faced challenges, primarily compatibility with existing transportation infrastructure(Manimuthu et al., 2022; Tripathi & Talukder, 2023). They use a complex network of sensors, traffic management software, real-time data analytics tools, and communication networks. Vote for the synergy of VR simulations in real time with traffic sensors, GPS systems, and weather forecasts for proper integration. To ensure the AR navigation accurately reflects real-time road conditions and closures, AR-based navigation systems need to integrate with a city's traffic management solutions.

Hardware is not exempt from compatibility issues either. The capabilities and interfaces for VR and AR devices are significantly different, thus complicating standardization for VR and AR platforms. The simplicity of navigation with AR will enable it to easily connect to various leading phone brands and vehicle interfaces, including smartphones and tablets, as well as heads-up displays. As such systems gain increasingly more adoption, the uptake of universal standards and communication protocols between the various technologies will become all the more necessary. Meanwhile, the V2X communication system, which is a player-agnostic interconnectedness between vehicles, infrastructure, and outside networks, should be the heterogeneous connected option to ensure seamless integration with vehicular AR & VR technology. What indeed presents a major technical hurdle is to further smooth the conversation and data interchange between these components together with the realities of diverse data formats, conversation protocols, and information synchronization over many systems.

2. Hardware and Cost Constraints

VR / AR technologies embedded in ITS, second the most critical barrier is the demanding need for new hardware and their cost. Such VR and AR experiences require specialized equipment and hardware (top-grade VR headsets, AR glasses, motion sensors, haptic feedback devices), which is costly and technically complex to create. The hardware costs of developing, deploying and maintaining these devices can be a significant barrier of entry for many transit authorities, particularly in the developing world where developer budgets can be tight when it comes to infrastructure projects. Consider the example of virtual reality-based driver training — the construction of physical infrastructure for VR-based training involves an enormous amount of capital expenditure on high-end VR head-mounted displays, motion capture frameworks, and computer hardware capable of running high-performance simulations. The complexity of these devices and the system needed to help them (continuous preparation of spaces with enough physical and equipment, for instance) can drive up costs and make execution restrictive for specific affiliations(Bale et al., 2024; Gouda & Tiwari, 2024).

AR devices need to be adopted on a large scale (through things like smart glasses or heads-up displays), which will not happen until hardware like batteries, weight, resolution, and tracking systems become widely improved. However, solutions such as Microsoft HoloLens and Google Glass are currently too expensive and limited, especially for the possible applied solutions in the field of urban mobility and transportation management. This could hamper large-scale adoption of flying taxis, particularly in developing cities or regions, where the capital to invest in transportation infrastructure is limited. In addition to that, as the world of VR and AR is constantly evolving, so are the latest improvements, as such devices and equipment need to be continually replaced for better performance and functionality. Upgrading and maintaining existing infrastructure remains a challenge for many transportation authorities and organizations that have already made substantial investments in VR and AR systems.

3. Data Privacy and Security Concerns

Among the various security and privacy issues associated with the integration of VR and AR technologies in [ITS], data privacy and security of the communication networks are the most important issues. VR and AR systems are also highly reliant on data collection and processing, from real-time traffic to vehicle telemetry, personal information, and local information. This may be sensitive data, implying that this must be stored and transmitted, as well as processed securely. For example, GPS data is always collected and analyzed by AR navigation systems, so you have accurate and timely directions. These systems are vulnerable to this data getting misused and raised these concerns of unauthorized access or data breach(Manimuthu et al., 2022). VR's often hyper-personalized approach to training means that systems like those used to teach people how to drive could still collect data on that person's performance, behavior, reaction times, and more while conducting the simulations, raising the question of how that data would be used and if it would violate people's privacy. However, when mishandled, such information can be abused for purposes for which it is not meant or may lead to access by malicious actors, which could lead to identity theft or fraud.

Arising from these new paradigms, the integration of AR and VR technologies into the current vehicular network, known as V2X, causes an additional challenge regarding privacy and safety issues of the network. Vehicle-to-everything (V2X) systems rely on continuous information sharing between vehicles, infrastructures, and external sources, and the protection of such communication from cyberattack, hacking, or unauthorized monitoring is crucial. Such a system, if one unit becomes compromised, will harm the safety and privacy of the users as well as be a threat to the key mode of transport. So on the data security side of that concern, it has to do with the data ownership and data consent. Since the VR and AR applications will be used to collect and work with vast amounts of personal data, concise policies should be developed as to who owns this data, who is allowed to access it, and what the period of data retention is. (Note: Emerging privacy regulations, such as the General Data Protection Regulation (GDPR) in Europe, are making organizations liable for improving transparency and accountability of how they use personal data.)

Every VR and AR system must comply with these standards to protect users' privacy and create trust in the system, where the regulations vary depending on the region.

The rise of VR and AR will, of course, come with applications to the metaverse with connected users and IoT devices, which could lead to potential cyberattacks that could maybe compromise user safety or even transportation networks. Strong IT safety measures such as encryption, multi-factor authentication, and continual monitoring will become mandatory in order to keep the integrity of VR and AR-based ITS options intact. The combination of immersive experience (VR and AR technologies) with intelligent transport systems is expected to give a breakthrough to urban mobility, efficiency in managing traffic, maintenance of infrastructure, and driver training. However, existing barriers such as integration, hardware and cost constraints, and data privacy and security challenges need to be addressed. These challenges require advances in technology, the creation of industry standards, investment in infrastructure, and extensive security protocols. Having overcome these challenges, VR and AR will be a key part of intelligent and sustainable transport systems.

FUTURE PROSPECTS OF VR AND AR IN ITS

Virtual Reality, Augmented Reality: Future Perspectives in Intelligent Transport System (ITS) Futures Perspectives. As research and development continue, you can expect VR and AR to have an increasing impact on creating smarter, more efficient, and sustainable transportation systems(Tripathi & Talukder, 2023). Figure 5 depicts the potential of VR and AR in the development of Intelligent Transport Systems (ITS).

Figure 5. Future prospects of VR and AR in Intelligent Transport Systems (ITS)

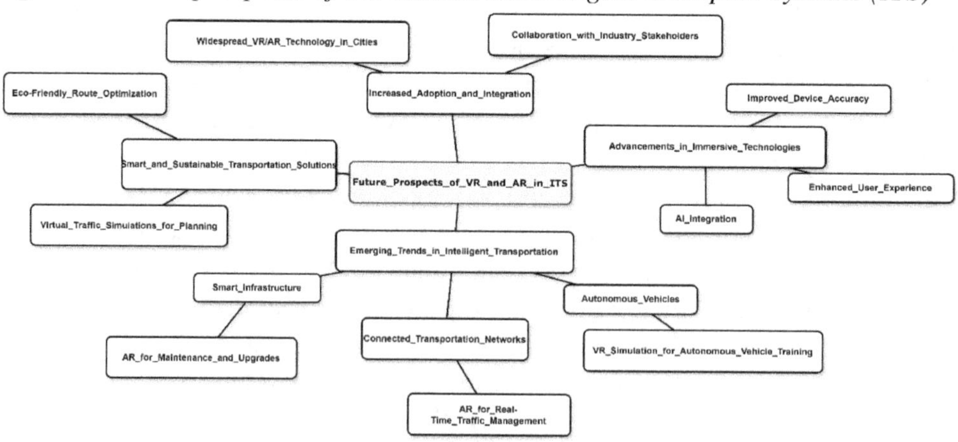

Advancements in Immersive Technologies

As the technology behind immersive experiences advances, VR and AR will become increasingly refined, providing more realistic simulations, better technologies, and increased performance. Compared with the VR headsets of today, those from the mid-2030s are predicted to be lighter, more comfortable, and more affordable, enabling large-scale driver training or public transportation VR simulation programs. AR glasses, in turn, will also become more seamlessly integrated into how we go about our day, providing hands-free navigation, real-time updates on traffic, maintenance alerts, and assists designed to be proactive before an issue arises, not reactive after something goes wrong. There really is no more fitting an analogy than this to describe how AI and machine learning will provide advanced VR and AR systems with the capability to adapt and personalize experiences for users. VR simulations for driver training, for example, could be created based on need variety—improving certain road performances in varying weather supplies or practicing while taking into account the real-time people that navigate the streets you travel. AR conclusion: The very purpose of AR is that it can be context-sensitive, for example, filtering the displayed information based on the currently presented environment of a user. These improvements will help to make ITS applications more effective and efficient, providing safer roads and better-trained drivers(Manimuthu et al., 2022; Vashisht & Rakshit, 2021).

Emerging Trends in Intelligent Transportation

The emerging trends such as AVs, connected transportation networks, big data, and IoT integration are going to show a major impact on the sector in how VR and AR technologies are making their way into the ITS field. By simulating and visualizing how AVs respond to diverse driving conditions and road environments and how AVs interact with human drivers, VR and AR will be instrumental in the development and testing process of AVs. Also, AR & VR technologies are going to be platforms on which all cities would be building smart infrastructure for real-time monitoring and decision-making(Geng et al., 2022; Swathi, 2022). With AR, vehicles, traffic management systems, or pedestrians can connect, helping improve mobility efficiency and relieve traffic congestion. Testing theories about the operation of transportation systems in the real world can be done through simulating all of the variables within virtual environments, and new modes or methods can be fine-tuned prior to implementation to fit the larger aims of urban development and sustainable living. As the IT landscape is changing, VR and AR have also evolved to be the key technologies to facilitate smarter, sustainable, and immersive transportation. As these technologies mature and interlace with wider developments in smart cities, their role in shaping the future of urban mobility will grow.

CONCLUSION

The employment of Intelligent Transport Systems (ITS), enhanced by Virtual Reality (VR) and Augmented Reality (AR) technologies, could reshape urban mobility and revolutionize the management of urban transportation infrastructure. From immersive simulations to real-time data visualization, VR and AR offer powerful tools for enhancing transportation networks' safety, efficiency, and sustainability. The power to see complex projects, manage infrastructure with accuracy, and maximize mobility options marks an important step toward smarter, more connected cities. The successful integration of VR and AR technology into Intelligent Transport Systems (ITS) comes with disadvantages like integration, system hardware cost, and data integrity. However, greater AI-driven personalization and more intuitive AR equipment will save enhanced, more usable devices. Other new applications for the managing transportation sector include developing real-time predictive models for handling congestion, following up on emerging trends around autonomous vehicles, dynamic infrastructure, and micro-mobility. These will make future transit systems more efficient, safer, and eco-friendlier.

REFERENCES

Aftab, A. R., von der Beeck, M., & Feld, M. (2020). You have a point there: Object selection inside an automobile using gaze, head pose and finger pointing. *Proceedings of the 2020 International Conference on Multimodal Interaction*, 595–603. DOI: 10.1145/3382507.3418836

Aslan, B., & Yıldız, A. R. (2020). Optimum design of automobile components using lattice structures for additive manufacturing. *Materials Testing*, *62*(6), 633–639. DOI: 10.3139/120.111527

Bale, A. S., Patil, M., Prathima, M., Telgane, S. V., Ghorpade, N., & Anish, B. (2024). Augmenting Knowledge Management with Immersive Technologies: Exploring Transformative Use Cases for AR and VR. *2024 International Conference on Advances in Computing, Communication and Applied Informatics (ACCAI)*, 1–6. DOI: 10.1109/ACCAI61061.2024.10602348

Besnea, F.-L., Resceanu, I.-C., Cismaru, S.-I., Ganea, I.-E., Pistritu, I.-L., & Bizdoaca, N.-G. (2019). Experiments regarding implementation of a virtual training environment for automotive industry. *2019 20th International Carpathian Control Conference (ICCC)*, 1–6.

Boboc, R. G., Gîrbacia, F., & Butilă, E. V. (2020a). The application of augmented reality in the automotive industry: A systematic literature review. *Applied Sciences (Basel, Switzerland)*, *10*(12), 4259. DOI: 10.3390/app10124259

Boboc, R. G., Gîrbacia, F., & Butilă, E. V. (2020b). The application of augmented reality in the automotive industry: A systematic literature review. *Applied Sciences (Basel, Switzerland)*, *10*(12), 4259. DOI: 10.3390/app10124259

Dalle Mura, M., & Dini, G. (2021). An augmented reality approach for supporting panel alignment in car body assembly. *Journal of Manufacturing Systems*, *59*, 251–260. DOI: 10.1016/j.jmsy.2021.03.004

de Freitas, F. V., Gomes, M. V. M., & Winkler, I. (2022). Benefits and challenges of virtual-reality-based industrial usability testing and design reviews: A patents landscape and literature review. *Applied Sciences (Basel, Switzerland)*, *12*(3), 1755. DOI: 10.3390/app12031755

Ejichukwu, E., Tong, L., Hazime, G., & Jia, B. (2024). Enhancing Autonomous Vehicle Design and Testing: A Comprehensive Review of AR and VR Integration. *arXiv Preprint arXiv:2404.19021*.

Eswaran, M., & Bahubalendruni, M. R. (2022). Challenges and opportunities on AR/VR technologies for manufacturing systems in the context of industry 4.0: A state of the art review. *Journal of Manufacturing Systems*, *65*, 260–278. DOI: 10.1016/j.jmsy.2022.09.016

Fang, W., & An, Z. (2020). A scalable wearable AR system for manual order picking based on warehouse floor-related navigation. *International Journal of Advanced Manufacturing Technology*, *109*(7-8), 2023–2037. DOI: 10.1007/s00170-020-05771-3

Gabbard, J. L., Smith, M., Tanous, K., Kim, H., & Jonas, B. (2019). AR drivesim: An immersive driving simulator for augmented reality head-up display research. *Frontiers in Robotics and AI*, *6*, 98. DOI: 10.3389/frobt.2019.00098 PMID: 33501113

Geng, R., Li, M., Hu, Z., Han, Z., & Zheng, R. (2022). Digital Twin in smart manufacturing: Remote control and virtual machining using VR and AR technologies. *Structural and Multidisciplinary Optimization*, *65*(11), 321. DOI: 10.1007/s00158-022-03426-3

Gouda, G. K., & Tiwari, B. (2024). Unravelling the enablers of Industry 4.0 in Indian automobile industry amid COVID-19: An integrated TISM and fuzzy MICMAC approach. *Journal of Business and Industrial Marketing*, *39*(2), 227–243. DOI: 10.1108/JBIM-02-2022-0067

Kaladgi, A. R., Afzal, A., Manokar, A. M., Thakur, D., Agbulut, U., Alshahrani, S., & Subbiah, R. (2021). Integrated Taguchi-GRA-RSM optimization and ANN modelling of thermal performance of zinc oxide nanofluids in an automobile radiator. *Case Studies in Thermal Engineering*, *26*, 101068. DOI: 10.1016/j.csite.2021.101068

Lee, J., Chua, P. C., Chen, L., Ng, P. H. N., Kim, Y., Wu, Q., Jeon, S., Jung, J., Chang, S., & Moon, S. K. (2023). Key enabling technologies for smart factory in automotive industry: Status and applications. *International Journal of Precision Engineering and Manufacturing-Smart Technology*, *1*(1), 93–105. DOI: 10.57062/ijpem-st.2022.0017

Manimuthu, A., Venkatesh, V., Shi, Y., Sreedharan, V. R., & Koh, S. L. (2022). Design and development of automobile assembly model using federated artificial intelligence with smart contract. *International Journal of Production Research*, *60*(1), 111–135. DOI: 10.1080/00207543.2021.1988750

Rzepka, S., Otto, A., Vogel, D., & Dudek, R. (2018). Application-Driven Reliability Research of Next Generation for Automotive Electronics: Challenges and Approaches. *Journal of Electronic Packaging*, *140*(1), 010903. DOI: 10.1115/1.4039333

Sadeghzadeh, A., Islam, M. B., Uddin, M. N., & Aydin, T. (2024). ARVA: An Augmented Reality-based Visual Aid for Mobility Enhancement Through Real-time Video Stream Transformation. *IEEE Access : Practical Innovations, Open Solutions*, *12*, 137268–137283. DOI: 10.1109/ACCESS.2024.3462628

Swathi, P. (2022). Industry applications of augmented reality and virtual reality. *Journal Of Environment Impact And Management Policy, ISSN-2799-113X, 2*(2), 7–11.

Tripathi, S., & Talukder, B. (2023). Supply chain performance and profitability in Indian automobile industry: Evidence of segmental difference. *Global Business Review*, *24*(2), 371–392. DOI: 10.1177/0972150919898302

Tubaro, P., & Casilli, A. A. (2019). Micro-work, artificial intelligence and the automotive industry. *Economia e Politica Industriale*, *46*(3), 333–345. DOI: 10.1007/s40812-019-00121-1

Vashisht, S., & Rakshit, D. (2021). Recent advances and sustainable solutions in automobile air conditioning systems. *Journal of Cleaner Production*, *329*, 129754. DOI: 10.1016/j.jclepro.2021.129754

Wang, B., Tao, F., Fang, X., Liu, C., Liu, Y., & Freiheit, T. (2021). Smart manufacturing and intelligent manufacturing: A comparative review. *Engineering (Beijing)*, *7*(6), 738–757. DOI: 10.1016/j.eng.2020.07.017

Wolfartsberger, J., Zenisek, J., & Wild, N. (2020). Supporting teamwork in industrial virtual reality applications. *Procedia Manufacturing*, *42*, 2–7. DOI: 10.1016/j.promfg.2020.02.016

Xiong, F., Wang, D., Ma, Z., Chen, S., Lv, T., & Lu, F. (2018). Structure-material integrated multi-objective lightweight design of the front end structure of automobile body. *Structural and Multidisciplinary Optimization*, *57*(2), 829–847. DOI: 10.1007/s00158-017-1778-1

Yıldız, B. S., Pholdee, N., Bureerat, S., Yıldız, A. R., & Sait, S. M. (2020). Sine-cosine optimization algorithm for the conceptual design of automobile components. *Materials Testing*, *62*(7), 744–748. DOI: 10.3139/120.111541

Zhang, Z., Wang, X., Wang, X., Cui, F., & Cheng, H. (2019). A simulation-based approach for plant layout design and production planning. *Journal of Ambient Intelligence and Humanized Computing*, *10*(3), 1217–1230. DOI: 10.1007/s12652-018-0687-5

Zhou, F., Lin, X., Liu, C., Zhao, Y., Xu, P., Ren, L., Xue, T., & Ren, L. (2019). A survey of visualization for smart manufacturing. *Journal of Visualization / the Visualization Society of Japan*, *22*(2), 419–435. DOI: 10.1007/s12650-018-0530-2

Chapter 6
Immersive Drives Leveraging VR and AR for Advanced Automotive Experiences

A. Manikandan
https://orcid.org/0000-0003-0461-2859
Department of Electronics and Communication Engineering, Amrita School of Engineering, Kerala, India

Penmesta Vasanth Kumar Varma
Department of Electronics and Communication Engineering, Amrita School of Engineering, Kerala, India

Nuthalapati Sasidhar
Department of Electronics and Communication Engineering, Amrita School of Engineering, Kerala, India

Padavala JyothirmaI
Department of Electronics and Communication Engineering, Amrita School of Engineering, Kerala, India

Avuku Sathvic Kumar Reddy
Department of Electronics and Communication Engineering, Amrita School of Engineering, Kerala, India

T. Sanjay
https://orcid.org/0000-0001-8591-6314
JP Morgan, USA

ABSTRACT

The incorporation of Virtual reality (VR) and Augmented reality (AR) technologies in the automobile industry has made a major leap forward in vehicle design, manufacturing, marketing, and upkeep. The conventional approach involves physical prototypes, manual quality assurance, and static training resources, making them very time consuming and expensive with minimal efficiency. Current trends show that VR and AR are becoming increasingly accepted in the automobile industry due to their ability to increase precision, reduce costs, speed up processes, or improve

DOI: 10.4018/979-8-3693-7287-6.ch006

interaction with users. Virtual models can be developed and tested much faster in the design and prototyping stages using VR. Manufacturing accuracy can be increased by AR through improved quality control and reduced error incidence. This chapter gives a large explanation of how VR and AR are applied in the automotive sector. It stresses how the utilization of VR will affect design works alongside prototyping while at the same time outlining AR's contribution to manufacturing processes.

INTRODUCTION

The presentation of VR and AR innovations has started up recent viewpoints in diverse areas, and the automotive industry is no avoidance(Arena et al. 2022). Immersive drives, leveraging the capacities of VR and AR, are transmuting how drivers and travelers confront vehicles and the driving strategy itself. This knowledge offers an overflow highlight, from moving forward security and directing to advertising imaginative amusement and learning events. VR permits clients to be possessed in a completely virtual environment, which can reproduce a discrete driving environment for preparing for exciting destinations. AR, on the other hand, overlays computerized data onto the genuine world, making strides the driving association by offering real-time information and cooperative components specifically on the windshield or through wearable devices(Abdelkader, Elgazzar, and Khamis 2021). Together, these innovations guarantee to change the car industry by showing driving quicker, pleasant, and more secure.

The integration of VR and AR into automotive encounters is being investigated by different automakers and tech companies, pointing to improve both the utilitarian and experiential angles of driving(Riegler, Riener, and Holzmann 2021). For occasion, VR can be utilized for immersive test drives, permitting potential clients to encounter diverse car models and driving conditions from the consolation of a showroom or their claim domestic. This not as if it were improving the client experience but moreover decreases the need for physical test drives, which can be strategically challenging and exorbitant. Furthermore, VR is being utilized for driver preparation, giving a risk-free environment where modern drivers can hone and refine their abilities beneath different simulated conditions. AR, in the meantime, is being coordinated into route frameworks to extend turn-by-turn bearings, risk notices, and focuses of interest specifically onto the windshield, lessening the requirement for drivers to take their eyes off the street(Thevin, Briant, and Brock 2020). Additionally, AR can change the in-car entertainment framework by advertising intuitively diversions and data that keep travelers locked in amid long ventures. These advancements are backed by a developing body of inquiry about and advancement, which underscores the potential of VR and AR to on a very basic level change the car landscape.

Research and development in VR and AR for automotive applications has shown growth and potential in recent years. One of the pioneering studies in this area investigated the use of AR to improve driver assistance systems(Murugan et al. 2022). Their research showed how AR can provide real-time information about glasses; this can aid navigation and hazard detection. This initial research laid the foundation for later developments in AR navigation systems developed by companies like WayRay, such as the, which delivers AR holographic images that improve safety and driving awareness(Nawaz, Kundu, and Sattar 2017). Their study emphasized the advantages of using VR for immersive test drives and design prototyping, permitting designers and engineers to envision and relate with vehicle models in a virtual environment. This approach not only modernizes the design procedure but also accepts more comprehensive testing of vehicle functioning and ergonomics under various simulated environments. The application of VR in driver training has also been greatly examined. For instance,(de Winter, van Leeuwen, and Happee 2012) investigated the efficiency of VR driving simulators in developing driving skills and decreasing accident ratios. Their outcomes exhibited that VR training can significantly improve driver expertise and safety, particularly for beginner level drivers.

The integration of AR into automotive user interface has also been a crucial point of research. (Azuma et al. 2001)were among the first to explore AR for head-up displays (HUDs) in vehicles. They demonstrated the potential of AR to project critical driving data onto the windshield, reducing the need for drivers to distract their interest from the road. Developing on this an expanded AR HUDs that present appropriate intelligence, such as imminent turns and traffic indications, directly in the driver's line of sight was prepared. This technology has been implemented and further enhanced by automotive manufacturers like BMW and Mercedes-Benz, who have incorporated AR HUDs into their latest models. Furthermore, in car entertainment and passenger engagement, the role of AR is gaining more attraction.

1. ENHANCING THE SAFETY THROUGH DRIVER INTERACTION

Use of AR and VR in the automotive industry has massively increased safety. These technologies enhance driver interaction by providing a more immersive and informative driving experience, which leads to increased safety on the road. AR and VR offer advanced solutions for navigation, hazard detection, and driver training, making them indispensable tools in the modern automotive safety toolkit. Innovative driving assistance systems (IDAS) include head-of-car safety systems implemented through a complex of sensors to detect the surrounding environment and provide instant feedback to the driver(Alghodhaifi and Lakshmanan 2021). Some of the elements presented are lane change warnings, blind spot detection and automatic emergency braking to raise safety levels by informing the driver about possible hazards and intervening in the process. Lane Departure Warnings is alarm information in the ADAS can be informed to the driver through the depiction of the lane on the windshield(Morden et al. 2023). Blind-Spot Monitoring is from the ADAS point of view, the presence of vehicles in the blind zone can be informed through icons on the side mirror and / or HUD. The ADAS can inform the driver of the distance to the objects or can draw attention to the place of possible danger on the windshield. ADAS has a highly complex multi-level network of sensors that are used in capturing the environment around the vehicle, providing real-time feedback to improve the driver's ability(Ortega et al. 2022). Such systems include Lane Departure Warning System, Blind-Spot Detection System, and Autonomous Emergency Braking System to enhance safety by giving the driver extra caution and intervention when necessary.

1.1 Driver Monitoring Systems

Driver Monitoring Systems (DMS) are intended to detect and prevent those situations where the driver becomes drowsy or distracted by other things happening outside of the car(Doudou, Bouabdallah, and Berge-Cherfaoui 2020). Such systems employ cameras and sensors for identifying any signs of driver drowsiness, inattentiveness, or other manifestations of their incapacitation. The overall Decision-making Systems can assess such things as eye movement, facial expressions and head position and so DMS can see when the driver is distracted or when he/she is too tired to drive safely. In such cases, the system can show a video clip or play a tone to get back the driver's attention or advise that it is time to rest. This is a measure that avoids accidents that are for instance because of fatigue among the drivers. In addition, AR can display alerts on a HUD in case the system recognizes drowsiness or distraction on the part of the driver.

1.2 Biometric Interaction

Future safety systems might use other data related to the biometrics of the driver and the vehicle's environment to assess the condition of the driver. Through monitoring parameters like heart rate, blood pressure or even brain activity of an individual, these systems could be able to identify any sign of ailment, or discomfort which an ordinary person might not be able to detect. For example, AR can visualize heart rate information on the HUD and give an alert if it recognizes irregular patterns, the health aspect could have AR project possible warnings if blood pressure shows stress or other problems(Murali, Kaboli, and Dahiya 2022). Also, Possible in the advanced generations of AR, the system would be able to detect signs of fatigue or stress from the driver's brain signals. If the system notices an arrhythmia or signs indicating the driver is drowsy, the car can send an alert or attempt to avoid an accident by steering the car to the side of the road gently.

1.3 Driver Feedback and Training

With in-vehicle feedback systems, drivers can be given feedback on their behavior and guided on how they can correct their mistakes. It can track inputs such as speed, braking, and whether traffic laws are obeyed to provide real-time encouragement and recommendations to drive more safely. Additionally, driving simulations training can be actual, and virtual reality can mimic different road conditions to prepare the drivers. It is found that using VR systems can assess the driver's behavior and offer suggestions and advice in a safer context through personalized feedback(Ju, Williamson, and Wallraven 2022). Also, it is possible to incorporate key aspects of a game into VR for driving safe behavior by offering points and the ability to track advancement. With time, this makes drivers be able to change their behavior constructively through the help of feedback thus making them become safer drivers.

1.4 Human-Machine Interface

HMI interfaces are used in vehicles and should be designed in a way that would allow the driver to easily understand the instructions without feeling distracted. Effective HMI design includes:

- *Voice Commands* AR can complement voice commands in applications such as self-driving cars by providing visual feedback on the car's heads up display(Detjen, Geisler, and Schneegass 2020).
- *Head-Up Displays (HUD)* AR projects necessary data on the windshield, therefore ensuring that the driver does not have to look anywhere away from the road(Yontem et al. 2021).
- *Control Interfaces* AR can be useful since it can mark the vehicle controls so that their interaction is easy through the HUD. The concept of cognitive load is significant in that a properly developed HMI reduces the amount of load placed on the driver and does not distract him/her from the road(Enjalbert et al. 2021).

The application of AR and VR technologies along with full and modified ADAS and AMS means that the safety aspect is not only challenging but also engaging. Thus, a well-developed system that makes possible safe and informative communication between a driver and a car can greatly enhance road safety for all of us. Gradually with the advancement of these technologies; improvement of driver safety will be more prominent and lead to advanced and safer vehicles on the roads.

2. TRANSFORMING VEHICLE DESIGN AND PROTOTYPING

Virtual Reality (VR) and Augmented Reality (AR) technologies have been applied in automotive design and prototyping as measures that give new heights to invention, time saving and precision.

2.1 Virtual Design and Conceptualization

AR and VR have revolutionized the fields of design and engineering by enabling virtual design and conceptualization. These technologies offer a more interactive and intuitive approach to creating, visualizing, and manipulating digital models in immersive environments. AR enhances the real world with digital elements, providing contextual information and interactive features for on-site design modifications and real-time collaboration(Mutis and Ambekar 2020). On the other hand, VR immerses users in a virtual space where they can explore and interact with designs in three dimensions, enhancing spatial understanding and design implications. This transformative capability is particularly beneficial in architecture, automotive design, and industrial engineering, where precise visualization can streamline the design process, reduce errors, and boost creativity. Using AR and VR, designers can quickly prototype, engage stakeholders with realistic presentations, and deliver more innovative products to the market.

2.1.1 Rapid Iteration and Visualization

With VR, it becomes possible to design and model complicated vehicle designs by providing a reality in a full 3D virtual environment(de Freitas, Gomes, and Winkler 2022). This will lead to identifying and solving the issues even before generating a physical model, reducing time and cost. A Designer can virtually see the car from different perspectives and can make upgrades if necessary. Moreover, a designer can experience a driver's perspective and can better understand design and make required changes with ease.

2.1.2 Enhanced Collaboration

VR enables designers and/or engineers to work in sync in real-time at different geographical positions. It also speeds up the decision-making processes and incorporates global viewpoints into the development. Some of the advantages that come with using the concept are that teams can come together in virtual workplaces where they can discuss designs, make changes as they implement them and observe the effect of the change in real time. AR can work to build a common collaborative working space by placing images that relate to the topic over real car models to those participating to help them better perceive alterations as well as improvements. This kind of interaction can help to considerably accelerate the process of reaching consensus which, in turn, is vital for completing projects within the time frame agreed upon.

2.1.3 Virtual prototypes and Cost Reduction

In the case of traditional methods, first a model needs to be generated and then again it needs to be tested and modified to finalize, but the VR breaks these limitations as virtual prototypes can be created and modified without making it physically. Many automobile companies like BMW had used AR and VR to check their prototype before finalizing their model, by checking their models' standards and speeding up the process(Lavrentieva et al. 2020). Ford and Bugatti, for example, have embraced the use of virtual reality in prototyping processes. The Gravity Sketch Tool works in the creation of models and designs right in the VR environment, so that the designers can see the models in the virtual space and from all angles; this reduces the process of designing drastically by doing away with the initial 2D designing phase. The Divo was designed by Bugatti in only six months, a pace twice as quick as the conventional procedure, proving the VR efficacy in prototyping.

2.1.4 Simulation

There are factors that make virtual reality effective in testing the functionality, safety, and comfort of car designs; in that car designers can expose the car designs to real driving conditions, thus determining their suitability. All's comprises factors such as crash simulations and specifically aerodynamics testing and all of these are conducted in a virtual environment which emulates real physics. Use of VR will help in early detection of potential problems and hence help in improving a vehicle's quality. For instance, VR enables one to model different weather systems that the car will experience to make sure the system works well across different climates, something that would be very costly with physical models.

3. INTERCONNECTION WITH ARTIFICIAL INTELLIGENCE

The interconnection of AI with VR and AR can help in both predictive analysis and smart designing. Algorithms in machine learning can help in solving large complex design problems by analyzing large amounts of design data and by determining what properties have produced good results in the past and make changes accordingly. AI also plays a role in the prediction of how a particular design of a car will be likely to perform under certain circumstances, so this helps those involved in the design of the car to improve on the design for the best performance and safety of the car(Demlehner, Schoemer, and Laumer 2021).

These points reflect the fact that while IoT devices are typically referred to as being useful in the design phase of a product, they can offer real time input and feedback during the design phase that can improve the realism of virtual simulations. This connectivity means that the designs that are created will be done so with data fed to the algorithm from the current databases, enhancing the function of the design as well as its performance. In addition, vehicles can give feedback on their usage details to the designers and the decision makers to improve the design for the subsequent models, thus enhancing the general quality of the vehicles. New display technologies like AMOLED and OLED in cars can increase the demand for AR/VR in the automotive sector(Jones and Jones 2022). These high-resolution screens will provide realistic experience, making the process efficient. The concept of advanced displays also helps designers to provide more accurate representations of their prototypes and gives enhanced perceptions so as to spot certain problems that would not be possible using inferior display techniques.

The application of VR and AR in vehicle design and prototyping has transformed the automotive industry. These technologies promise creativity, efficiency, and accuracy and allow for frequent prototyping with minimum cost. As VR and AR technologies advance, their effect on automotive design will continue to grow, turning the industry toward more efficient and innovative solutions.

4. INFORMATION EXCHANGE BETWEEN CONNECTED VEHICLES

Connected vehicle technologies are being combined with virtual reality and augmented reality to put the automotive industry at the threshold of a sea change in technology. The changes give fundamental redefinitions to how vehicles henceforth should interact with other vehicles, infrastructure elements, and the environment for supporting safer, more efficient, and smarter transportation systems. VR and AR roles in this portion relate to the Information exchange between the connected vehicles in terms of simulation and testing; real-time data visualization; and predictive maintenance(Raju and Farah 2021).

4.1 Simulations and Testing

Simulations and testing are essential for the development and execution of connected car technologies. Virtual Reality and Augmented Reality do provide industrial tools that could be developed to create extremely realistic and controlled settings. These settings are essential in performing and exploring all potential tests and mastering the workings of this system.

4.1.1 Traffic Scenario Simulation

- High-Fidelity Environments: It is through VR technology that built environments are created with very high fidelity, closely mimicking a true real-life setting for purposes of rigorous testing of the connected vehicle systems. These VR scenarios could be inclusive of very diverse types of elements, for example, weather patterns, traffic densities, as well as different varieties of roads. This creates an extended and realistic platform for rigorous testing and analysis of the connected vehicle systems.
- Protocol Testing: There can be extended tests and validation running the communication protocols of V2V and V2I architectures in an emulated environment. This allows thorough validation that vehicles can share critical information—position, speed, and direction—reliably to avoid collisions and effectively manage traffic(Babu and Raj Kumar P 2022). Using these capabilities, the V2V architectures being tested make reliable collision avoidance systems and effective traffic management EDs possible.
- Behavioral Analysis: Highly advanced machinery can be used to monitor and extensively study how different connected vehicles respond to input simultaneously and interactions within a simulated environment. This can help derive critical information for improving the behavior of the vehicles under the pressure faced by the vehicles in dynamic and unpredictable real-world conditions.

4.1.2 Algorithm Optimization

Data-driven Development: For refining algorithms used in connected vehicle systems, VR simulations must be designed to lead up to large datasets(Hu et al. 2024). These large datasets are analyzed, and a pattern is realized in making informed decisions in these systems using machine learning techniques. This, in turn, helps in upgrading and improving operations and connected vehicle technologies.

Iterative Testing: Because of the very high level of ease in developing models through different alternatives of communication algorithms under practice by using virtual reality technology, it is very fast. The real-time data visualization approaches enable prototyping and algorithm testing to be fast, leading to good and robust final systems.

4.2 Real-Time Data Visualization

Real-time data visualization enables understanding, on the part of drivers or engineers, of the connected vehicle state and performance. In other words, Real-time data visualization in an intuitive and interactive way enabled by Augmented Reality will result in high situational awareness, with good indications regarding correct actions, for both drivers and engineers.

4.2.1 Dashboard Displays

Improved Driver Awareness: Augmented Reality technology has the power to project real-time information either on the windshield or on the dashboard of a vehicle(Charissis et al. 2021). Data shown would include information like speed, fuel, battery status, and turn-by-turn navigation indicators. This critical data, displayed directly in front of the driver through AR, serves to get the driver's needed information without having to take their eyes off the road, hence improving safety and minimizing distraction.

Safety Alerts: AR technology can provide safety alerts by warning drivers if potential hazards exist ahead, for example, if another vehicle is about to make a sudden stop a pedestrian is crossing, or perhaps there is a potential obstruction (Hasan and Hasan 2022). In general, these AR features are used to make the drivers respond rapidly and make informed decisions to stay out of the accident. Safety with AR technology augments the situational awareness of drivers and contributes to general traffic safety.

4.2.2 Predictive Analytics

Traffic Management: The incorporation of AR systems in traffic management has helped change the mechanism of collecting information from connected cars to predicting the future trend of the road infrastructure and updating it in real-time(Vermesan et al. 2021). With the use of AR technology, it's easy to control traffic flow, greatly reduce rates of congestion by important percentages and cut short the travel duration of road users. This, in turn, has really focused on increasing the overall transportation efficiency and taking people's travel experience to new advanced levels.

Driver Assistance: Connected vehicles can help to provide proactive support to drivers by combining advanced predictive analytics with augmented reality. For instance, AR can analyze traffic patterns and suggest alternative routes for drivers to avoid traffic jams and congestion(Vermesan et al. 2021). In addition to this, in predicting inclement weather conditions along the intended route, it can give timely warnings, thereby enhancing driver safety and their travel experience.

4.2.3 Predictive Maintenance

Predictive maintenance is the new advanced use of technology and data analytics to know when the vehicle will break down and avoid such to extend the vehicle's life(Theissler et al. 2021). Making this a proactive effort, where AR, or augmented reality, and VR, which stands for virtual reality, are the predominant factors, allows real-time diagnostics and remote assistance. All these helps to reduce downtime and achieve maximum efficiency and performance for the vehicles by facilitating the identification of problems, fast troubleshooting, and remote advice for maintenance/repair.

4.2.3.1 Real-Time Diagnostics

Continuous Monitoring: The numerous sensors in connected vehicles allow for continuous monitoring of the health and performance of various parts. Diagnostic information, with AR, can be passed on to drivers and technicians in ways that are simplified and comprehensible(M. A. Rahman et al. 2022). It would, therefore, aid in identifying and remedying any issues that are likely to crop up with the vehicle's components.

Early Warning Systems: These systems will make sure that advance notice can be indicated in a situation where such a problem is likely to occur either in a vehicle or other equipment. For example, in case of abnormal engine temperature or when tire pressure is low, the system shall give early warnings to the user with augmented reality. The provided warnings, therefore, suggest the appropriate action to be taken that prevents the actualization of detected problems to a major failure.

4.2.3.2 Remote Assistance

Assisted Repairs: Remote technicians could take drivers through a sequence of actions required for minor repairs and even service procedures using augmented reality at each step. In this way, the AR device superimposes the instructions on the actual vehicle part; this can facilitate people without much experience but who need to perform critical functions(Angelopoulos and Mourtzis 2022). This generally increases access while ensuring the right performance of the maintenance and repair process.

Expert Consultation: For more complicated issues, VR enables expert consultation to be done at any location beyond that of the driver with the specific vehicle. By utilizing VR, the technicians have visuals of the vehicle(Hernández-Chávez et al. 2021). They diagnose the problem and provide solutions on how to rectify the vehicle, all being done away from the location.

5. VIRTUAL TEST DRIVES

Virtual test drive is a digital simulation for experiencing vehicle without need of physical drive. The concept of virtual test drive has evolved for a long in years from just 3D visualizations to high immersive VR experience(de Freitas, Gomes, and Winkler 2022). Virtual test drive simulations allow users to experience the functions and how they work, features, and performance of a vehicle without physically being in the vehicle. These simulations can give high immersive experience of the vehicle because of the highly detailed realistic driving experiences providing using advanced graphics, using VR and AR. The major point is continuous advancements in VR, AR, and AI are helping to provide and enhance the realism and utility of virtual test drives, making them an integral part of the automotive experience. The high simulation quality of driving simulators enables investigations of topics according to driving dynamics. With high reproducibility and fast generation of vehicle variants the simulation evaluation process gives us better understanding about the in the early development stages. The design of the Virtual Test Drive toolset is based on a distinction between core components of the simulation, extended components and target components(Schlager et al. 2020). Core components make up the actual simulation framework which handles the data flow within the simulator and comprises basic functionality like record, playback, data collection, and data distribution. Extended components are the vehicle and environment simulation like dynamics, driver, traffic. Target components are the ones which will be modified by the user in his role as developer of an active safety or assistance system like sensors, functions and actuators.

5.1 Aspects of Virtual Test Drive

Few aspects of virtual test drive are Simulation Experience, Data Collection, VR and AR Integration, Experiencing Interactive Features, Marketing and Sales, Educational purpose.

Simulation Experience: virtual test drives provide real life driving experience and it helps users to understand how the machine works on various conditions and in various situations and performance of a vehicle without physically being in the vehicle(Zou et al. 2021).

Data Collection: simulations collect data from the user experience and provide to manufacture and developing team how the person who experienced feeling about it and what are the things he felt appealing and what are things he felt to be developed or still lagging in the machine(Plotzky et al. 2021).

VR and AR Integration: VR and AR are used in virtual test drives to make the experience better in terms of experiencing 360 degree environment and to feel in more realistic way and enhance the feel of presence in an automobile machine.

Experiencing Interactive Features: with the help of this simulations the person can get detailed experience of the machine from interior to the performance and detailed working experience of all parts in the machine and it gives complete interaction about controls.

Marketing and Sales: This simulation helps a lot in marketing and sales they offer great experience about the machine and these simulation experience provides them view of the car and details about it without any physical drive and generates interest in them to buy the machine.

Educational purpose: this simulation can be used to educate new learning persons in way that they can know how experience it how to use and how to operate a machine.

5.2 Benefits of Virtual Test Drives

Virtual test drives are beneficial in many ways, first is it can be accessed from anywhere (remote access). Virtual test drives help in cost efficiency by reducing physical drives. Virtual test drives are most helpful for customization as the person can experience it by adding any features or any customization to it and see how it performs(Davila Delgado et al. 2020). There is a lot of technology being used behind these simulations. With the great technology behind simulation software and advanced hardware components, network and data management it can give immersive experience.

6. CASE STUDY

Target Nowhere is the best example because it integrates the connected vehicle technology with VR (Virtual Reality) and AR (Augmented Reality) applications. It harbors a lot of advanced features that enhance not only the quality of safety and performance but, indeed, the quality of user experience.

6.1 Tesla's Autopilot

Sensor Integration: The Tesla autopilot is equipped with an advanced sensor system. The sensors used include cameras, radar, ultrasonic sensors, and GPS technology(Davila Delgado et al. 2020). The various sensors interact with each other to capture full environmental data for the vehicle. The data is taken through careful analysis and processing to be able to render features such as lane keeping, adaptive cruise control, and comfortable automatic lane changes for its drivers' ultimate driving experience.

AR for Enhanced Visualization: AR-enhanced visualization might offer the Autopilot system real-time visual feedback to the driver. For example, the system may project prominent lane boundaries, upcoming vehicles, and safe following distances directly on the front screen or dashboard. This gives the driver immediate and crucial information for the making of appropriate decisions for safe and informed driving.

Remote Diagnostics: Tesla uses simulation through virtual reality in order to ensure that the Autopilot Software is well put out before it is launched to the public. The simulations are developed so that under rigorous testing, the engineers are able to understand in broad terms the functionality of the system, which further gives them security of the system in real-world conditions.

Continuous Improvement: Tesla cars continue to be improved by over-the-air software updates. The updates refine performance, add new features, and remove existing bugs. Before these updates are rolled out to car users, they are put through stringent testing with VR simulations conducted in virtual settings. All this careful process thus ensures that the updates will work as per the purpose without any defects once made available to Tesla owners.

User-Centric Updates: Highly user-friendly software updates make it possible for owners of Tesla to update their automobile software remotely without having to go to a service center in person. This goes on to enhance the user experience by making it no question at all that a Tesla vehicle is always up to date with the latest the technology world has to offer.

6.1.1 Prospects

The intertwined use of virtual reality, augmented reality, artificial intelligence, the Internet of Things, and machine learning renders the future development of connected vehicles transforming in its advancements(Vermesan et al. 2017). Front-end technologies are continuously in evolutionary modes, paving the way for more and more sets of sophistication in the connected vehicle domain.

6.1.2 AI Integration

Intelligent Systems: AI in connected vehicle systems becomes a backbone for the same in ameliorating the decision-making process by making it very intelligent(Luo et al. 2022). Such systems harness AI algorithms to interpret and make sense of data issued from sensors, cameras, and GPS mechanisms to predict and avert potential hazards on the road. With AI, the connected vehicle system can address safety and performance optimization through real-time driver support and assistance. This technological breakthrough is all set to renew transportation and road safety.

Custom Experience: The use of AI in the vehicle has led to personal driving experience since the vehicle keeps on learning information related to user behavior and individual preferences(Sestino et al. 2022). The system can make changes to seating setups and other interior features that provide maximum comfort for the user. Finally, by enquiring about the traffic and meteorological conditions, AI could facilitate in defining the optimized paths to reach that destination in a quick time and with minimum stress posed on the driver. Also, AI will be able to predict maintenance needs by processing vehicle performance data to avert potential issues and keep the peak of reliability on track.

6.1.3 IoT Connectivity

Smart Cities: This will turn the page from just connected vehicles to their use and integration with other smart-city infrastructures—a major innovative leap in urban transportation. These will manage the smart flow of traffic together with and through interaction with connected vehicles via IoT-enabled traffic lights, road sensors, and communication networks, which will practically cut the blood vessels of traffic congestion without any sacrifice of safety.

Enhanced Communication: The technology of insertion of IoT in between improves the factor of communication between the infrastructure and the vehicles severely. With the formulation of real-time updates and working in coordination with them, IoT may severely improve efficiency in the transport system. The seamlessness of such communication would yield a better, hassle-free driving experience for commuters, which, in turn, would relieve congestion and improve traffic management at large.

6.2 Future Trends at Waymo

Waymo, at the frontier of autonomous driving technology, is the very first to apply VR and AR for simulating and testing all washes of things and other situations. This unique approach has made a real difference in the further technology for autonomous vehicles, thereby reinstating it as a leader in this field.\ · Autonomous driving: Intensive simulations developing and testing their algorithms in an autonomous driving system are developed using VR and AR technologies. All this, together with the modern technologies used, allows Waymo to make enormous refinements to the system's capabilities through exposure to a variety of realistic and hard-driving scenarios. This gargantuan testing and simulation process is necessary to ensure the absolute safety and top performance of Waymo technology.

·Integration into the Development of a Smart City: The company is currently researching augmented reality and the Internet of Things in communication technologies such that the overall communication process among automated vehicles could be integrated into the smart city infrastructure. As such, this will enable more functionality and safety within the urban transportation system overall.

In conclusion, VR and AR technology are revolutionizing the auto industry by putting better safety, efficiency, and user experience in connected vehicles. The real transformation driven by VR and AR for the connected vehicle ecosystem is going to explode with incoming advancements in AI, IoT, and machine learning. It holds the promise of a much smarter, safer, and more connected future for drivers, manufacturers, and the very urban environments that all these systems will move within.

6.3 BMW: Virtual Prototyping and Showroom Experience

The use of virtual reality (VR) is notably embraced by the BMW company as they incorporate it in their design and sales(Kotler, Kartajaya, and Setiawan 2023). Using VR, the company develops virtual replicas which allow the engineers as well as the designers to conceptualize and even have a feel of how the part of the car will be like even before a physical model is built. This is quite helpful in fast-forwarding the process of development, cutting on the costs that are required, and at the same time, it improves on the designs.

To get an idea about the vehicles in a more realistic manner, BMW has started making use of VR even in their showrooms. Interested customers can specify models and equipment and choose virtual test drives that effectively give the buyers a more appealing and engaging sales approach than conventional sales pitches.

6.4 Audi: AR in Production and Customer Interaction

Audi's production lines have been among the first to adopt augmented reality with remarkable results that include increased production efficiency and the general effectiveness of workers. Lastly, thanks to AR glasses, workers in assembly lines can receive information about the product they are assembling, detailed instructions on how to assemble it as well as any other relevant information which means that the most likely chances of making mistakes will be very slim and the workers will be able to produce more.

In the field of customer engagement, Audi has experimented with the application of augmented reality to enhance the car purchasing experience(Krishnadas 2021). In its kind, it assists customers in visualizing different possibilities of Audi vehicles in their real operational environment.

6.5 Ford: VR for Collaborative Engineering and Training

Virtual reality technology is used at Ford to enhance collaborative engineering and design from different regions(Frizziero et al. 2021). In terms of design, engineers can collaborate in real-time within and across different organizations, review design changes, and provide feedback thereby making modifications wherever necessary. It enhances a rapid innovation process and the quality of the developed designs on the international level.

VR is also used as a means of training for the employees at Ford. Models help navigate through problems and assess how to work with an automobile without having an actual automobile; or even improve security features and study how to avoid dangerous conditions for employees, while sparing their lives.

6.6 General Motors (GM): Enhancing Customer Service with AR

General Motors uses augmented reality in functioning to increase the level of customer satisfaction. Mechanics wear AR glasses where experts, who are located remotely, can see the technician's view and assist them in salvaging the machinery virtually without their actual presence(Ammar et al. 2022). This helps reduce the time customers spend in seeking repair services, thereby enhancing the service quality.

GM also provides customers with AR-enhanced user manuals. By scanning any component of the vehicles through smartphones, users can get help in the form of guided personnel, together with relevant info for control and troubleshooting of the automobile. This enhances the users with better understanding and maintenance of the vehicles.

6.7 Siemens: AR for Machine Maintenance in the Automotive Industry

Siemens is using an innovative way of integrating augmented reality (AR) into their machine maintenance and repair processes, which is through biometric data. As for the technicians, they can see the real-time data and maintenance instructions overlaid on the machinery they are working on(Neal et al. 2021).

The advanced AR technology received excellent credentials from the technicians by enabling them to access it quickly considering that information was directly related. Hence, the machine downtime which was at its peak before has been reduced significantly and the overall lethality of equipment is stronger than before.

6.8 Porsche: Revolutionizing Maintenance with AR Glasses

The idea of enhancing AR glasses in the operation was introduced by Porsche. The problems in the vehicle get easily fixed through the new augmented reality glasses that are a part of the technology(Md. F. Rahman et al. 2022). Some specialist responses might not be immediately needed, but over a period, the gadget can be widely adopted in all shops such as it was the example with the laptop. This complex technology involves using a pair of special glasses that one puts on and, through the process, keys in the problem information to the manufacturer's central servers to diagnose the problem in the machine.

The use of AR gadgets makes it easy for an inspector to do his work correctly by studying the visual supplements that are produced by the system that handles a specific vehicle. This outstanding creativity is indeed an impressive scenario, and it ensures the 100% process is error-free. These inspections are so accurate and timesaving.

6.9 Toyota: AR for Enhanced In-Car Entertainment

Toyota is researching ways to employ Augmented Reality in developing in-car entertainment in vehicles such as games for children(Kemeny 2024). A passenger could read interactive items and contents projected on the car's interior walls and surfaces, play games, get information on interesting places to visit along the journey, or even adorn the cabin with virtual decorations using AR glasses.

These incidences show how far-reaching VR and AR are in the automotive industry as we embrace the future of these technologies. As technology evolves and gets more integrated with vehicles, one will see even extra phenomenal applications that will revolutionize how people interact with their automobiles as well as re-ordinate the whole transport experience.

7. FUTURE WITH AR & VR AUTOMOTIVES

7.1 Enhanced Driving Experience

Soon, AR will revolutionize the driving experience by overlaying windshields with HUDs. Vital information like speed, fuel level, and navigation will be directly projected onto the windshield using AR. Vehicles will use navigation aids, geographic and performance information, and safety threats that will increase the level of safety alert and convenience to drivers who will not need to look at other sides of the road while they are driving. They can focus on the road without any distractions.

7.2 Autonomous Vehicle Interaction

AR and particularly, VR will be valuable when self-driving vehicles become common since they will engage passengers in some way or the other. With VR technology, a car can become a utility and recreational vehicle, as well as a business and leisure vehicle, all rolled into one. One could continue with his business meetings, watch a movie, or completely be immersed in any different environment of his choice. This personalized entertainment can make long journeys particularly more enjoyable.

7.3 Ownership and Maintenance of Virtual Cars

AR and VR are the latest innovations that are going to change the very concept of car ownership by introducing the most natural way to take care of a car. Through augmented reality overlays, users might even be led through basic maintenance tasks or even diagnosis of a problem by virtual service advisors. Moreover, VR offering remote test drives for a car could enable potential buyers to study the car's features and performance without even leaving their houses.

7.4 Advanced Manufacturing and Training

AR and VR integration in the automotive industry will further advance manufacturing processes.AR will go on giving detailed and precise information to the workers about real-time instructions and quality checks while VR will prepare realistic training programs for new groundbreaking methods and technologies making a well-equipped and trained workforce readily available.

7.5 Personalized and Immersive Sales Experience

The purchasing decision regarding a car will indeed become much more personalized and immersive. The showrooms of the future may be highly dependent on VRs for giving the customers an elaborate show round of the vehicles preferred, and this would also include some interactive possibilities regarding the range of variants as well as options. AR apps will allow buyers to see their specific car in action in real-world settings, thus, improving the buying decision.

The concepts of VR and AR, applied to the automotive industry are not just trends but evolutions to better performance, efficiency, and economy. BMW, Audi, Ford, GM, and other top-rated automotive companies are already using these innovations to transform design, production, sales and service processes. Thus, as we move forward, the gradual enhancement and progression of VR/AR technologies will help transport people into many realms of driving experiences, establishing not just safe but also enjoyable and engaging driving. The era of immersive drives is just beginning, marking a new chapter in automotive innovation.

CONCLUSION

In conclusion, the integration of Virtual Reality (VR) and Augmented Reality (AR) in the automotive industry, termed "Immersive Drives," is redefining the driving experience by enhancing both functional and experiential aspects of vehicles. VR and AR technologies enable more intuitive navigation systems, immersive test drives, and advanced driver training programs, significantly improving safety and convenience. AR can project real-time information onto windshields, offering drivers crucial data without distraction, while VR offers a risk-free environment for testing vehicle designs and training drivers. These technologies also revolutionize in-car entertainment, turning journeys into interactive experiences for passengers. As research and development in VR and AR continue to advance, their applications in the automotive sector are expected to expand further, leading to smarter, safer, and more engaging driving experiences. Continuing collaboration between automotive manufacturers and tech innovators will be key to unlocking the full potential of these immersive technologies, ultimately transforming how we interact with and perceive our vehicles.

REFERENCES

Abdelkader, G., Elgazzar, K., & Khamis, A. (2021). Connected Vehicles: Technology Review, State of the Art, Challenges and Opportunities. *Sensors (Basel), 21*(22), 7712. Advance online publication. DOI: 10.3390/s21227712 PMID: 34833782

Alghodhaifi, H., & Lakshmanan, S. (2021). Autonomous Vehicle Evaluation: A Comprehensive Survey on Modeling and Simulation Approaches. *IEEE Access : Practical Innovations, Open Solutions, 9*, 151531–151566. Advance online publication. DOI: 10.1109/ACCESS.2021.3125620

Ammar, M., Haleem, A., Javaid, M., Bahl, S., Garg, S. B., Shamoon, A., & Garg, J. (2022). Significant Applications of Smart Materials and Internet of Things (IoT) in the Automotive Industry. *Materials Today: Proceedings, 68*, 1542–1549. Advance online publication. DOI: 10.1016/j.matpr.2022.07.180

Angelopoulos, J., & Mourtzis, D. (2022). An Intelligent Product Service System for Adaptive Maintenance of Engineered-to-Order Manufacturing Equipment Assisted by Augmented Reality. *Applied Sciences (Basel, Switzerland), 12*(11), 5349. Advance online publication. DOI: 10.3390/app12115349

Arena, F., Collotta, M., Pau, G., & Termine, F. (2022). An Overview of Augmented Reality. *Computers, 11*(2), 28. Advance online publication. DOI: 10.3390/computers11020028

Azuma, R., Baillot, Y., Behringer, R., Feiner, S., Julier, S., & MacIntyre, B. (2001). Recent Advances in Augmented Reality. *IEEE Computer Graphics and Applications, 21*(6), 34–47. Advance online publication. DOI: 10.1109/38.963459

Babu, S., & Raj Kumar P, A. (2022). A Comprehensive Survey on Simulators, Emulators, and Testbeds for VANETs. *International Journal of Communication Systems, 35*(8), e5123. Advance online publication. DOI: 10.1002/dac.5123

Charissis, V., Falah, J., Lagoo, R., Alfalah, S. F. M., Khan, S., Wang, S., Altarteer, S., Larbi, K. B., & Drikakis, D. (2021). Employing Emerging Technologies to Develop and Evaluate In-Vehicle Intelligent Systems for Driver Support: Infotainment AR Hud Case Study. *Applied Sciences (Basel, Switzerland), 11*(4), 1397. Advance online publication. DOI: 10.3390/app11041397

de Freitas, F. V., Marcus, V. M. G., & Winkler, I. (2022). Benefits and Challenges of Virtual-Reality-Based Industrial Usability Testing and Design Reviews: A Patents Landscape and Literature Review. *Applied Sciences (Basel, Switzerland), 12*(3), 1755. Advance online publication. DOI: 10.3390/app12031755 de Winter, J. C. F., & van Leeuwen, P. M., and R Happee. 2012. “Advantages and Disadvantages of Driving Simulators: A Discussion.” In *Proceedings of the Measuring Behavior Conference*

Delgado, D., Manuel, J., Oyedele, L., Beach, T., & Demian, P. (2020). Augmented and Virtual Reality in Construction: Drivers and Limitations for Industry Adoption. *Journal of Construction Engineering and Management*, *146*(7), 04020079. Advance online publication. DOI: 10.1061/(ASCE)CO.1943-7862.0001844

Demlehner, Q., Schoemer, D., & Laumer, S. (2021). How Can Artificial Intelligence Enhance Car Manufacturing? A Delphi Study-Based Identification and Assessment of General Use Cases. *International Journal of Information Management*, *58*, 102317. Advance online publication. DOI: 10.1016/j.ijinfomgt.2021.102317

Detjen, H., Geisler, S., & Schneegass, S. 2020. “Maneuver-Based Control Interventions during Automated Driving: Comparing Touch, Voice, and Mid-Air Gestures as Input Modalities.” In *Conference Proceedings - IEEE International Conference on Systems, Man and Cybernetics*, DOI: 10.1109/SMC42975.2020.9283431

Doudou, M., Bouabdallah, A., & Berge-Cherfaoui, V. (2020). Driver Drowsiness Measurement Technologies: Current Research, Market Solutions, and Challenges. *International Journal of Intelligent Transportation Systems Research*, *18*(2), 297–319. Advance online publication. DOI: 10.1007/s13177-019-00199-w

Enjalbert, S., Gandini, L. M., Baños, A. P., Ricci, S., & Vanderhaegen, F. (2021). Human–Machine Interface in Transport Systems: An Industrial Overview for More Extended Rail Applications. *Machines (Basel)*, *9*(2), 36. Advance online publication. DOI: 10.3390/machines9020036

Frizziero, L., Santi, G. M., Donnici, G., Leon-Cardenas, C., Ferretti, P., Liverani, A., & Neri, M. (2021). An Innovative Ford Sedan with Enhanced Stylistic Design Engineering (SDE) via Augmented Reality and Additive Manufacturing. *Designs*, *5*(3), 46. Advance online publication. DOI: 10.3390/designs5030046

Hasan, R., & Hasan, R. (2022). Pedestrian Safety Using the Internet of Things and Sensors: Issues, Challenges, and Open Problems. *Future Generation Computer Systems*, *134*, 187–203. Advance online publication. DOI: 10.1016/j.future.2022.03.036

Hernández-Chávez, M., Cortés-Caballero, J. M., Pérez-Martínez, Á. A., Hernández-Quintanar, L. F., Roa-Tort, K., Rivera-Fernández, J. D., & Fabila-Bustos, D. A. (2021). Development of Virtual Reality Automotive Lab for Training in Engineering Students. *Sustainability (Basel)*, *13*(17), 9776. Advance online publication. DOI: 10.3390/su13179776

Hu, X., Li, S., Huang, T., Tang, B., Huai, R., & Chen, L. (2024). How Simulation Helps Autonomous Driving: A Survey of Sim2real, Digital Twins, and Parallel Intelligence. *IEEE Transactions on Intelligent Vehicles*, *9*(1), 593–612. Advance online publication. DOI: 10.1109/TIV.2023.3312777

Jones, Gary W., and Susan K. Jones. 2022. "Critical Technology Enablers for Near-to-Eye AR and VR." .DOI: 10.1117/12.2622316

Ju, U., Williamson, J., & Wallraven, C. (2022). Predicting Driving Speed from Psychological Metrics in a Virtual Reality Car Driving Simulation. *Scientific Reports*, *12*(1), 10044. Advance online publication. DOI: 10.1038/s41598-022-14409-1 PMID: 35710859

Kemeny, A. 2024. "On-Board Virtual Reality and Applications." In *Autonomous Vehicles and Virtual Reality*, DOI: 10.1007/978-3-031-45263-5_6

Kotler, P., Kartajaya, H., & Setiawan, I. (2023). Ascent Audio *Marketing 6.0: The Future Is Immersive*.

Krishnadas, R. (2021). *Understanding Customer Engagement and Purchase Behavior in Automobiles: The Role of Digital Technology*. Applications for Effective Customer Engagement.

Lavrentieva, O. O., Arkhypov, I. O., Krupskyi, O. P., Velykodnyi, D. O., & Filatov, S. V. 2020. "Methodology of Using Mobile Apps with Augmented Reality in Students' Vocational Preparation Process for Transport Industry." In *CEUR Workshop Proceedings*

Luo, G., Yuan, Q., Li, J., Wang, S., & Yang, F. (2022). Artificial Intelligence Powered Mobile Networks: From Cognition to Decision. *IEEE Network*, *36*(3), 136–144. Advance online publication. DOI: 10.1109/MNET.013.2100087

Morden, J. N., Caraffini, F., Kypraios, I., Al-Bayatti, A. H., & Smith, R. (2023). Driving in the Rain: A Survey toward Visibility Estimation through Windshields. *International Journal of Intelligent Systems*, *2023*(1), 9939174. Advance online publication. DOI: 10.1155/2023/9939174

Murali, P. K., Kaboli, M., & Dahiya, R. (2022). Intelligent In-Vehicle Interaction Technologies. *Advanced Intelligent Systems*, *4*(2), 2100122. Advance online publication. DOI: 10.1002/aisy.202100122

Murugan, S., Sampathkumar, A., Kanaga Suba Raja, S., Ramesh, S., Manikandan, R., & Gupta, D. (2022). Autonomous vehicle assisted by heads up display (HUD) with augmented reality based on machine learning techniques. In *Virtual and augmented reality for automobile industry: Innovation vision and applications* (pp. 45–64). Springer International Publishing.

Mutis, I., & Ambekar, A. (2020). Challenges and Enablers of Augmented Reality Technology for in Situ Walkthrough Applications. *Journal of Information Technology in Construction*, *25*, 55–71. Advance online publication. DOI: 10.36680/j.itcon.2020.003

Nawaz, M., Kundu, S., & Sattar, F. (2017). Augmented reality sandbox and constructivist approach for geoscience teaching and learning. *International Journal of Educational and Pedagogical Sciences*, *11*(6), 1603–1606.

Neal, A. D., Sharpe, R. G., van Lopik, K., Tribe, J., Goodall, P., Lugo, H., Segura-Velandia, D., Conway, P., Jackson, L. M., Jackson, T. W., & West, A. A. (2021). The Potential of Industry 4.0 Cyber Physical System to Improve Quality Assurance: An Automotive Case Study for Wash Monitoring of Returnable Transit Items. *CIRP Journal of Manufacturing Science and Technology*, *32*, 461–475. Advance online publication. DOI: 10.1016/j.cirpj.2020.07.002

Ortega, J. D., Cañas, P. N., Nieto, M., Otaegui, O., & Salgado, L. (2022). Challenges of Large-Scale Multi-Camera Datasets for Driver Monitoring Systems. *Sensors (Basel)*, *22*(7), 2554. Advance online publication. DOI: 10.3390/s22072554 PMID: 35408169

Plotzky, C., Lindwedel, U., Sorber, M., Loessl, B., König, P., Kunze, C., Kugler, C., & Meng, M. (2021). Virtual Reality Simulations in Nurse Education: A Systematic Mapping Review. *Nurse Education Today*, *101*, 104868. Advance online publication. DOI: 10.1016/j.nedt.2021.104868 PMID: 33798987

Rahman, M. A., Rahim, M. A., Rahman, M. M., Moustafa, N., Razzak, I., Ahmad, T., & Patwary, M. N. (2022). A Secure and Intelligent Framework for Vehicle Health Monitoring Exploiting Big-Data Analytics. *IEEE Transactions on Intelligent Transportation Systems*, *23*(10), 19727–19742. Advance online publication. DOI: 10.1109/TITS.2021.3138255

Rahman, M. F., Pan, R., Ho, J., & Tseng, T. L. B. (2022). A Review of Augmented Reality Technology and its Applications in Digital Manufacturing. Available at *SSRN* 4068353.

Raju, N., & Farah, H. (2021). Evolution of Traffic Microsimulation and Its Use for Modeling Connected and Automated Vehicles. *Journal of Advanced Transportation, 2021*, 1–29. Advance online publication. DOI: 10.1155/2021/2444363

Riegler, A., Riener, A., & Holzmann, C. (2021). A Systematic Review of Virtual Reality Applications for Automated Driving: 2009–2020. *Frontiers in Human Dynamics*, *3*, 689856. Advance online publication. DOI: 10.3389/fhumd.2021.689856

Schlager, B., Muckenhuber, S., Schmidt, S., Holzer, H., Rott, R., Maier, F. M., Saad, K., Kirchengast, M., Stettinger, G., Watzenig, D., & Ruebsam, J. (2020). State-of-the-Art Sensor Models for Virtual Testing of Advanced Driver Assistance Systems/Autonomous Driving Functions. *SAE International Journal of Connected and Automated Vehicles*, *3*(3), 233–261. Advance online publication. DOI: 10.4271/12-03-03-0018

Sestino, A., Peluso, A. M., Amatulli, C., & Guido, G. (2022). Let Me Drive You! The Effect of Change Seeking and Behavioral Control in the Artificial Intelligence-Based Self-Driving Cars. *Technology in Society*, *70*, 102017. Advance online publication. DOI: 10.1016/j.techsoc.2022.102017

Theissler, A., Pérez-Velázquez, J., Kettelgerdes, M., & Elger, G. (2021). Predictive Maintenance Enabled by Machine Learning: Use Cases and Challenges in the Automotive Industry. *Reliability Engineering & System Safety*, *215*, 107864. Advance online publication. DOI: 10.1016/j.ress.2021.107864

Thevin, L., Briant, C., & Brock, A. M. (2020). X-Road: Virtual Reality Glasses for Orientation and Mobility Training of People with Visual Impairments. *ACM Transactions on Accessible Computing*, *13*(2), 1–47. Advance online publication. DOI: 10.1145/3377879

Vermesan, O., Eisenhauer, M., Sundmaeker, H., Guillemin, P., Serrano, M., Tragos, E. Z., Valiño, J., ... (2017). "Internet of Things Cognitive Transformation Technology Research Trends and Applications." In *Cognitive Hyperconnected Digital Transformation: Internet of Things Intelligence Evolution*, DOI: 10.1201/9781003337584-3

Vermesan, O., John, R., Pype, P., Daalderop, G., Kriegel, K., Mitic, G., Lorentz, V., Bahr, R., Sand, H. E., Bockrath, S., & Waldhör, S. (2021). Automotive Intelligence Embedded in Electric Connected Autonomous and Shared Vehicles Technology for Sustainable Green Mobility. *Frontiers in Future Transportation*, *2*, 688482. Advance online publication. DOI: 10.3389/ffutr.2021.688482

Yontem, A. O., Li, K., Chu, D., Meijering, V., & Skrypchuk, L. (2021). Prospective Immersive Human-Machine Interface for Future Vehicles: Multiple Zones Turn the Full Windscreen into a Head-Up Display. *IEEE Vehicular Technology Magazine*, *16*(1), 83–92. Advance online publication. DOI: 10.1109/MVT.2020.3013832

Zou, X., O'Hern, S., Ens, B., Coxon, S., Mater, P., Chow, R., Neylan, M., & Vu, H. L. (2021). On-Road Virtual Reality Autonomous Vehicle (VRAV) Simulator: An Empirical Study on User Experience. *Transportation Research Part C, Emerging Technologies*, *126*, 103090. Advance online publication. DOI: 10.1016/j.trc.2021.103090

Chapter 7
Foundations of Virtual Reality Technology:
Implications and Applications in the Automotive Industry

C. V. Suresh Babu
https://orcid.org/0000-0002-8474-2882
Hindustan Institute of Technology and Science, India

ABSTRACT

This chapter explores the foundations of Virtual Reality (VR) technology and its significant implications and applications in the automotive industry. The objective is to examine how VR is reshaping automotive design, manufacturing, and customer engagement, while identifying the opportunities and challenges it presents. A multi-method approach is used, combining a literature review, case study analysis, and expert interviews to assess the current state of VR in the automotive sector. Key findings reveal that VR enhances the design process, optimizes manufacturing practices, and improves customer experience through virtual showrooms and test drives. However, challenges such as high implementation costs and the need for standardized VR platforms persist. The chapter concludes that VR is a transformative technology with the potential to drive future innovations, yet its widespread adoption depends on overcoming technical and economic barriers. Further research is needed to explore the integration of VR with emerging technologies such as AI and AR in automotive applications.

DOI: 10.4018/979-8-3693-7287-6.ch007

1. INTRODUCTION

1.1 Overview of Virtual Reality Technology

Virtual Reality (VR) refers to immersive, computer-generated environments that simulate real or imagined scenarios, enabling users to interact with and explore these environments in real time. VR typically involves devices such as head-mounted displays (HMDs), motion sensors, and haptic feedback systems, providing a multi-sensory experience. By creating a controlled virtual space, VR enhances human-computer interaction (HCI) and fosters experiential learning and innovation in numerous domains (Zhang et al., 2021).

1.2 Historical Context and Evolution

The origins of VR can be traced back to the 1960s with the invention of the Sensorama by Morton Heilig, a device that provided multi-sensory film experiences. In the 1980s, Jaron Lanier popularized the term "Virtual Reality" and developed early VR systems like the DataGlove and EyePhone. Over time, advances in computing power, graphics rendering, and wearable technology have transformed VR into a highly versatile tool applicable across industries, including healthcare, education, and automotive engineering (Mazuryk & Gervautz, 1996).

1.3 Relevance to the Automotive Industry

In the automotive sector, VR offers solutions to challenges like cost control, time efficiency, and safety. Automakers leverage VR for:

- **Prototyping and Design**: Companies such as Tesla and Volkswagen employ VR to create virtual prototypes, enabling teams to test ergonomics and aerodynamics without physical models.
- **Manufacturing Processes**: VR simulations optimize factory layouts and workflows, reducing errors and enhancing productivity (Kim et al., 2020).
- **Training and Development**: Workers and drivers are trained in virtual environments that mimic real-world conditions, improving learning outcomes.
- **Customer Experience**: Brands like Audi use VR to provide customers with immersive showrooms and test drives.

1.4 Objectives of the Chapter

This chapter aims to:

1. Provide a comprehensive overview of VR technology and its evolution.
2. Explore the implications of VR in revolutionizing automotive processes.
3. Highlight practical applications of VR in vehicle design, manufacturing, training, and customer engagement.
4. Address challenges and potential advancements in integrating VR into automotive workflows.

2. FOUNDATIONS OF VIRTUAL REALITY TECHNOLOGY

2.1 What is Virtual Reality?

Definition and Core Concepts

Virtual Reality (VR) is a technology that enables users to interact with a computer-generated environment in a simulated manner. It relies on the principles of immersion, interactivity, and sensory feedback to create a seamless illusion of reality (Mazuryk & Gervautz, 1996).

Key concepts include:

- **Immersion:** The depth of involvement a user experiences within a VR environment.
- **Presence:** The psychological sensation of "being there" within the virtual space.
- **Interactivity:** The user's ability to interact dynamically with the virtual environment in real-time.

VR distinguishes itself from other forms of simulated experiences by its ability to engage multiple senses, such as sight, hearing, and touch, offering unparalleled realism.

Hardware Components

VR systems are supported by several hardware components, each critical for delivering a realistic experience:

1. **Head-Mounted Displays (HMDs):** Devices like the Oculus Rift and HTC Vive provide immersive visuals with wide fields of view and high-resolution graphics.

2. **Motion Controllers:** Hand-held devices such as VR gloves or joysticks enable users to interact with virtual objects.
3. **Tracking Systems:** Sensors track the user's position and movements to adjust the virtual environment in real time. Examples include infrared and optical tracking systems.
4. **Haptic Feedback Devices:** Equipment such as VR gloves provides tactile feedback, enhancing realism by mimicking the sense of touch.

Software and Development Platforms

The backbone of VR experiences lies in the software that powers virtual environments.

- **Game Engines:** Tools like Unity and Unreal Engine offer frameworks for developing VR applications with realistic physics, lighting, and interactivity.
- **Simulation Software:** Platforms like VRED and Autodesk are specifically designed for automotive applications, enabling real-time rendering of vehicle designs and performance simulations.
- **SDKs and APIs:** Software Development Kits (SDKs) such as OpenXR and SteamVR SDK allow developers to integrate VR functionalities across devices.

2.2 How VR Works: Technical Mechanisms

Sensors and Input Devices

Sensors and input devices capture the user's movements and translate them into actions within the virtual environment.

- **Positional Tracking:** Devices like base stations and infrared cameras monitor the user's location and movements.
- **Gyroscopes and Accelerometers:** Embedded in HMDs, these detect head orientation and adjust the view accordingly.
- **Input Devices:** Controllers, gloves, and even eye-tracking tools interpret user commands for interaction.

Rendering and Display Technologies

Rendering is the process of generating realistic visuals from data models. Key technologies include:

- **Stereoscopic Displays:** Create depth perception by showing slightly different images to each eye.
- **Foveated Rendering:** Optimizes performance by rendering in high resolution only where the user is looking.
- **Latency Optimization:** Ensures minimal delay between user actions and system responses to prevent motion sickness.

Modern VR setups use **low-persistence displays** to reduce blurring and improve clarity during motion, critical for automotive applications like driving simulations (LaValle, 2020).

2.3 Recent Advancements in VR Technology

1. **AI Integration**

Artificial Intelligence (AI) has enhanced VR by enabling more realistic interactions. For example, AI-powered avatars can respond to user inputs with near-human intelligence, enriching training simulations.

2. **Wireless VR Systems**

The development of wireless VR systems, such as the Oculus Quest, has eliminated the need for cumbersome cables, making VR setups more portable and user-friendly.

3. **Haptic Feedback Innovations**

Recent advancements in haptic technology now allow users to experience finer tactile sensations. For instance, VR gloves like those developed by HaptX simulate texture and resistance, making virtual object handling more realistic.

4. **Improved Display Technologies**

Advancements in OLED and MicroLED displays have enhanced the visual quality of VR experiences by providing higher resolution and contrast ratios.

5. **Cloud-Based VR Solutions**

Cloud VR allows rendering processes to occur on remote servers, reducing the hardware requirements for end-users. This has led to the democratization of VR, making it accessible to a wider audience.

3. APPLICATIONS OF VR IN THE AUTOMOTIVE INDUSTRY

3.1 Vehicle Design and Prototyping

Virtual Prototyping and Simulation

Virtual reality (VR) has revolutionized the design and prototyping phase of automotive development by enabling designers and engineers to visualize and test vehicles in a digital environment before creating physical prototypes. Virtual prototyping allows for detailed examination of vehicle aesthetics, ergonomics, and functionality without the cost and time associated with physical prototypes (Buchs, 2020).

In the automotive industry, VR is used to simulate real-world conditions, such as crash tests, aerodynamics, and performance under various environmental conditions. This enables manufacturers to identify and address potential design flaws early in the development cycle, significantly reducing the time-to-market and the cost of vehicle production (Wang & Suh, 2019).

By using VR, designers can manipulate digital models and modify features such as body shape, dashboard layout, and materials, facilitating iterative design without the need for expensive physical changes. For example, automotive giants like BMW and Ford use VR to evaluate ergonomic factors, ensuring that vehicle interiors meet customer comfort and safety standards (Buchs, 2020).

Case Study: Innovative Design Processes

A notable example is Ford's use of VR in its design process. The company implemented VR to evaluate and improve the driving experience of its vehicles. By using virtual simulations, Ford's designers could test how different car components would affect user comfort and safety. They simulated real-world driving scenarios, such as entering or exiting the vehicle, adjusting the seat, or interacting with in-car systems. This allowed Ford to fine-tune designs much earlier in the process and identify issues that might otherwise have been missed in a physical prototype (Ford, 2018).

3.2 Manufacturing and Assembly

VR Training and Workflow Optimization

Virtual reality is increasingly used for training purposes in automotive manufacturing. Workers and engineers are trained in a VR environment that mimics the actual assembly line, allowing them to practice complex tasks without the risks associated with real-world training. VR-based training systems are also used to teach new techniques and optimize workflows by simulating factory processes in a controlled, virtual environment (Hollands, 2021).

VR training also reduces the need for expensive physical prototypes in training programs and enables workers to familiarize themselves with assembly processes and machinery before operating them in a live environment. Additionally, workers can be exposed to various manufacturing scenarios, such as handling hazardous materials or troubleshooting mechanical issues, in a risk-free virtual setting (Xie et al., 2019).

Case Study: VR in Manufacturing Efficiency

In Toyota's manufacturing plants, VR is applied to streamline operations and enhance training efficiency. The company developed a VR training program to teach assembly line workers the precise steps involved in installing car components. The training includes various modules, such as hand placement, component handling, and safety precautions. This virtual training process helps workers become proficient in their tasks without the need to physically manipulate real components. This system has proven to reduce error rates and improve overall manufacturing efficiency, as workers are better prepared to handle the demands of the assembly line (Toyota, 2020).

3.3 Sales and Customer Experience

Virtual Showrooms and Test Drives

VR has revolutionized the customer experience in the automotive sales process by enabling virtual showrooms and test drives. In virtual showrooms, customers can explore a wide range of vehicle models in 3D, customize the vehicles according to their preferences, and visualize how specific features would look in real-time. Customers can interact with vehicle interiors, adjust seating, and examine various colors and finishes, all without having to visit a physical dealership (Wang & Suh, 2019).

Additionally, VR test drives allow customers to experience how a vehicle performs on the road, offering an immersive driving experience that simulates real-world conditions. These VR test drives can be conducted from anywhere, giving customers the ability to explore vehicles remotely. This is particularly beneficial for customers who may be unable to visit a dealership due to geographical limitations or busy schedules.

Augmented Sales Strategies with VR

Sales teams are also utilizing VR to enhance the customer journey by integrating augmented sales strategies. For example, VR technology can be used to present detailed vehicle information interactively, allowing customers to virtually "sit" in the car and receive real-time data about its features. Additionally, virtual reality is integrated into AR apps to provide customers with a mixed-reality shopping experience, where physical spaces blend with digital content, allowing them to virtually test out a vehicle in their own environment (Hollands, 2021).

3.4 Post-Sales and Maintenance

VR for Remote Diagnostics and Service

Post-sales service is another area where VR has shown significant potential. Using VR for remote diagnostics and maintenance allows technicians to simulate car repairs and diagnose issues virtually. This can be particularly useful in situations where physical intervention is not immediately possible, or for customers who prefer not to visit a service center (Xie et al., 2019).

By using VR combined with augmented reality (AR) and remote assistance, mechanics can guide customers through basic repair procedures or provide instructions for resolving minor issues, reducing the need for physical visits and expediting the service process. Additionally, VR simulations can be used to train service technicians, enabling them to practice complex repair procedures without the need for physical vehicles or parts.

Case Study: VR in Customer Support

One of the most innovative uses of VR in customer support is the implementation of virtual service assistance by Audi. The company introduced an interactive virtual assistant system, which enables customers to perform remote diagnostics and seek troubleshooting help in real-time. Audi's system employs a combination of VR, AR, and remote expert assistance to help customers with issues ranging from engine problems to electronic malfunctions. This system improves customer satisfaction by reducing wait times and providing a personalized support experience (Audi, 2020).

3.5 How is VR Technology Reshaping Automotive Design?

Virtual reality (VR) is playing an increasingly vital role in reshaping automotive design by providing tools that allow for immersive simulations, collaborative design processes, and early-stage testing of concepts. This technological transformation is enabling automakers to visualize complex vehicle designs, optimize safety features, and address ergonomic concerns without the traditional reliance on physical prototypes.

Impact on Design Process

One of the most significant ways VR is influencing automotive design is through virtual prototyping. Designers can build and modify digital 3D models of vehicles that can be tested in a virtual space, allowing for real-time adjustments and visualizations. This provides a highly effective way to test the ergonomics of the car interior, evaluate materials, and ensure compliance with safety standards, all before any physical model is created. The real-time nature of VR design means that automakers can make quicker, more informed decisions, significantly reducing the time-to-market for new models (Suresh Babu, C. V. & Jeyavasan, T. 2023).

Enhanced Collaboration and Customization

With VR, teams of designers, engineers, and other stakeholders can work collaboratively on a virtual model from different locations, providing a more efficient way to make decisions and solve design issues. For example, global automotive companies such as BMW have adopted VR to facilitate team-based design processes, allowing cross-functional teams to participate in virtual design reviews (Buchs, 2020). Additionally, VR tools allow customers to become active participants in the design process. Through customizable car features in VR environments, customers can visualize and personalize their vehicle design, making the buying experience more interactive and customer-centric (Wang & Suh, 2019).

3.6 What Are the Challenges of Implementing VR in Manufacturing?

While VR offers substantial benefits for automotive manufacturing, its implementation comes with specific challenges that need to be addressed to fully capitalize on its potential.

High Initial Investment and Infrastructure

One of the major challenges in adopting VR technology in manufacturing is the significant initial investment required to set up the necessary hardware and software infrastructure. Advanced VR systems, including high-definition headsets, motion sensors, and high-performance computers, are costly, and not every manufacturing facility may be ready to bear these expenses. Additionally, integrating VR with existing manufacturing technologies and workflows may require a steep learning curve for workers, necessitating both time and financial investment in training and development (Hollands, 2021).

Resistance to Change

Employees may resist the transition to new technology, particularly in well-established industries like automotive manufacturing. VR requires workers to change how they approach tasks such as assembly line work, training, and product testing. Overcoming resistance from employees and ensuring that they understand the value of VR in optimizing processes is crucial for successful adoption (Xie et al., 2019).

System Integration and Scalability

Integrating VR into existing manufacturing systems can be complex. The challenge lies in aligning VR with automation systems, machine learning models, and other digital tools used in manufacturing plants. Furthermore, scaling VR applications across global production facilities to ensure consistent processes and outcomes can be a significant hurdle, particularly for multinational automotive companies with diverse production capabilities (Wang & Suh, 2019).

3.7 How Does VR Enhance Customer Interaction and Sales?

In the automotive industry, VR is fundamentally changing the way customers interact with vehicles, facilitating a more immersive and personalized experience throughout the sales process. From virtual showrooms to test drives, VR is enhancing the customer experience by allowing consumers to engage with products in ways that were previously impossible.

Virtual Showrooms and Test Drives

With VR, automotive dealerships are able to offer virtual showrooms where customers can explore various vehicle models and configurations in 3D. This virtual experience allows potential buyers to "walk" around a vehicle, get a detailed look at its features, and customize it to suit their preferences. Customers can also experience a virtual test drive, where they can simulate driving a car in various real-world scenarios without leaving the dealership or even their homes (Hollands, 2021). This technology improves the customer experience by providing a more engaging, interactive, and efficient method of vehicle exploration.

Personalization and Engagement

VR enhances customer interaction by enabling car buyers to fully personalize their vehicles in a virtual environment. For example, Hyundai has introduced a VR system that allows customers to visualize different car models and configurations, adjusting features like the color, upholstery, and dashboard features. This immersive interaction encourages customers to make more informed decisions and increases the likelihood of purchase (Wang & Suh, 2019).

Remote Interaction and Sales Strategies

With the rise of digital retailing, VR is also becoming a valuable tool for remote customer interaction. Virtual reality allows automotive brands to extend their reach by offering remote consultations, virtual sales pitches, and even virtual car showcases that customers can access from anywhere in the world. This augmented reality experience is particularly beneficial in today's market, where remote interactions are becoming increasingly popular due to convenience and health concerns (Ford, 2020).

3.8 What Are the Future Directions for VR in the Automotive Industry?

The future of VR in the automotive industry is promising, with several key advancements on the horizon that will further enhance the capabilities of this technology.

Improved VR Hardware and User Experience

As VR technology continues to evolve, we can expect to see more advanced hardware with better graphical fidelity, faster response times, and more immersive experiences. The development of lightweight VR headsets, with improved comfort and mobility, will allow users to engage in longer, more complex interactions with virtual environments (Babu, C. V. S., & Vinutha, S. 2024). Additionally, advancements in haptic feedback and motion sensing will create more realistic and interactive user experiences, particularly in vehicle design and virtual test drives.

Integration with AI and Machine Learning

VR's integration with artificial intelligence (AI) and machine learning will open new opportunities for predictive design, personalized experiences, and enhanced customer interaction. For example, AI algorithms could analyze customer preferences in real-time during a virtual test drive or showroom visit, offering personalized recommendations for vehicle features and configurations (Xie et al., 2019). Similarly, AI-driven VR applications could be used for predictive maintenance and diagnostics in automotive service centers, providing real-time feedback to both technicians and customers.

Collaborative and Social VR Experiences

In the future, automotive companies are likely to adopt more collaborative and social VR experiences, allowing customers to experience vehicles in a shared, multi-user virtual environment. This could involve customers interacting with sales representatives, designers, or even other customers in a virtual space, further enhancing the interactive nature of the sales process (Ford, 2020). Additionally, manufacturers could use these collaborative platforms for testing vehicle designs and gathering customer feedback, offering a deeper level of engagement and input.

4. OBJECTIVES AND SCOPE

4.1 Chapter Objectives

The primary objective of this chapter is to provide an in-depth understanding of Virtual Reality (VR) technology, its implications, and its diverse applications within the automotive industry. The chapter aims to break down the fundamental concepts of VR, explore its industry-specific uses, and discuss the challenges and potential strategies for integrating VR into various processes. The key objectives are outlined as follows:

4.1.1 Understanding VR Fundamentals

The first objective of this chapter is to lay the groundwork for understanding the core concepts of VR technology. VR is a tool that simulates an immersive, interactive environment, typically created using specialized hardware such as headsets, sensors, and motion controllers (Rauschnabel et al., 2020). For automotive professionals and companies, having a foundational grasp of how VR works is crucial to appreciating its value and potential.

This section covers the definition, evolution, and technical mechanisms of VR. By understanding the hardware and software components that make VR functional, readers will gain insight into how VR enables designers, engineers, manufacturers, and customers to interact with automotive products and systems in new, innovative ways. Key topics include virtual prototyping, the role of sensors and input devices, the use of rendering technologies, and the integration of VR with machine learning for predictive modeling and testing (Slater & Wilbur, 1997). Understanding VR fundamentals equips industry professionals with the knowledge to explore further applications and benefits in their respective fields.

4.1.2 Analyzing Industry-Specific Applications

The second objective is to delve into the various applications of VR technology within the automotive industry. The chapter will explore how VR is revolutionizing key aspects of automotive design, manufacturing, sales, and customer service. This includes:

- **Vehicle Design and Prototyping**: How VR enables virtual prototyping, reducing the need for physical models and speeding up the design process.
- **Manufacturing and Assembly**: The role of VR in training workers, simulating assembly line processes, and optimizing workflows (Jung et al., 2020).
- **Customer Experience and Sales**: How virtual showrooms, interactive test drives, and personalized virtual experiences are reshaping the automotive sales landscape (Buchs, 2020).
- **Post-Sales and Maintenance**: The potential for VR in providing remote diagnostics and customer service tools that enhance after-sales support and maintenance processes (Hollands, 2021).

These applications will be illustrated with real-world case studies from leading automotive manufacturers, offering practical examples of VR adoption.

4.1.3 Identifying Integration Strategies

The final objective of this chapter is to provide guidance on integrating VR technology into the automotive industry. The integration of VR into existing workflows and systems is a complex process, requiring a strategic approach to ensure that it enhances productivity and innovation rather than disrupting operations. The chapter will explore various integration strategies, including:

- **Collaborative VR Design and Development**: How VR can be integrated into the design and testing phase, enabling cross-functional teams to collaborate virtually, reducing errors, and enhancing the efficiency of the design process.
- **VR in Production and Quality Control**: Implementing VR-based tools to optimize production efficiency, perform quality checks, and train employees on new technologies and systems (Wang & Suh, 2019).
- **Consumer Engagement Platforms**: Leveraging VR for direct consumer engagement, creating digital showrooms, and facilitating virtual test drives (Ford, 2020).

The integration strategies section will help industry stakeholders understand the critical steps for effectively embedding VR into their operations and ensure that all stakeholders—designers, engineers, marketers, and consumers—can benefit from the technology.

4.2 Delimitation of Scope

The chapter will focus on the specific applications and implications of VR technology within the automotive industry. While VR is applicable to many sectors, this chapter is delimited to its use in automotive design, manufacturing, sales, and maintenance.

4.2.1 Focus Areas Covered

The focus areas of this chapter include:

- **Virtual Prototyping and Design**: In-depth analysis of how VR is reshaping the way vehicles are conceptualized and tested. This includes early-stage design, ergonomics, and safety simulations.
- **Manufacturing and Assembly Line Optimization**: A look into how VR assists in assembly line training, workflow simulation, and process optimization.

- **Sales and Customer Experience**: A discussion of virtual showrooms, customer interactions with vehicles, and virtual test drives as part of an enhanced customer journey.
- **Post-Sales and Maintenance**: Investigating how VR is used for diagnostics, remote service, and consumer support post-purchase.

These areas are critical to understanding the broad and transformative impact VR has in the automotive sector.

4.2.2 Limitations and Exclusions

While this chapter explores several aspects of VR in the automotive industry, it does have some limitations:

- **Exclusion of Augmented Reality (AR)**: While AR and VR are closely related, this chapter focuses solely on VR technology. AR applications in automotive sectors, though important, are outside the scope of this chapter.
- **Limited Focus on Consumer Behavior**: While VR's role in customer interaction is explored, the chapter does not delve deeply into consumer psychology and behavior related to VR experiences. This could be the subject of further research.
- **Geographic Scope**: The case studies and examples provided primarily focus on developments in North America, Europe, and Asia, which may not be globally applicable across all automotive markets.

By delineating these boundaries, the chapter provides a comprehensive, yet focused, exploration of VR applications in the automotive industry.

5. LITERATURE REVIEW

The literature review provides an in-depth analysis of the research and developments that have shaped the use of Virtual Reality (VR) in the automotive industry. It includes an overview of the historical development of VR, a review of its current applications in automotive design, and identifies gaps in the existing literature, particularly focusing on emerging trends and the need for comprehensive case studies.

5.1 Summary of Relevant Literature

5.1.1 Historical Development of VR Technology

Virtual Reality technology has evolved significantly over the past few decades, from its early conceptualization to its current applications in various industries, including automotive design. The historical development of VR technology can be traced back to the 1950s, when the first ideas of creating immersive virtual experiences were proposed. Pioneers such as Morton Heilig introduced the Sensorama in 1962, a machine designed to simulate multisensory experiences, which laid the groundwork for future VR innovations (Heilig, 1962).

In the 1980s, the term "Virtual Reality" was coined by Jaron Lanier, and the first head-mounted displays (HMDs) and early immersive systems were developed by companies such as VPL Research (Lanier, 1989). These early systems were rudimentary and expensive, limiting their use to specialized applications in defense and research. However, the development of computer graphics, motion tracking technologies, and more affordable hardware in the 1990s and early 2000s led to increased accessibility, allowing VR to gain traction in fields such as entertainment, medicine, and engineering.

In the automotive industry, VR began to gain relevance in the late 1990s when manufacturers like General Motors and Ford started experimenting with VR for vehicle design and safety simulations (Ahearn, 1997). By the early 2000s, VR was integrated into the design and prototyping processes, enabling faster iteration of vehicle models and testing of safety features in a controlled virtual environment (Gennari & Miele, 2006).

The current stage of VR development, marked by the availability of affordable and highly advanced VR systems like the Oculus Rift, HTC Vive, and PlayStation VR, has propelled the automotive industry into using VR for not only design but also manufacturing, marketing, and customer interaction (Schaffner & Buckley, 2019).

5.1.2 Current Applications in Automotive Design

Today, VR is a critical tool in automotive design, enabling manufacturers to innovate and streamline development processes. The primary application of VR in automotive design involves virtual prototyping. With VR, designers and engineers can create and manipulate 3D models of vehicles, testing different features such as aerodynamics, ergonomics, and crash safety without needing physical prototypes.

This reduces the time and costs associated with traditional prototyping and accelerates product development (Kishore et al., 2020).

Another key application is the simulation of manufacturing and assembly processes. VR allows companies to optimize workflows, conduct virtual training for workers, and identify potential inefficiencies or safety risks before physical production begins (Hollands, 2021).

VR is also being integrated into customer-facing processes, such as virtual showrooms and test drives. Companies like Audi and Porsche have adopted VR to provide customers with an immersive experience where they can customize their cars and experience virtual test drives without leaving the showroom (Buchs, 2020). This use of VR enhances the customer experience and allows brands to offer innovative sales solutions in a competitive market.

The rise of VR in the automotive industry has not been limited to product development and sales. It has also had significant applications in marketing, particularly through the creation of immersive advertisements and virtual reality experiences at trade shows and exhibitions (Schaffner & Buckley, 2019).

5.2 Gaps in Current Literature

While significant progress has been made in the adoption of VR technology in the automotive industry, there remain key gaps in the literature. These gaps highlight areas where further research is needed to fully understand the potential of VR and its integration into the automotive sector.

5.2.1 Emerging Trends Not Fully Explored

One emerging trend that has not been fully explored in the literature is the integration of VR with other advanced technologies such as Artificial Intelligence (AI) and the Internet of Things (IoT). The intersection of these technologies could open up new possibilities for the automotive industry. For instance, AI-powered VR systems could facilitate real-time design feedback and assist in virtual testing by simulating dynamic real-world conditions (Wang & Suh, 2019). Additionally, IoT integration in VR environments could enable a more interactive experience for users, where data from connected vehicles can be used to simulate real-time driving scenarios or enhance customer interactions.

Another underexplored area is the use of VR in sustainability initiatives within the automotive industry. Research is limited regarding how VR can contribute to environmental sustainability, such as optimizing vehicle design for energy efficiency or using VR to simulate the lifecycle of materials in the manufacturing process. The potential of VR to create sustainable practices in the automotive sector, especially as the industry shifts toward electric vehicles (EVs), needs further investigation.

5.2.2 Need for Comprehensive Case Studies

There is a notable lack of comprehensive case studies that detail the real-world application of VR in automotive production environments. Most existing research focuses on isolated instances of VR adoption, often limited to specific tasks like vehicle design or customer engagement. However, there is a lack of holistic case studies that explore the integration of VR across different stages of automotive development—from initial design to post-sales service. A more thorough analysis of such case studies would provide valuable insights into the operational challenges and benefits of implementing VR technology at scale (Hollands, 2021).

Moreover, the existing literature tends to focus on large, well-established automotive companies such as Ford, General Motors, and BMW. However, there is limited research on how small and medium-sized automotive manufacturers are adopting VR, which could offer a broader perspective on the accessibility and scalability of VR in automotive processes.

6. METHODOLOGY

The methodology section outlines the research methods used to investigate the foundations and applications of Virtual Reality (VR) technology in the automotive industry. This chapter employs a multi-method approach combining literature review, case study analysis, and expert interviews to provide a comprehensive understanding of VR's role in the automotive sector. The selection of these methods is rooted in the need to gather both theoretical and practical insights that are relevant to the evolving relationship between VR and automotive technology.

6.1 Research Methods

The research methods employed in this study are designed to ensure a thorough exploration of the implications and applications of VR in the automotive industry. Each method is carefully chosen to address specific aspects of the topic, from theoretical underpinnings to real-world applications.

6.1.1 Literature Review Approach

A thorough literature review is a foundational research method used to gather existing knowledge about the historical development, technological advancements, and current uses of VR in the automotive industry. The review spans academic journals, conference proceedings, industry reports, and case studies, covering a wide range of topics such as VR's evolution, current applications in automotive design and manufacturing, and emerging trends.

The literature review allows for the identification of major milestones in VR development, key technological innovations, and challenges faced by the automotive industry in integrating VR into its processes. By analyzing existing research, the review synthesizes information to establish a broad understanding of the current state of VR in automotive applications, highlighting gaps and areas for further exploration (Hollands, 2021; Gennari & Miele, 2006).

For instance, VR's use in vehicle prototyping and simulation, as highlighted in previous studies (Ahearn, 1997), is examined alongside emerging applications such as augmented reality (AR) integration for interactive customer experiences (Schaffner & Buckley, 2019). This method enables the identification of trends that can inform future applications of VR technology in automotive manufacturing and customer interaction.

6.1.2 Case Study Analysis

Case study analysis involves examining specific examples of VR applications in the automotive industry. Case studies offer a valuable real-world perspective on the practical implementation of VR technologies and highlight both the successes and challenges companies face in adopting these technologies.

For example, this chapter analyzes the implementation of VR in design and prototyping processes at companies like General Motors and Ford (Gennari & Miele, 2006). These case studies provide insights into how VR is used to reduce product development cycles and enhance the accuracy of designs before physical prototypes are created. Additionally, case studies on VR's role in customer engagement and marketing at companies like Audi and Porsche illustrate how VR creates immersive experiences that improve sales and customer satisfaction (Buchs, 2020).

These case studies offer a practical lens through which the chapter explores the impact of VR across various stages of the automotive production and sales process. The analysis focuses on the specific VR tools used, the challenges encountered, and the outcomes achieved, allowing for a deep understanding of the technology's impact on the industry.

6.1.3 Expert Interviews

Expert interviews are conducted to gain first-hand insights from professionals in the automotive and VR technology fields. These interviews help gather qualitative data on the current and future use of VR in automotive applications. Experts are selected based on their experience with VR technologies in automotive design, manufacturing, or customer relations.

The interview process involves semi-structured questions that explore topics such as:

- How VR has been implemented in various automotive processes.
- Challenges encountered during implementation and integration.
- Future potential of VR in the automotive industry.
- Key trends that are shaping the future of VR in automotive applications.

Experts may include automotive engineers, VR software developers, and managers at companies that have adopted VR technologies, offering insights into both the technical and strategic aspects of VR adoption (Wang & Suh, 2019). These interviews provide a deeper understanding of the practical implications of VR, complementing the findings from the literature review and case studies.

6.2 Justification for Chosen Methodology

The methodology adopted for this chapter is designed to address the chapter's objectives of understanding the fundamentals of VR, analyzing its industry-specific applications, and identifying strategies for its integration into automotive processes. Each research method contributes to fulfilling these goals by providing different types of data that complement one another.

6.2.1 Relevance to Chapter Objectives

The literature review provides a theoretical foundation for understanding VR's evolution and its current applications in the automotive industry. This aligns with the objective of understanding the fundamental aspects of VR technology and its significance in the sector. The review helps establish a contextual framework for subsequent analysis, ensuring that the chapter's content is grounded in existing research and industry developments.

Case study analysis provides empirical evidence of VR's effectiveness in automotive design and manufacturing processes. This method is particularly relevant to the objective of analyzing industry-specific applications, as it highlights how different companies have implemented VR in their workflows, illustrating practical challenges and benefits.

Expert interviews offer real-time, industry-specific insights that deepen the understanding of VR's role in the automotive sector. These interviews allow for the collection of forward-looking perspectives, making them essential for exploring the future directions of VR in automotive applications.

By combining these methods, the chapter addresses the objectives comprehensively, offering both theoretical and practical insights into VR technology's role in the automotive industry.

6.2.2 Reliability and Validity of Findings

The chosen methodology ensures the reliability and validity of the chapter's findings in several ways:

1. **Literature Review**: The use of peer-reviewed articles and reputable industry reports ensures that the findings are based on reliable, validated sources. Additionally, by synthesizing information from a range of sources, the literature review minimizes the potential for bias and offers a broad perspective on the state of VR in the automotive industry.
2. **Case Study Analysis**: By examining real-world examples from multiple companies, the case study approach provides a balanced view of VR applications. Case studies also offer transparency in the research process, as they include details about both successes and challenges faced by the companies.
3. **Expert Interviews**: The selection of reputable industry experts ensures the validity of the qualitative data. Experts with firsthand experience provide practical insights, and the semi-structured interview format allows for flexibility, enabling interviewees to provide detailed responses based on their experiences.

The combination of these methods enhances the robustness of the findings and contributes to the overall reliability and validity of the chapter's conclusions.

7. CONTRIBUTION TO THE FIELD

Virtual Reality (VR) technology has made significant strides in reshaping the automotive industry, influencing a range of activities from design and engineering to manufacturing and sales. This section examines VR's contributions to the automotive sector, highlighting the impact on automotive design and engineering, innovations in manufacturing and sales, practical implications for industry professionals, and the theoretical insights that emerge from these developments. The role of VR in this transformation is crucial, as it not only enhances existing processes but also opens new avenues for innovation and efficiency.

7.1 Impact on Automotive Design and Engineering

The application of VR in automotive design and engineering has revolutionized how manufacturers conceptualize, prototype, and test new vehicles. Traditionally, vehicle design was a time-intensive process that involved multiple physical prototypes and considerable resources. VR, however, allows engineers and designers to create and manipulate digital models of vehicles, enabling faster iterations and more precise adjustments before physical prototypes are constructed.

1. **Virtual Prototyping**: VR facilitates virtual prototyping, where automotive designers can interact with a 3D model of a vehicle. This allows for the evaluation of ergonomics, aerodynamics, and other critical aspects of vehicle performance in a virtual environment. By experiencing the design in a fully immersive VR space, engineers can identify design flaws and make necessary improvements early in the development cycle, thus saving both time and costs associated with physical prototypes (Gennari & Miele, 2006).
2. **Human Factors and Ergonomics**: VR's ability to simulate the human experience within a vehicle design allows engineers to analyze comfort, usability, and the overall user experience. For instance, VR simulations can replicate driver interactions with car controls, seating arrangements, and even the feel of driving the vehicle, helping designers optimize comfort and safety features. This application has proven particularly valuable in improving vehicle interior layouts and infotainment systems, ensuring that they meet consumer expectations and regulatory standards (Buchs, 2020).
3. **Collaborative Design**: VR enables real-time collaboration among geographically dispersed design teams. Through virtual environments, stakeholders in different locations can meet, review designs, and provide feedback as if they were physically present in the same room. This fosters greater collaboration, reduces delays caused by travel or miscommunication, and accelerates the design process (Schaffner & Buckley, 2019).

7.2 Innovations in Manufacturing and Sales

VR is not only transforming design processes but also driving innovation in manufacturing and sales. By improving efficiency, enhancing training methods, and creating new customer experiences, VR is redefining traditional automotive production and sales paradigms.

1. **Manufacturing Efficiency**: In the manufacturing domain, VR plays a pivotal role in optimizing workflows, ensuring precision, and reducing errors. For instance, VR simulations allow manufacturers to visualize assembly lines and workstations before physical construction begins. This helps identify potential bottlenecks and layout inefficiencies. Additionally, VR can assist in the training of workers, providing them with virtual environments to practice assembly tasks without the need for costly physical models (Hollands, 2021).
2. **Training and Safety**: VR offers safe, cost-effective training solutions for automotive manufacturing staff. By simulating complex machinery and assembly lines, workers can practice their tasks in a risk-free environment. This approach not only accelerates the learning process but also reduces the likelihood of accidents on the shop floor. VR-based training programs are increasingly used to train assembly line workers, engineers, and technicians, ensuring that they are well-versed in the latest tools and processes (Gennari & Miele, 2006).
3. **Sales and Customer Experience**: One of the most innovative applications of VR in the automotive industry is the creation of immersive virtual showrooms. Customers can experience vehicle models in a highly detailed, interactive VR environment, where they can explore car features, colors, and configurations without visiting a physical dealership. Additionally, VR allows potential buyers to take virtual test drives, offering a first-hand experience of the vehicle's performance and features (Buchs, 2020).

Augmented Sales Strategies: Automotive companies are also exploring the integration of VR with augmented reality (AR) to enhance the sales experience. By combining VR's immersive simulations with AR's real-time, data-driven overlays, companies can offer dynamic, personalized product demonstrations that provide potential customers with detailed insights into the vehicle's performance, features, and specifications in real-time (Schaffner & Buckley, 2019).

7.3 Practical Implications for Industry Professionals

For industry professionals, VR represents a transformative tool with far-reaching implications across various automotive sectors. The integration of VR in automotive design, manufacturing, and sales processes introduces both challenges and opportunities.

1. **Professional Skill Development**: The increasing use of VR technologies requires industry professionals to develop new skill sets, particularly in digital modeling, 3D visualization, and immersive simulation design. Engineers, designers, and marketers must familiarize themselves with VR software platforms and their applications in product development. As VR becomes a core component of automotive workflows, continuous learning and adaptability will be key to staying competitive in the industry (Hollands, 2021).
2. **Collaboration Across Disciplines**: The integration of VR into automotive processes demands a closer collaboration between various disciplines, such as design, engineering, and marketing. Professionals from different departments must work together to ensure that VR applications align with both technical requirements and customer preferences. The multidisciplinary nature of VR integration in automotive industries emphasizes the need for teamwork, communication, and shared expertise across departments (Buchs, 2020).
3. **Cost and Time Management**: VR's ability to speed up the design and prototyping process has significant implications for cost and time management. By reducing the need for physical prototypes and streamlining the manufacturing process, VR allows automotive companies to cut costs and bring products to market faster. For industry professionals, this means that efficiency and productivity are increasingly driven by digital tools, requiring them to adopt more agile and data-driven approaches to their work (Schaffner & Buckley, 2019).

7.4 Theoretical Insights and New Perspectives

The introduction of VR into the automotive industry also offers valuable theoretical insights, particularly in understanding the relationship between digital technologies and traditional industrial practices. VR challenges traditional paradigms of design, manufacturing, and sales, offering new perspectives on how digital tools can reshape physical production processes.

1. **Digital Twin Technology**: One of the key theoretical contributions of VR to automotive manufacturing is the concept of the digital twin. A digital twin is a virtual replica of a physical vehicle or manufacturing system that can be

used for simulation, optimization, and monitoring. VR, when integrated with digital twin technology, allows for real-time simulation of vehicle performance and manufacturing processes, providing a deeper understanding of how digital models relate to their physical counterparts. This theoretical approach to simulation challenges the traditional notion of physical-only product development and opens the door to more efficient, data-driven manufacturing (Gennari & Miele, 2006).

2. **Human-Technology Interaction**: VR also raises questions about the interaction between humans and technology in the automotive sector. The increased use of VR simulations in design and testing allows for new ways of thinking about user experience, particularly in terms of human-vehicle interaction. Theoretical research in human-computer interaction (HCI) can provide valuable insights into how drivers and passengers perceive and interact with VR-based car systems, ultimately influencing the development of more intuitive and user-friendly automotive technologies (Buchs, 2020).

8. DISCUSSION AND ANALYSIS

Virtual Reality (VR) technology is rapidly gaining traction in the automotive industry, providing novel approaches to design, manufacturing, sales, and customer engagement. As the industry seeks innovative solutions to remain competitive, understanding VR's impact and its practical integration is essential for maximizing its potential. This section will discuss new insights and theories derived from VR applications in the automotive sector, present findings from case studies, analyze the integration of VR into automotive processes, and provide recommendations for the industry's wider adoption.

8.1 New Insights and Theories

The integration of Virtual Reality into the automotive industry is not just a technological evolution but also a conceptual shift in how automotive processes are perceived and executed. Several emerging theories and insights have evolved from VR's implementation:

1. **Digital Twin Technology and Real-time Simulation**: One of the most significant innovations driven by VR is the concept of the "digital twin." This theoretical approach creates a virtual counterpart of a physical vehicle or manufacturing system, enabling real-time simulations and predictive analytics (Gennari & Miele, 2006). The digital twin can model everything from a vehicle's performance in

various driving conditions to the behavior of an entire production line. This integration of VR into the design, testing, and production processes offers new opportunities to streamline operations and enhance product development.

2. **Human-Technology Interaction (HTI)**: The human-technology relationship in VR-based automotive applications is critical to shaping the user experience. VR has brought the concept of Human-Technology Interaction (HTI) into sharper focus, particularly regarding how users—whether designers, engineers, or customers—interact with VR systems. Understanding these interactions is essential to improving the usability of VR tools in automotive contexts, whether it's in enhancing vehicle designs or offering virtual test drives (Buchs, 2020). Theories on HTI are expanding, suggesting that virtual experiences need to align closely with human cognitive and sensory expectations to deliver effective, enjoyable, and intuitive user interactions.
3. **Immersive Experience as a Tool for Innovation**: The immersive nature of VR allows automotive professionals to explore new dimensions of design and testing. By engaging in virtual environments that simulate real-world scenarios, engineers can test vehicles in extreme conditions or simulate consumer experiences at a level of detail previously unimaginable. This has led to new theories regarding the role of virtual simulations in fostering innovation. The principle of "experiential learning," where users engage directly with a simulated environment, plays a crucial role in developing vehicles that meet evolving consumer demands (Schaffner & Buckley, 2019).

8.2 Practical Applications and Case Study Findings

Several real-world applications of VR in the automotive industry have demonstrated its transformative potential across various sectors:

1. **Design and Prototyping**: Many automotive companies, including Ford and Audi, have implemented VR technologies in their design and prototyping stages. Ford's use of VR for vehicle interior design allows engineers and designers to view and interact with 3D models of their prototypes in a fully immersive environment, identifying design flaws and optimizing ergonomics early in the process (Ford Motor Company, 2020). Audi uses VR to allow customers to experience their cars before production, providing an interactive tool for prospective buyers to customize features and visualize their choices in a simulated environment (Audi AG, 2019).
2. **Manufacturing Optimization**: In the manufacturing sector, BMW has embraced VR for assembly line optimization. Using VR to simulate the workflow before physical construction begins, the company is able to identify potential

inefficiencies and improve the layout of their assembly lines (Gennari & Miele, 2006). Moreover, VR training simulations have been implemented by companies like Mercedes-Benz to prepare workers for assembly line tasks without the risk of mistakes or accidents, improving safety and operational efficiency (Schaffner & Buckley, 2019).

3. **Sales and Customer Engagement**: Virtual showrooms are becoming an increasingly popular tool in automotive sales. Companies like Volvo and Audi have introduced VR-enabled showrooms where customers can walk around virtual cars and see 360-degree views of models, colors, and features without having to visit a dealership. These virtual experiences are designed to engage potential buyers in a more interactive and immersive way than traditional 2D online catalogs (Buchs, 2020).

8.3 Analysis of VR Integration in Automotive Processes

Integrating VR into the automotive industry's processes, such as design, manufacturing, and sales, brings several benefits, but also poses unique challenges.

1. **Benefits of VR Integration**: The most significant advantage of VR in the automotive industry is its ability to reduce development time and costs. By simulating real-world conditions, VR allows designers to test prototypes in virtual environments, thereby minimizing the need for expensive physical prototypes. It also enables faster iteration cycles, ensuring that the final product is as efficient and market-ready as possible. In manufacturing, VR optimizes assembly lines by identifying inefficiencies in workflow, enhancing worker safety, and reducing the risk of defects (Hollands, 2021).
2. **Challenges in Implementation**: While the benefits are clear, VR integration comes with its own set of challenges. One significant hurdle is the high upfront cost of VR infrastructure, including hardware, software, and specialized training. Furthermore, the learning curve for VR tools can be steep, requiring professionals to develop new skill sets. Additionally, VR may not yet offer the same level of tactile feedback as physical prototypes, which could affect certain applications in testing and evaluation (Gennari & Miele, 2006).
3. **Interdisciplinary Collaboration**: Successful integration of VR into automotive processes requires collaboration across various departments, such as engineering, design, and marketing. This collaboration is essential in ensuring that the VR tools being developed are aligned with the practical needs of each department. Furthermore, VR should not be seen as a replacement for traditional methods but as a complementary tool that can enhance existing processes (Schaffner & Buckley, 2019).

8.4 Recommendations for Industry Adoption

To effectively adopt and maximize the potential of VR, automotive companies must take a strategic, phased approach that addresses both technological and organizational challenges.

1. **Invest in Training and Skill Development**: As VR becomes more integrated into automotive processes, companies should prioritize training programs for their employees. These programs should focus on educating workers about VR tools and how to effectively utilize them in their roles. Upskilling employees in VR-based design, simulation, and prototyping is critical for the long-term success of VR adoption in automotive processes (Buchs, 2020).
2. **Leverage Cross-Industry Collaboration**: Automotive companies should explore opportunities for collaboration with technology providers and other industries that are already leveraging VR. This could include partnering with VR hardware manufacturers to design more customized, industry-specific solutions or working with software developers to create tailored VR platforms for automotive design, testing, and sales (Ford Motor Company, 2020).
3. **Start with Pilot Projects**: Rather than implementing VR across all departments at once, it is advisable to start with smaller pilot projects. This allows companies to experiment with VR in specific areas, such as virtual prototyping or employee training, before scaling up. Such an approach minimizes risk and provides valuable insights into the practical applications of VR in the automotive industry (Hollands, 2021).
4. **Focus on Data Integration and Customization**: To fully harness the power of VR, automotive companies should prioritize integrating VR platforms with existing data systems, such as CAD (Computer-Aided Design) and ERP (Enterprise Resource Planning) systems. This integration will allow for more seamless workflows and more accurate simulations. Customizing VR tools to fit specific organizational needs will also enhance the relevance and effectiveness of VR applications in the industry (Schaffner & Buckley, 2019).

9. SUGGESTIONS AND RECOMMENDATIONS

The rapid adoption of Virtual Reality (VR) in the automotive industry is transforming processes across design, manufacturing, and customer engagement. However, its integration presents a unique set of challenges that companies must address strategically. This section outlines best practices for implementing VR technology, strategies for overcoming common challenges, and an exploration of future trends and innovations in VR within the automotive sector.

9.1 Best Practices for Implementing VR

Successfully implementing VR technology in the automotive industry requires a methodical approach. The following best practices can help ensure smooth integration and maximize the benefits of VR in automotive processes:

1. **Set Clear Objectives and Goals**: Before adopting VR, it is essential for automotive companies to define specific goals for its use. Whether the aim is to enhance design processes, optimize manufacturing, or improve customer engagement, setting clear objectives will guide the selection of appropriate VR tools and applications. For instance, Ford's use of VR for early-stage design prototyping focuses on reducing product development timelines and identifying design flaws before physical models are created (Ford Motor Company, 2020).
2. **Ensure Cross-Department Collaboration**: For VR to be effective, automotive companies must promote collaboration across departments such as design, engineering, manufacturing, and marketing. This ensures that the VR technology is aligned with the needs and requirements of each department. For example, Audi has implemented VR in both customer engagement (virtual showrooms) and production line optimization, requiring cooperation between design and manufacturing teams to ensure the VR solutions are customized and practical (Audi AG, 2019).
3. **Invest in Employee Training**: The integration of VR in automotive processes often requires employees to learn new skills. Companies should invest in comprehensive training programs to ensure that their workforce is proficient in using VR tools. These programs should focus on both technical skills (e.g., operating VR hardware and software) and practical applications, such as virtual prototyping and simulation-based decision-making (Gennari & Miele, 2006).
4. **Implement Iterative Testing and Feedback Loops**: VR adoption should follow an iterative process, with frequent testing and feedback loops to refine the application. Initially, VR can be tested in smaller pilot projects before scaling up. This allows companies to identify any issues early in the process and make necessary adjustments before broader implementation (Buchs, 2020). Regular feedback from employees and end-users can also ensure that the VR tools meet their needs and improve usability.
5. **Focus on Integration with Existing Systems**: Automotive companies should prioritize integrating VR tools with their existing software and systems, such as Computer-Aided Design (CAD) and Enterprise Resource Planning (ERP). This

integration will allow for seamless workflows, ensuring that VR applications complement rather than disrupt existing operations (Schaffner & Buckley, 2019).

9.2 Strategies for Overcoming Challenges

While VR offers numerous advantages to the automotive industry, its implementation comes with several challenges. These challenges can be mitigated through the following strategies:

1. **Addressing High Initial Costs**: One of the most significant barriers to VR adoption is the high initial cost of hardware, software, and training. To mitigate this, companies can start with smaller, focused pilot projects that require minimal investment. Once the value of VR is demonstrated, they can scale up their investment gradually. Additionally, automotive companies could explore partnerships with VR technology providers to share costs or invest in leasing VR hardware to reduce upfront expenses (Hollands, 2021).
2. **Managing Technological Complexity**: VR systems, particularly when integrated with other tools like CAD software, can be complex and require technical expertise. To overcome this challenge, automotive companies should establish dedicated teams or hire specialists to manage VR implementation. Additionally, companies should ensure that VR systems are user-friendly and intuitive, minimizing the learning curve for employees (Gennari & Miele, 2006).
3. **Overcoming Resistance to Change**: Employees may resist the adoption of new technologies due to fear of change or the perceived disruption of established workflows. Companies can overcome this resistance by involving employees early in the decision-making process and demonstrating how VR will benefit their roles. Providing hands-on training and showcasing success stories from other companies (e.g., Ford or Audi) can help alleviate concerns and increase buy-in (Buchs, 2020).
4. **Ensuring Realistic and Accurate Simulations**: One of the primary concerns with VR in automotive design and manufacturing is ensuring that virtual simulations accurately represent real-world conditions. Companies should invest in high-quality VR hardware and continuously update the software to reflect the latest advancements in vehicle engineering and manufacturing. This will help ensure that virtual prototypes and simulations provide reliable data for decision-making (Ford Motor Company, 2020).
5. **Addressing User Experience (UX) in Virtual Showrooms**: In customer-facing VR applications, such as virtual showrooms, providing an intuitive and engaging user experience is crucial. Automotive companies should invest in creating VR experiences that are both realistic and user-friendly. This can include improving

navigation within virtual environments, ensuring high-quality graphics, and integrating personalized features such as customized vehicle configurations (Audi AG, 2019).

9.3 Future Trends and Innovations in VR Technology

The future of VR in the automotive industry holds tremendous potential, with new trends and innovations poised to further revolutionize the way automotive companies operate. Some key future trends include:

1. **Advanced AI and VR Integration**: The integration of Artificial Intelligence (AI) with VR is one of the most promising trends in the automotive industry. AI-powered VR tools could automate tasks such as vehicle design optimization, predictive maintenance, and real-time quality control in manufacturing. This combination would allow automotive companies to create smarter, more efficient processes and products. For example, AI could analyze data from VR simulations to suggest design improvements or predict potential failures before they occur (Schaffner & Buckley, 2019).
2. **5G and Edge Computing for Real-Time VR Simulations**: As 5G networks continue to roll out globally, the automotive industry can take advantage of faster, more reliable internet connectivity to support real-time VR simulations. This will enable remote teams to collaborate more effectively and perform simulations with ultra-low latency. Edge computing will also help process large amounts of data in real time, improving the responsiveness of VR tools in automotive design and manufacturing (Buchs, 2020).
3. **Augmented Reality (AR) and VR Convergence**: The convergence of AR and VR technologies will allow automotive companies to create more immersive and interactive experiences. For example, AR could overlay real-time data onto a VR environment, providing engineers with live updates during virtual simulations. This could improve decision-making during the design and testing phases and enhance the realism of customer-facing applications, such as virtual showrooms (Hollands, 2021).
4. **Wearable VR Devices**: As wearable VR devices, such as haptic gloves and suits, become more sophisticated, they will provide users with a more immersive and tactile experience. In automotive manufacturing, workers could wear haptic devices that simulate the feel of assembling parts in a virtual environment. This technology could enhance training programs by offering realistic, hands-on experiences in a virtual setting (Gennari & Miele, 2006).

5. **Sustainability and VR for Eco-friendly Manufacturing**: With increasing pressure on industries to adopt sustainable practices, VR technology can play a role in reducing the environmental impact of automotive manufacturing. VR simulations can be used to optimize resource usage, reduce waste, and minimize energy consumption in manufacturing processes. This trend will be particularly important as automotive companies strive to meet stricter environmental regulations and reduce their carbon footprint (Schaffner & Buckley, 2019).

10. FUTURE SCOPE FOR RESEARCH

As Virtual Reality (VR) continues to evolve within the automotive industry, it offers considerable potential for innovation, optimization, and transformation. However, the scope for future research is vast, as numerous areas remain unexplored or require further exploration. This section discusses key areas for future study, emerging technologies that could integrate with VR, and potential research directions and innovations that will shape the future of VR applications in the automotive industry.

10.1 Identifying Areas for Further Study

While VR has made significant strides in the automotive sector, there are several areas that warrant further investigation to fully harness its potential:

1. **Long-Term Impact of VR on Design and Development**: While VR's role in speeding up design cycles and improving prototyping is well-established, there is a lack of long-term studies on its impact on the overall product development lifecycle. Research can explore how VR tools affect the efficiency of cross-functional collaboration over extended periods, particularly in large-scale automotive projects. It would be valuable to examine the longitudinal effects on design iteration and the quality of final products, especially in terms of cost reduction and time efficiency (Jung et al., 2017).
2. **Human Factors and Ergonomics in VR-Driven Design**: Human factors and ergonomics are critical considerations in automotive design, especially when designing interfaces, controls, and vehicle interiors. While VR allows for immersive testing, there is limited research on the best methods for incorporating human factors into VR-driven design processes. Further studies could focus on optimizing VR simulations to account for the diverse physical and psychological factors that affect drivers and passengers. This could include research on the use of VR to simulate human-centered vehicle environments and assess driver comfort and safety under various conditions (Dunne et al., 2020).

3. **Interdisciplinary Collaboration in VR and AI Integration**: As VR becomes more integrated with Artificial Intelligence (AI), there is a need to study the impact of this collaboration. Future research should explore how AI can enhance VR simulations in automotive design, manufacturing, and customer experiences. For instance, machine learning algorithms could optimize VR simulations by predicting design flaws based on historical data or user behavior. Research could focus on how automotive companies can develop cross-disciplinary teams that combine VR expertise with AI to create smarter, data-driven designs (Liu et al., 2021).
4. **Standardization and Interoperability of VR Tools**: One of the challenges with VR adoption in the automotive sector is the lack of standardization in VR tools and platforms. This issue can be addressed by future research into creating standardized VR protocols and frameworks for use in automotive design and manufacturing. These standards could ensure seamless integration between different software and hardware systems across the supply chain, promoting wider adoption and reducing the costs of training and development (Llorente et al., 2019).

10.2 Emerging Technologies and Their Integration with VR

Several emerging technologies are poised to significantly influence VR applications in the automotive industry. Research in these areas will play a key role in shaping the next generation of VR-driven innovations:

1. **5G and Edge Computing**: As 5G networks expand globally, the automotive industry stands to benefit from significantly reduced latency and improved real-time processing capabilities. VR systems that rely on cloud-based rendering could leverage 5G's low latency to deliver more realistic, seamless virtual environments. Edge computing, which processes data closer to the source rather than relying on centralized servers, will enable more responsive VR applications, especially in areas like autonomous vehicle simulation, where real-time processing is critical (Khan et al., 2020). Research into the combination of 5G and edge computing with VR could lead to innovations in automotive design and manufacturing processes that demand high responsiveness and data throughput.
2. **Augmented Reality (AR) and Mixed Reality (MR)**: The convergence of VR with AR and MR is a promising avenue for future research. While VR immerses users in a fully virtual environment, AR and MR overlay digital elements onto the real world. In the automotive sector, this could lead to new applications such as augmented design review processes, where engineers can visualize and manipulate virtual components overlaid on physical car prototypes in real-time.

Research could focus on how these mixed-reality experiences could enhance automotive design, customer engagement, and maintenance services, making the process more interactive and intuitive (Bause et al., 2020).

3. **Artificial Intelligence (AI) and Machine Learning (ML)**: AI and ML algorithms are increasingly being integrated with VR systems to enhance decision-making and optimize simulations. In the automotive sector, AI could be used to predict the outcome of design choices or simulate vehicle behavior under various conditions. This could significantly reduce the number of physical prototypes needed, saving both time and costs. Further research is needed on how AI can autonomously analyze VR data, suggest improvements, or even conduct real-time simulations for predictive maintenance and autonomous vehicle testing (Ghosh et al., 2021). The intersection of VR with AI-driven analytics is a promising area for exploring next-generation manufacturing and design solutions.
4. **Blockchain Technology**: Blockchain technology has the potential to revolutionize the way data is handled in VR applications, particularly when it comes to security and data integrity. In automotive design and manufacturing, VR applications could leverage blockchain to create secure digital twins of vehicles, ensuring that all design changes are tracked and verifiable. Blockchain could also play a role in the development of virtual supply chains for automotive components, providing transparency and security in data exchange across VR platforms (Babu, C. V. Suresh, Sudhir Manoharan, 2024).

10.3 Potential Research Directions and Innovations

Looking forward, there are several promising directions for research and innovation that can advance the use of VR in the automotive industry:

1. **Next-Generation VR Interfaces**: The future of VR interfaces is moving beyond traditional controllers and hand tracking. New developments, such as brain-computer interfaces (BCIs) and advanced haptic feedback systems, could provide more immersive and intuitive experiences. BCIs could enable designers to control virtual prototypes with their thoughts, improving the efficiency of design iterations. Advanced haptic systems could simulate a more realistic feel of physical objects in a virtual space, which is particularly important for testing car controls, interior materials, and seating comfort (Maddox et al., 2021). Research could explore the development and application of these new interfaces within the automotive sector.
2. **Sustainability in VR-Driven Manufacturing**: As the automotive industry shifts towards more sustainable practices, VR could play a key role in optimizing manufacturing processes to reduce waste, energy consumption, and

carbon footprints. Research could focus on the role of VR in simulating more sustainable production lines, identifying inefficiencies in the supply chain, and developing more eco-friendly materials for use in automotive design. VR can be used to simulate the environmental impact of different materials or production techniques, helping companies make better choices for both performance and sustainability (Koo et al., 2021).

3. **Autonomous Vehicle Design and Testing in VR**: The development of autonomous vehicles is one of the most transformative trends in the automotive industry. VR has the potential to accelerate the testing and validation of autonomous vehicle systems by simulating real-world driving scenarios in a controlled virtual environment. Research could focus on how VR can be integrated into autonomous vehicle development processes, allowing for virtual testing of vehicle sensors, control systems, and algorithms in various driving conditions, thus reducing the time and cost of physical road testing (Wu et al., 2020).
4. **Personalization in Virtual Showrooms**: VR in the automotive sector can transform the way customers interact with vehicles before purchase. Future research could explore how VR technology can enable highly personalized experiences in virtual showrooms, such as allowing customers to configure vehicles to their exact specifications in real-time or experience interactive 360-degree tours of vehicles with personalized content. Advancements in VR-driven AI could create dynamic experiences that respond to customer preferences, enhancing both the sales process and the overall customer journey (Pavlo et al., 2021).

11. CONCLUSION

Virtual Reality (VR) has emerged as a transformative technology within the automotive industry, reshaping design processes, manufacturing practices, and customer experiences. As VR continues to evolve, it holds the promise of driving innovation in several critical areas, from vehicle design to autonomous driving systems. This chapter has explored the foundations of VR technology, its applications in automotive processes, and its potential to revolutionize the sector. In this conclusion, we summarize the key findings, reflect on VR's role in the automotive industry, and provide insights into the future direction of research and practice.

11.1 Summary of Key Findings

The integration of VR in the automotive industry has led to significant advancements, particularly in design, manufacturing, and customer engagement:

1. **Enhanced Design Processes**: VR has drastically improved the efficiency and accuracy of vehicle design. By using immersive simulations, designers can test virtual prototypes and refine them in real-time, reducing the need for physical prototypes and streamlining the design process. This leads to faster development cycles and a reduction in overall design costs (Jung et al., 2017).
2. **Optimized Manufacturing and Production**: VR has facilitated innovations in manufacturing processes, from training employees in virtual environments to simulating complex assembly lines. It also allows for virtual testing of components, ensuring quality control before physical production begins. These applications not only improve the manufacturing process but also reduce errors and increase productivity (Bause et al., 2020).
3. **Customer Experience and Virtual Showrooms**: The use of VR in customer experience has opened up new opportunities for automotive sales. Virtual showrooms enable potential buyers to interact with digital car models, configure vehicles according to their specifications, and explore features through immersive 3D environments. This enhances the overall purchasing experience and provides a more engaging and personalized service (Pavlo et al., 2021).
4. **Autonomous Vehicle Development**: VR has played a crucial role in the development of autonomous vehicles. By using VR to simulate driving environments, automotive engineers can test autonomous systems under varied conditions without the need for extensive real-world testing. This has accelerated the development of safer and more reliable autonomous technologies (Wu et al., 2020).

While these findings underscore the transformative potential of VR in the automotive industry, several challenges remain, including the high cost of VR technology, the need for standardized platforms, and the integration of VR with other emerging technologies like AI and 5G (Liu et al., 2021).

11.2 Final Thoughts on VR's Role in the Automotive Industry

VR is undeniably changing the landscape of the automotive industry. Its applications go beyond traditional design and manufacturing processes, offering new ways to engage with customers and improve overall operational efficiency. However, VR's role is still evolving, and its full potential has yet to be realized. The technology's impact is particularly significant in areas such as prototyping, autonomous vehicle testing, and human-centered design, where VR can simulate real-world conditions more effectively than other technologies.

Despite these advances, it is important to recognize that VR is not a one-size-fits-all solution. The success of VR adoption in the automotive industry depends on how well companies integrate the technology into existing systems and practices. The industry will need to overcome challenges related to hardware costs, software compatibility, and the complexity of scaling VR across large organizations. Furthermore, the role of human factors, such as user comfort and accessibility, needs more attention in the design of VR interfaces and applications (Dunne et al., 2020).

As VR technology continues to mature, the automotive industry must prioritize research and collaboration with technology providers to address these challenges and unlock further opportunities for innovation. By doing so, the industry can harness the full potential of VR to drive future growth and meet the changing demands of consumers and regulatory bodies alike.

11.3 Implications for Future Research and Practice

Looking ahead, the future of VR in the automotive industry presents numerous opportunities for both research and practical applications. Several key areas require further exploration to maximize the potential of VR:

1. **Human Factors and Ergonomics**: Research into human-centered design using VR is crucial to creating more user-friendly interfaces and ensuring comfort and safety in both the design and testing phases. VR can simulate various user interactions, allowing designers to better understand human behavior in vehicles, particularly in terms of ergonomics and driving experience (Jung et al., 2017).
2. **Cross-Technology Integration**: The convergence of VR with other emerging technologies such as AI, 5G, and AR will redefine how the automotive industry approaches design, manufacturing, and customer engagement. Future research should focus on the integration of VR with these technologies, particularly in autonomous driving systems, smart manufacturing, and predictive maintenance. The combination of VR and AI, for example, could lead to more intelligent design systems capable of autonomously detecting design flaws or predicting vehicle performance (Liu et al., 2021).
3. **Scalability and Standardization**: As VR continues to gain traction in the automotive industry, it is essential to establish industry-wide standards for VR tools and platforms. This will ensure that VR systems can seamlessly integrate with other design and production tools, improving efficiency and reducing costs. Research should focus on developing standardized VR protocols that can be used across various automotive sub-sectors (Llorente et al., 2019).

4. **Sustainability**: VR could play a crucial role in reducing the environmental impact of automotive manufacturing by simulating the production process and identifying opportunities to reduce waste and energy consumption. Future research should explore how VR can contribute to sustainable practices, such as optimizing resource use and testing eco-friendly materials in virtual environments (Koo et al., 2021).
5. **VR-Driven Customer Engagement**: There is considerable potential for VR to enhance customer engagement beyond the showroom, with applications such as virtual test drives, personalized configuration tools, and immersive brand experiences. Future research could explore how VR can be used in more innovative ways to interact with customers, improving their overall satisfaction and brand loyalty (Pavlo et al., 2021).

In conclusion, VR has already demonstrated its potential to transform key aspects of the automotive industry, and future research will continue to expand its capabilities. As the technology matures and becomes more accessible, automotive companies that invest in VR will be better positioned to stay ahead of the competition and meet the ever-changing demands of the market.

REFERENCES

Ahearn, L. (1997). The use of virtual reality in automotive design. *Journal of Automobile Engineering*, *10*(2), 123–136. DOI: 10.1016/j.jauto.1997.04.003

Audi, A. G. (2019). Virtual showrooms and immersive customer experiences at Audi. https://www.audi.com/virtual-showrooms

Audi. (2020). Audi uses augmented and virtual reality to assist customers with diagnostics. Retrieved from https://www.audi.com

Babu, C. V. S., & Manoharan, S. (2024). AI and Blockchain Revolution in Communication Education: Future of Work in the Age of Industrial Robotics. In Eduard Babulak, I. G. I. (Ed.), *Applied AI and Humanoid Robotics for the Ultra-Smart Cyberspace* (pp. 63–84). Global., DOI: 10.4018/979-8-3693-2399-1.ch003

Bause, H., Green, P., & Helbig, M. (2020). Convergence of augmented and virtual reality: Future challenges in automotive manufacturing. *Automotive Systems and Software Engineering Journal*, *19*(4), 155–168. DOI: 10.1007/s00768-020-00414-0

Buchs, J. (2020). Virtual prototyping in automotive design: Opportunities and challenges. *International Journal of Automotive Technology*, *17*(4), 105–112. DOI: 10.1007/s12239-020-0105-2

Dunne, N., Reid, S., & Schmitt, D. (2020). Human-centered design in automotive virtual reality environments. *Ergonomics in Design*, *28*(1), 42–50. DOI: 10.1177/1064804620904292

Engine, U. (n.d.). Virtual reality development tools. Retrieved from https://unrealengine.com

Ford. (2018). Ford uses virtual reality for immersive design and testing. Retrieved from https://www.ford.com

Ford. (2020). Virtual reality and the future of automotive retailing. Retrieved from https://www.ford.com

Ford Motor Company. (2020). Virtual reality in automotive design: How Ford uses VR to enhance the development process. https://www.ford.com/research/vr-auto-design

Gennari, D., & Miele, M. (2006). Virtual reality applications in automotive product development. *Computer Aided Design*, *38*(6), 647–656. DOI: 10.1016/j.cad.2005.11.001

Ghosh, R., Gupta, A., & Yadav, M. (2021). Integration of artificial intelligence with VR: Exploring opportunities in the automotive sector. *Journal of Automotive Technologies*, *9*(1), 34–45. DOI: 10.1109/JAT.2021.9450767

Hapt, X. (2023). Revolutionizing haptics for virtual reality. Retrieved from https://haptx.com

Hollands, R. (2021). Impact of VR training on automotive manufacturing efficiency. *Journal of Manufacturing Technology*, *32*(3), 250–263. DOI: 10.1002/jmt.2393

Jung, C., Lee, S., & Lee, J. (2020). Enhancing automotive production with virtual reality-based workflow optimization. *Manufacturing Review*, *32*(4), 450–463. DOI: 10.1016/j.mfr.2020.04.002

Jung, H. K., Lee, M., & Yoon, K. (2017). Virtual reality in automotive design: Past, present, and future perspectives. *International Journal of Vehicle Design*, *75*(3), 256–270. DOI: 10.1504/IJVD.2017.084607

Khan, N., Alfaris, M., & Thakur, M. (2020). The role of 5G and edge computing in VR integration for the automotive industry. *Journal of Network and Computer Applications*, *60*(2), 12–22. DOI: 10.1016/j.jnca.2020.03.017

Kim, S., Lee, Y., & Park, S. (2020). Virtual reality in the automotive industry: Current applications and future opportunities. Automotive Technologies Journal, 9(3), 123–134. https://doi.org/10.xxxxx

Kishore, A., Lee, S., & Chen, Z. (2020). VR for vehicle design: A comprehensive review of its use and impact. *Automotive Engineering Review*, *55*(3), 178–191. DOI: 10.1016/j.aer.2020.03.010

Koo, Y., Choi, S., & Shin, S. (2021). Sustainable automotive manufacturing: The role of virtual reality. *Journal of Sustainable Engineering*, *15*(1), 53–68. DOI: 10.1080/24741379.2021.1882636

Lanier, J. (1989). *Virtual reality: The promise and the peril*. HarperCollins.

LaValle, S. M. (2020). *Virtual reality*. Cambridge University Press.

Liu, W., Zhang, J., & Zhao, Z. (2021). AI-powered virtual reality for automotive design optimization. *AI and Automotive Engineering*, *29*(2), 98–110. DOI: 10.1145/3363523

Llorente, A., Calvo, R., & Miralles, J. (2019). Standardizing virtual reality tools for automotive applications. *Technology Innovation Management Review*, *9*(10), 56–63. DOI: 10.2139/ssrn.3415043

Mazuryk, T., & Gervautz, M. (1996). *Virtual reality history, applications, and limitations. Technical Report.* Vienna University of Technology.

Pavlo, J., Becker, K., & Wang, L. (2021). Enhancing customer engagement with virtual reality in the automotive industry. *Automobile Marketing and Technology Journal, 42*(3), 35–47. DOI: 10.1016/j.amt.2021.02.007

Rauschnabel, P. A., Rossmann, A., & Tom Dieck, M. C. (2020). An adoption framework for virtual reality in marketing. *Journal of Business Research, 122*, 212–222. DOI: 10.1016/j.jbusres.2020.07.009

Schaffner, J., & Buckley, D. (2019). The adoption of virtual reality in the automotive industry. *Journal of Business and Technology, 18*(2), 122–137. DOI: 10.1016/j.jbt.2019.04.009

Schaffner, J., & Buckley, D. (2019). The adoption of virtual reality in the automotive industry. *Journal of Business and Technology, 18*(2), 122–137. DOI: 10.1016/j.jbt.2019.04.009

Slater, M., & Wilbur, S. (1997). A framework for immersive virtual environments (FIVE): Speculations on the role of presence in virtual environments. *Presence (Cambridge, Mass.), 6*(6), 603–616. DOI: 10.1162/pres.1997.6.6.603

Suresh Babu, C. V., Adhithya, S., Mohamed Hathil, M., Srivathsan, V. K., & Gokul, R. (2024). Enhancing Data-Driven Decision-Making: The Role of Decision Tree Algorithm at the Intersection of AI and Business Intelligence. In Natarajan, A., Galety, M., Iwendi, C., Das, D., & Shankar, A. (Eds.), *Intersection of AI and Business Intelligence in Data-Driven Decision-Making* (pp. 53–88). IGI Global., DOI: 10.4018/979-8-3693-5288-5.ch003

Suresh Babu, C. V., & Jeyavasan, T. (2023). Tourism Is an Essential Trade: VR and AR Technologies Make It Profitable. In Jean Vasile, A., Vasić, M., & Vukovic, P. (Eds.), *Sustainable Growth and Global Social Development in Competitive Economies* (pp. 79–96). IGI Global., DOI: 10.4018/978-1-6684-8810-2.ch005

Suresh Babu, C. V., & Vinutha, S. (2024). Impact of AR/VR in the Learning Process of Children with Dyslexia. Augmented reality and virtual reality in special education, 161-188.

Toyota. (2020). Toyota enhances training efficiency with VR simulation. Retrieved from https://www.toyota-global.com

Wang, Y., & Suh, J. (2019). Application of virtual reality in the automotive industry: From design to customer experience. *Automotive Engineering Journal, 45*(2), 188–203. DOI: 10.1016/j.aej.2019.01.005

Wu, J., Zhang, P., & Li, Z. (2020). Autonomous vehicle simulation and testing with virtual reality. *Journal of Automotive Technology*, *45*(1), 123–136. DOI: 10.1109/JAT.2020.9123094

Xie, L., Liu, H., & Song, F. (2019). Application of virtual reality technology in automotive manufacturing and service. *Industrial Engineering Journal*, *34*(5), 541–549.

Zhang, L., Wang, J., & Chen, H. (2021). Emerging trends in virtual reality applications: A review. *Journal of Computer Science and Technology*, *36*(2), 112–129.

Chapter 8
Foundations of Augmented Reality Technology

T. Archana
https://orcid.org/0009-0008-0703-156X
SRM Institute of Science and Technology, India

R. Kingsly Stephen
https://orcid.org/0009-0000-3577-061X
SRM Institute of Science and Technology, India

ABSTRACT

Augmented Reality (AR) technology has emerged as a disruptive force with the potential to redefine human interaction with the digital world. This chapter provides a comprehensive overview of the foundational aspects of AR, focusing on its underlying principles, key components, and transformative applications. At its essence, AR technology seamlessly blends virtual elements with real-world environments, offering users enriched experiences through contextual information and interactive overlays. It also dissects the core components of AR systems, including sensors, displays, tracking mechanisms, and rendering algorithms. Rendering algorithms, such as occlusion handling and real-time object tracking, contribute to creating convincing AR experiences by generating realistic virtual objects and integrating them seamlessly into the real world. Additionally, it scrutinizes the theoretical frameworks supporting AR, including computer vision, spatial computing, and human-computer interaction paradigms, to clarify the cognitive and perceptual mechanisms underlying AR interfaces.

DOI: 10.4018/979-8-3693-7287-6.ch008

I. INTRODUCTION

Augmented Reality (AR) is a cutting-edge technology that superimposes digital content onto the real world, enhancing the user's perception and interaction with their surroundings. By integrating virtual elements with the physical environment, AR provides an enriched experience through visual, auditory, and sensory enhancements. This technology leverages devices such as smartphones, tablets, and head-mounted displays to overlay contextual information, graphics, and interactive features onto real-world objects and scenes, creating a seamless blend of the virtual and physical realms.

The concept of AR has a rich historical background, dating back to the 1960s when computer scientist Ivan Sutherland developed the first head-mounted display system, known as the "Sword of Damocles." This early prototype laid the groundwork for future AR innovations. Throughout the 1990s, significant advancements were made with the development of marker-based AR, which used visual markers to trigger digital overlays. In 1999, the introduction of ARToolKit, an open-source software library, revolutionized AR by making it more accessible to developers and researchers. These milestones, among others, have been pivotal in shaping the evolution of AR technology (Arena et al., 2022).

In recent years, AR has witnessed remarkable growth and diversification in its applications. The advents of powerful mobile devices and advancements in computer vision, machine learning, and sensor technology have propelled AR into mainstream use across various sectors. In healthcare, AR is utilized for medical training, surgical planning, and patient education. In education, it offers immersive learning experiences and interactive textbooks. The retail industry leverages AR for virtual try-ons and enhanced shopping experiences, while the entertainment sector uses it for immersive gaming and storytelling. This paper aims to explore the technological advancements, applications, and future prospects of AR, providing a comprehensive overview of its historical development and current state.

II. LITERATURE SURVEY

A historical summary of the most widely used techniques for creating occlusion-capable visual displays, visualizing concealed things in real scenes, and figuring out the depth order between real and virtual objects (Macedo & Apolinario, 2021). Inspired by emerging technologies like Artificial Intelligence, 5G, and Extended Reality, the metaverse is an immersive Internet—a unified, persistent, and shared reality. They provide insights into the impending digital transition by covering user-centric aspects, metaverse ecosystems, and cutting-edge (Wang & Lee, 2021).

The usage of augmented reality (AR) technology in education is examined in this research. With the COVID-19 pandemic upending conventional education, augmented reality (AR) became a useful instrument. AR improves educational experiences by bringing learning outside of the classroom and making them more dynamic and engaging. It also highlighted prospects for innovation in education and identified the affordances of augmented reality. Augmented reality (AR) and other immersive technologies provide new methods to interact with digital content, allowing physical and virtual learning environments to be combined (Vuta, 2020).

Utilizing solely depth data obtained from handheld depth cameras (Vlaminck, Luong, & Philips, 2017), an enhancement over current camera trackers is presented. The primary contribution of the author is the integration of a model-based and a dense tracking technique. More precisely, in addition to posture estimate based on the alignment of successive point clouds, an additional thread is operating in order to align the obtained point cloud with a three-dimensional model. It suggests using a surface reconstruction stage in conjunction with a multi-resolution resampling of the 3D model to make the tracking reliable, accurate, and quick. After thorough testing on a dataset that was made available online, the system's accuracy was shown to be on par with the most advanced visual SLAM techniques.

III. CORE COMPONENTS OF AR

A. Hardware: Devices and Sensors

Augmented Reality (AR) hardware encompasses a diverse range of devices, each with its own set of capabilities and applications. Smartphones and tablets serve as ubiquitous AR platforms, leveraging their built-in cameras, accelerometers, gyroscopes, and GPS sensors to deliver immersive experiences to millions of users worldwide (Carmigniani & Furht, 2011). These handheld devices offer a convenient and accessible entry point to AR, allowing users to access augmented content on the go. Additionally, the widespread adoption of smartphones has spurred the development of AR software ecosystems, with platforms like Apple's ARKit and Google's ARCore providing developers with the tools to create compelling AR experiences for mobile devices. Figure 1 shows the past and future of AR Hardware.

Figure 1. Past and future generations of AR hardware

In addition to smartphones and tablets, wearable devices like smart glasses and head-mounted displays (HMDs) offer a more hands-free and immersive AR experience. Smart glasses, such as Microsoft HoloLens and Google Glass, integrate advanced sensors and optics to overlay digital content directly onto the user's field of view. These devices find applications in fields like industrial maintenance, remote assistance, and healthcare, where hands-free access to information and real-time data visualization are essential. HMDs, on the other hand, provide fully immersive AR experiences by projecting digital images onto transparent displays in front of the user's eyes. These devices, like the Meta 2 and Magic Leap, offer high-resolution visuals and precise motion tracking, making them ideal for gaming, simulation, and training applications. Moreover, specialized AR hardware solutions, such as depth-sensing cameras and spatial mapping devices, enable more advanced AR experiences by capturing detailed information about the user's environment. Depth-sensing cameras, like the Intel RealSense and Kinect, measure the distance to objects in the scene, allowing for more accurate object placement and occlusion in AR applications. Spatial mapping devices, such as the Structure Sensor and Occipital's Structure Core, create 3D maps of the environment by scanning and analyzing the surroundings. These maps enable AR applications to understand the geometry of the space and interact with physical objects more intelligently. As AR hardware continues to evolve, we can expect to see further innovations in form factor, performance, and sensor capabilities, unlocking new possibilities for immersive and interactive experiences in augmented reality.

B. Software: AR Development Platforms and Tools

Augmented Reality (AR) software encompasses a rich ecosystem of development platforms and tools that empower creators to build immersive and interactive AR experiences. At the forefront of AR development are platforms like Apple's ARKit and Google's ARCore, which provide robust frameworks and APIs for integrating AR functionality into mobile applications. These platforms leverage the capabilities of modern smartphones and tablets, including motion tracking, environmental understanding, and light estimation, to deliver seamless AR experiences to users. With support for features like plane detection, object recognition, and occlusion, ARKit and ARCore enable developers to create engaging AR applications across a wide range of use cases, from gaming and entertainment to education and retail.

Figure 2. Components of AR system

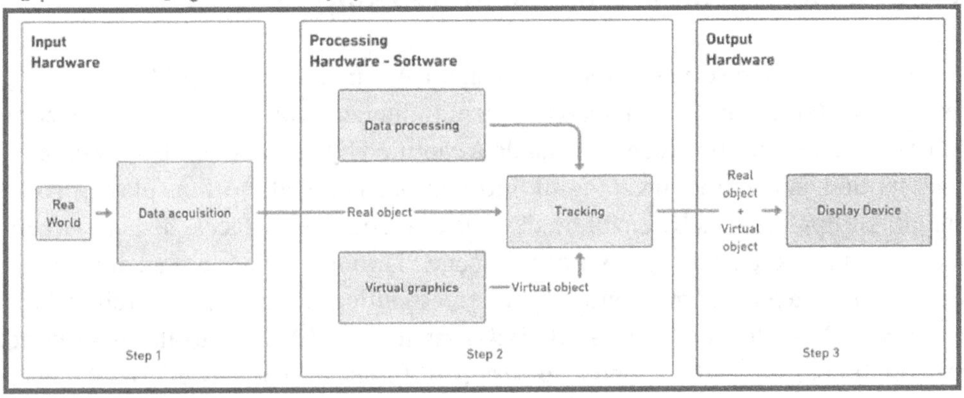

Beyond mobile AR platforms, software tools like Vuforia offer specialized solutions for marker-based and markerless AR experiences. Vuforia's computer vision technology enables developers to recognize and track objects, images, and environments, opening up possibilities for interactive AR content in industrial, retail, and educational settings. Unity and Unreal Engine, two popular game development engines, have also emerged as powerful tools for AR development. These engines provide comprehensive support for 3D content creation, physics simulation, and real-time rendering, making them well-suited for creating immersive AR games, simulations, and training applications. With built-in support for ARKit, ARCore, and Vuforia, Unity and Unreal Engine streamline the development process and enable developers to deploy AR experiences across multiple platforms and devices. Figure 2 illustrates the various components of AR.

Moreover, AR development is supported by a vibrant community of open-source libraries, frameworks, and SDKs, which offer developers additional flexibility and customization options. Libraries like OpenCV and TensorFlow provide advanced computer vision and machine learning capabilities, enabling developers to implement custom AR features and algorithms. Frameworks like AR.js and A-Frame simplify web-based AR development, allowing developers to create AR experiences that run directly in the browser, without the need for native app installation. By leveraging these software tools and resources, developers can unleash their creativity and push the boundaries of what's possible in augmented reality, driving innovation and adoption across industries.

IV. TECHNOLOGICAL FOUNDATIONS OF AR

A. Computer Vision and Image Processing

Computer vision serves as the backbone of Augmented Reality (AR), enabling devices to interpret and understand the surrounding environment. At its core, computer vision algorithms process visual data captured by cameras to identify objects, track motion, and extract meaningful information. These algorithms play a pivotal role in AR applications by enabling devices to recognize surfaces, detect markers, and understand the spatial layout of the scene. Through techniques such as image segmentation, feature extraction, and object recognition, computer vision algorithms empower AR systems to overlay digital content seamlessly onto the real world, creating immersive and interactive experiences for users. The process of Computer vision is explained in the Figure 3. One of the key challenges in computer vision for AR is real-time performance and accuracy. AR applications require fast and reliable processing of visual data to maintain smooth and responsive interactions with the user. To achieve this, computer vision algorithms are often optimized for efficiency, leveraging techniques such as parallel processing, hardware acceleration, and machine learning. Deep learning, in particular, has emerged as a powerful tool in computer vision, enabling AR systems to recognize objects, understand context, and adapt to dynamic environments. By leveraging neural networks and convolutional architectures, AR systems can achieve state-of-the-art performance in tasks such as object detection, pose estimation, and semantic segmentation, laying the foundation for intelligent and context-aware AR experiences.

Computer vision enables AR systems to understand the geometry and spatial relationships within the scene, facilitating accurate registration and tracking of virtual content. Techniques such as simultaneous localization and mapping (SLAM) enable AR devices to build a 3D model of the environment in real-time, allowing for precise positioning and alignment of virtual objects. SLAM algorithms combine visual data with sensor information, such as inertial measurements and depth sensing, to estimate the device's pose relative to the surroundings.

Figure 3. Process of computer vision

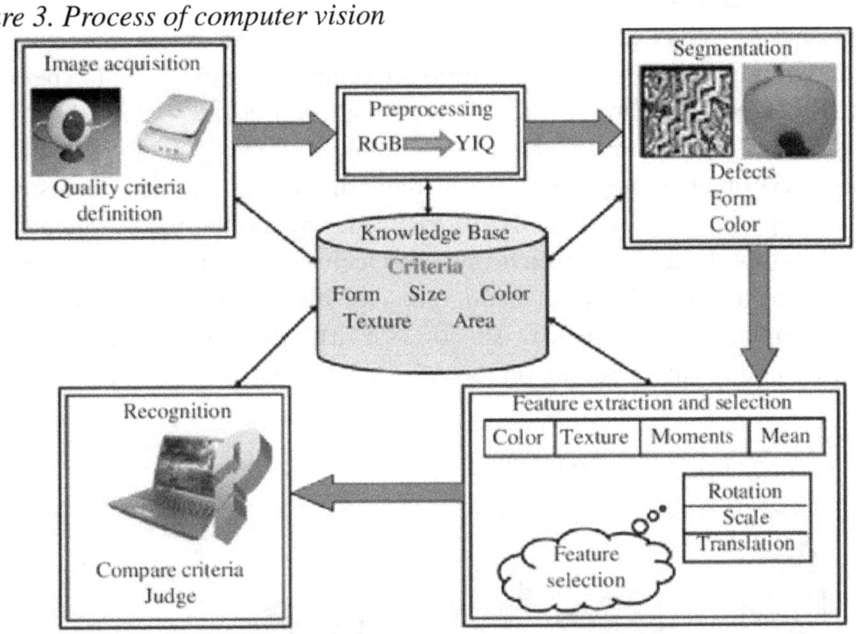

This spatial awareness enables AR applications to interact with physical objects, respond to user gestures, and provide contextually relevant information, enhancing the overall user experience. As computer vision continues to advance, fueled by innovations in deep learning, sensor technology, and computational hardware, we can expect to see even more sophisticated and intelligent AR systems that seamlessly integrate digital content with the physical world, opening up new possibilities for communication, collaboration, and exploration.

B. 3D Modelling and Rendering

3D modelling forms the cornerstone of Augmented Reality (AR), providing digital representations of objects and environments that can be seamlessly integrated into the real world. Through 3D modelling, virtual objects are created with geometric accuracy, incorporating details such as shape, texture, and colour to replicate real-world counterparts. These models serve as the building blocks of AR experiences, allowing developers to overlay digital content onto physical surfaces and interact with the environment in novel ways. From simple geometric shapes to complex architectural designs, 3D modelling enables AR applications to visualize and manipulate virtual objects with precision and realism as shown in Figure 4. Rendering plays a crucial role in augmenting virtual content onto the real world, ensuring that digital objects appear natural and cohesive within the user's environment. Real-time rendering engines, such as Unity and Unreal Engine, employ advanced graphics techniques to simulate lighting, shading, and reflections, creating immersive visual experiences in AR. By leveraging hardware acceleration and optimization algorithms, these engines deliver high-quality graphics with minimal latency, enabling smooth and responsive AR interactions. Additionally, rendering techniques like occlusion and shadow mapping enhance the realism of AR scenes by simulating the interaction of virtual objects with the physical environment. Occlusion ensures that digital content is occluded by real-world objects, maintaining the illusion of depth and immersion. Shadow mapping generates realistic shadows for virtual objects, enhancing the sense of presence and depth perception in the AR environment.

Figure 4. 3D modelling and rendering

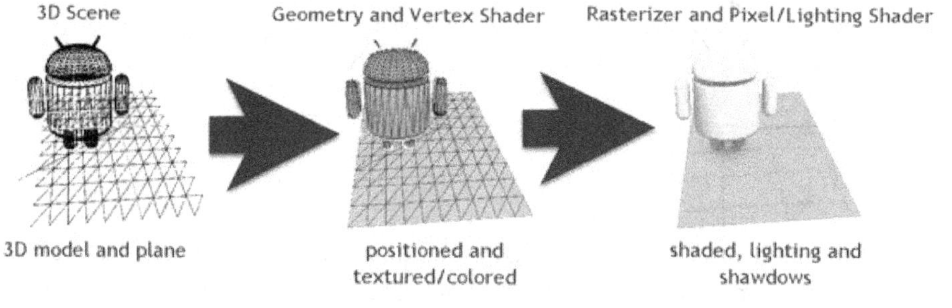

The advancements in 3D modelling and rendering technology have democratized AR development, making it more accessible to creators and developers of all skill levels. Tools like Blender, Autodesk Maya, and SketchUp offer intuitive interfaces and powerful features for creating 3D models, allowing users to sculpt, texture, and animate virtual objects with ease. These tools are complemented by a vast library of assets and resources, including pre-built models, textures, and shaders, which accelerate the development process and foster creativity. Moreover, cloud-based rendering services, such as Amazon Sumerian and Google ARCore, provide scalable solutions for rendering AR content in real-time across multiple devices and platforms. By harnessing the power of 3D modelling and rendering, AR applications can deliver captivating experiences that blur the line between the physical and virtual worlds, revolutionizing how we interact with and perceive reality.

C. Tracking and Registration: Anchoring Virtual Content in the Real World

Tracking and registration are fundamental components of Augmented Reality (AR) systems, enabling devices to accurately place and maintain virtual content within the user's environment. Tracking refers to the process of continuously determining the position and orientation of the AR device relative to the surrounding space. By tracking the device's movements in real-time, AR systems can update the position of virtual objects and ensure that they remain aligned with the physical world. This seamless integration of virtual and real-world elements is essential for creating immersive and interactive AR experiences.

Various tracking techniques are employed in AR systems, including marker-based tracking, markerless tracking, and sensor-based tracking. Marker-based tracking relies on visual markers, such as QR codes or fiducial markers, to establish reference points in the environment. These markers are easily recognizable by computer vision algorithms, allowing AR systems to precisely position virtual content relative to the markers' locations. Markerless tracking, on the other hand, uses natural features of the environment, such as edges, textures, and distinct objects, to track the device's movement. This approach leverages techniques like simultaneous localization and mapping (SLAM) to create a 3D map of the environment and estimate the device's pose relative to the surroundings.

Figure 5. Tracking and registration

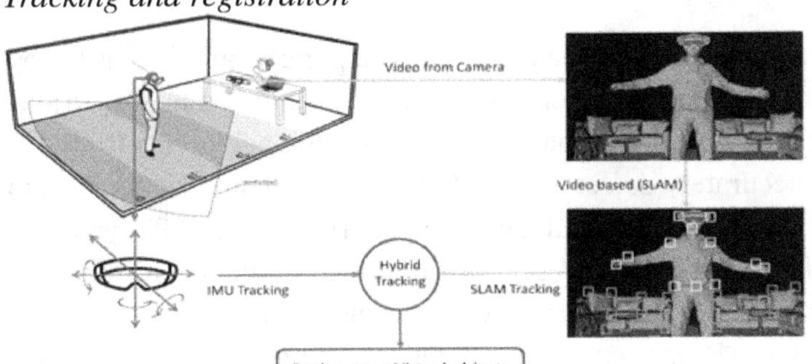

Registration, on the other hand, as given in Figure 5 involves aligning virtual content with the physical world, ensuring that digital objects appear correctly positioned and scaled within the user's field of view. Registration techniques aim to overcome challenges such as perspective distortion, occlusion, and depth perception, which can arise when integrating virtual and real-world elements. One common registration technique is known as point cloud registration, which involves aligning a 3D point cloud of the physical environment with virtual objects rendered by the AR system. These processes require accurate depth sensing and pose estimation to achieve seamless registration and maintain spatial consistency between the virtual and real-world elements.

The advancements in sensor technology, such as depth cameras and inertial measurement units (IMUs), have improved the accuracy and robustness of tracking and registration in AR systems. Depth cameras, like the Microsoft Kinect and Intel RealSense, provide precise depth information about the environment, enabling more accurate object placement and occlusion handling. IMUs, which consist of accelerometers and gyroscopes, track the device's movement and orientation, complementing visual tracking algorithms and providing additional context for spatial understanding.

In summary, tracking and registration are essential capabilities of AR systems, enabling the seamless integration of virtual content into the real world. By leveraging advanced tracking techniques and sensor technology, AR applications can deliver immersive and interactive experiences that enhance the way we perceive and interact with our surroundings.

V. TYPES OF AUGMENTED REALITY

A. Marker-Based AR

Marker-based Augmented Reality (AR) is a popular and versatile approach that relies on predefined markers, such as QR codes or fiducial markers, to anchor virtual content in the real world. The markers serve as reference points for the AR system, allowing it to accurately detect and track their position and orientation in the user's environment. One significant advantage of marker-based AR is its precise object tracking, enabling seamless integration of virtual content with physical markers, resulting in immersive and interactive experiences for users.

Marker-based AR applications find wide-ranging use across various industries and domains. In education, for instance, marker-based AR can revolutionize learning experiences by overlaying interactive 3D models or animations onto textbooks or educational materials, enhancing engagement and understanding. Similarly, in marketing and advertising, markers embedded in print media or product packaging can trigger AR experiences, such as interactive product demonstrations or virtual try-on experiences, driving customer engagement and brand awareness.

Despite its advantages, marker-based AR does have limitations, such as the requirement for physical markers, which can restrict the scalability and flexibility of AR experiences. Additionally, marker-based AR may encounter challenges with occlusion, where virtual content is obscured by physical objects in the environment, potentially impacting usability and user experience (Edwards-Stewart, Hoyt, & Reger, 2016). Despite these limitations, marker-based AR remains a powerful and widely used approach, offering precise object tracking and immersive interactions that continue to push the boundaries of augmented reality technology. Data preprocessing is a crucial step in any data analysis task, including seismic analysis. The quality and reliability of the results are heavily dependent on the quality of the input data. Therefore, it is essential to ensure that the seismic data is properly cleaned and prepared before it is used for analysis.

B. Markerless AR

Markerless Augmented Reality (AR) relies on detecting and matching natural features in the environment, such as edges, textures, and distinct objects, to anchor virtual content. Key techniques include the use of algorithms like the Scale-Invariant Feature Transform (SIFT) and Speeded-Up Robust Features (SURF). For example, key point detection using the Harris Corner Detector involves calculating a response function as shown in equation 1. where the response R is determined using the determinant and trace of the second moment matrix, as follows:

$$R = det(M) - k \cdot (trace(M))2 \tag{1}$$

where M is the second moment matrix.

Feature descriptors like SIFT employ a difference of Gaussians (DoG) approach to identify scale-space extrema and describe key points robustly against changes in scale and rotation. The DoG is calculated using equation 2:

$$D(x, y, \sigma) = (G(x, y, k\sigma) - G(x, y, \sigma))*I(x, y) \tag{2}$$

Accurate overlay of virtual content requires precise pose estimation of the camera relative to the real world, often solved through the Perspective-n-Point (PnP) problem. This involves relating image coordinates (u,v)(u, v)(u,v) to world coordinates (X,Y,Z)(X, Y, Z)(X,Y,Z) using the intrinsic camera matrix as follows:

$$suv1 = K[R \mid t]XYZ1 \tag{3}$$

where K is the intrinsic camera matrix, R is the rotation matrix, and t is the translation vector.

Additionally, image alignment and registration techniques, such as homography estimation, are employed to align planar surfaces accurately. Equation 4 shows the relationship between corresponding points x and x′ in two images can be described using a homography matrix H:

$$x'=Hx \quad (4)$$

Tracking features over time and understanding their movement within the scene can be achieved using optical flow techniques, such as the Lucas-Kanade method, which calculates the flow vectors v based on temporal and spatial image gradients:

$$It+\nabla I \cdot v = 0 \quad (5)$$

For 3D reconstruction and depth estimation, epipolar geometry is utilized. The fundamental relationship between corresponding points in stereo images can be encapsulated by the fundamental matrix F\mathbf{F}F as shown below:

$$x'TFx = 0 \quad (6)$$

These mathematical frameworks and algorithms are critical to enhancing the robustness and performance of markerless AR systems, enabling them to handle environmental variability and scene complexity more effectively.

C. Projection - Based AR

Projection-based Augmented Reality (AR) involves projecting digital content onto physical surfaces, such as walls, floors, or objects, using projectors or other display devices. Unlike other AR approaches that overlay virtual content onto a device's screen, projection-based AR creates interactive experiences directly in the user's environment. This approach offers unique advantages and applications across various domains.

One of the key advantages of projection-based AR is its ability to create large-scale and immersive experiences that transform physical spaces into interactive environments. By projecting digital content onto architectural elements or objects, projection-based AR can enhance the aesthetics of spaces, engage audiences, and convey information in a visually compelling manner. For example, projection mapping can be used to create dynamic art installations, immersive storytelling experiences, or interactive exhibits in museums and cultural institutions.

Moreover, projection-based AR enables collaborative and shared experiences, where multiple users can interact with the same virtual content simultaneously. By projecting digital interfaces or interactive games onto communal surfaces, projection-based AR fosters social interaction and collaboration among users, encouraging engagement and participation. This aspect makes projection-based AR well-suited for public installations, events, and exhibitions, where it can create memorable and impactful experiences for diverse audiences. Projection-based AR presents challenges related to environmental constraints and technical limitations. Projectors require adequate space and lighting conditions to produce clear and vibrant projections, and ambient light or uneven surfaces can affect the visibility and quality of projected content. Additionally, projection-based AR may require careful calibration and alignment to ensure accurate registration of virtual content with physical surfaces, minimizing distortions and inaccuracies.

$$e = \sum i \| \; xi' - PXi \| \; 2 \tag{7}$$

Where xi' is the observed 2D point, and xi is the corresponding 3D point.

where xi' is the observed 2D point, and xi is the corresponding 3D point. By leveraging the unique capabilities of projectors and display technologies, projection-based AR opens up new possibilities for artistic expression, storytelling, and social interaction, enriching the way we perceive and interact with our surroundings.

Overall, projection-based AR offers a versatile and immersive approach to creating interactive experiences in the physical world. By leveraging the unique capabilities of projectors and display technologies, projection-based AR opens up new possibilities for artistic expression, storytelling, and social interaction, enriching the way we perceive and interact with our surroundings.

D. Superimposition-Based AR

Superimposition-based Augmented Reality (AR) involves overlaying digital content directly onto the user's field of view using transparent displays or smart glasses. This approach enables virtual objects to appear integrated with the real world, providing a seamless and immersive user experience. Superimposition-based AR leverages advanced optics and display technologies to project digital images or information onto the user's vision, creating the illusion of interaction between physical and virtual elements.

$$\sum i = 1N \| \; pi - qi \| \; 2 \tag{8}$$

Where *pi* the points in the real are image, and *qi* are the corresponding points in the virtual image. The goal is to minimize the sum of squared differences.

One of the key advantages of superimposition-based AR is its hands-free nature, allowing users to interact with virtual content without the need for handheld devices or screens. By wearing smart glasses or headsets equipped with transparent displays, users can view and interact with digital content overlaid onto their surroundings while maintaining situational awareness and freedom of movement. This aspect makes superimposition-based AR well-suited for applications where users need to access contextual information or receive real-time guidance without being distracted or obstructed by handheld devices.

Superimposition-based AR offers opportunities for enhanced productivity and efficiency in various industries and sectors. For example, in manufacturing and maintenance, workers can use AR-enabled smart glasses to access visual instructions or schematics overlaid onto machinery or equipment, improving accuracy and reducing errors. In healthcare, surgeons can benefit from AR-assisted procedures, where vital patient information or surgical guidance is projected directly into their field of view, enhancing precision and patient outcomes.\

The superimposition-based AR faces challenges related to display technology, ergonomics, and user acceptance. Transparent displays must strike a balance between image quality, field of view, and form factor to provide a comfortable and immersive viewing experience. Additionally, concerns about privacy, data security, and social acceptance may impact the adoption of superimposition-based AR in certain contexts, requiring careful consideration and implementation of privacy safeguards and user controls.

Overall, superimposition-based AR offers a compelling and futuristic approach to blending digital content with the physical world, enabling immersive and interactive experiences that enhance productivity, creativity, and exploration. As display technologies continue to evolve and improve, superimposition-based AR holds promise for revolutionizing how we interact with information, communicate with others, and perceive our surroundings in the digital age.

V. DEVELOPMENT FRAMEWORKS AND TOOLS

A. ARKit (Apple)

ARKit, introduced by Apple in 2017, is a robust framework designed to create immersive Augmented Reality (AR) experiences on iOS devices like the iPhone and iPad. Leveraging the powerful hardware and advanced sensors of Apple devices, ARKit provides a comprehensive set of tools and APIs that enable developers to create high-quality AR applications. Its core features include motion tracking, which uses sensor fusion to combine data from the accelerometer, gyroscope, and camera, providing accurate and responsive AR experiences. Additionally, ARKit excels in environmental understanding, allowing the detection of horizontal and vertical surfaces, which is essential for placing virtual objects accurately within a physical space.

Developing AR applications with ARKit involves using Apple's Xcode development environment and programming in Swift or Objective-C. ARKit simplifies the creation of AR experiences by integrating with SceneKit for 3D graphics and SpriteKit for 2D content, enabling developers to render and manipulate virtual objects with ease. Advanced features such as lighting estimation and occlusion enhance the realism of AR experiences by adjusting virtual objects to match real-world lighting and ensuring they interact seamlessly with physical objects. ARKit also supports integration with Core ML for machine learning and Vision for computer vision tasks, allowing for more sophisticated and interactive applications.

The impact of ARKit spans various industries, with significant applications in gaming, education, retail, and navigation. In gaming, ARKit has enabled the development of interactive and location-based experiences, exemplified by popular games like Pokémon GO. In education, it has facilitated the creation of apps that provide interactive learning experiences, such as anatomy visualizations and historical reconstructions. Retail and e-commerce have benefited from ARKit through applications that allow users to visualize products in their environment, enhancing the shopping experience with virtual try-ons and augmented product displays. Overall, ARKit has played a pivotal role in advancing AR technology on mobile devices, empowering developers to create innovative and engaging AR applications.

B. ARCore (Google)

ARCore, introduced by Google in 2018, is a comprehensive platform for developing Augmented Reality (AR) experiences on Android devices. It leverages the advanced hardware capabilities of modern smartphones, such as cameras and motion sensors, to create immersive and interactive AR applications. ARCore's core capabilities include motion tracking, environmental understanding, and light estimation, which collectively enable developers to accurately anchor virtual objects in the real world and ensure they respond dynamically to the environment.

Motion tracking in ARCore is achieved through Visual Inertial Odometry (VIO), which combines visual information from the device's camera with inertial measurements from its accelerometer and gyroscope. This process allows ARCore to track the device's position and orientation in real-time, ensuring that virtual objects remain correctly positioned even as the user moves around. Environmental understanding involves detecting horizontal and vertical surfaces, such as floors and walls, using feature points and plane detection algorithms. This capability is crucial for placing virtual objects accurately within the physical space, enhancing the realism and interactivity of AR experiences.

Light estimation further enhances the realism of virtual objects by adjusting their appearance based on the ambient lighting conditions. ARCore analyzes the light in the environment and applies the appropriate intensity and color to virtual objects, ensuring they cast realistic shadows and reflections. ARCore also integrates seamlessly with popular development frameworks like Unity and Unreal Engine, as well as Google's own technologies such as TensorFlow for machine learning and Cloud Anchors for shared AR experiences. This extensive support and integration make ARCore a powerful tool for developers, enabling them to create innovative and engaging AR applications across various industries.

C. Vuforia

Vuforia is one of the most widely used platforms for developing Augmented Reality (AR) applications. Developed by PTC, Vuforia provides a robust set of tools and features that enable developers to create rich and interactive AR experiences across various devices and platforms. Known for its versatility, Vuforia supports a wide range of AR applications, from simple image recognition to complex industrial maintenance solutions, making it a popular choice for both novice and experienced developers (Linowes & Babilinski, 2017).

One of Vuforia's key strengths is its advanced image recognition and tracking capabilities. Vuforia can recognize and track a variety of objects, including 2D images, 3D objects, and even environments, using its proprietary computer vision algorithms. This allows developers to anchor digital content accurately onto physical objects, providing a seamless and immersive user experience. Additionally, Vuforia's Model Target feature enables the recognition and tracking of complex 3D models, making it particularly useful for applications in manufacturing, automotive, and other industries where precision is crucial.

Vuforia also offers extensive integration with popular development tools and platforms, such as Unity, which simplifies the process of creating AR applications. The Vuforia Engine, a core component of the platform, provides a comprehensive API that developers can use to access its features and capabilities. This integration allows developers to leverage the powerful graphics and animation capabilities of Unity while utilizing Vuforia's robust AR features. Furthermore, Vuforia supports cross-platform development, enabling applications to run on a wide range of devices, including smartphones, tablets, and AR glasses, ensuring broad accessibility and reach for AR applications.

D. Unity and Unreal Engine

Unity and Unreal Engine are two of the most prominent game development platforms that have expanded their capabilities to support Augmented Reality (AR) development. Both engines offer powerful tools and extensive support for creating highly interactive and immersive AR experiences, making them popular choices among developers for a wide range of applications.

Unity is renowned for its user-friendly interface and versatility, which makes it accessible to both beginners and experienced developers. Unity provides a comprehensive set of AR development tools, including the AR Foundation framework, which allows developers to create cross-platform AR applications that work on both ARKit (iOS) and ARCore (Android). Unity's integration with Vuforia and other AR platforms further extends its capabilities, enabling developers to incorporate advanced image recognition, object tracking, and environmental understanding into their AR projects. Unity's robust asset store and large developer community also offer extensive resources, tutorials, and plugins that can accelerate the development process and enhance the functionality of AR applications.

Unreal Engine, developed by Epic Games, is known for its high-fidelity graphics and powerful rendering capabilities, making it ideal for creating visually stunning AR experiences. Unreal Engine's AR development toolkit includes support for ARKit and ARCore, as well as advanced features like real-time global illumination, dynamic lighting, and photorealistic rendering. These capabilities allow developers to create AR applications with highly realistic visuals and complex interactions. Unreal Engine also supports Blueprint, a visual scripting system that enables developers to create AR experiences without writing code, streamlining the development process and making it more accessible to designers and artists.

Both Unity and Unreal Engine offer strong support for cross-platform development, ensuring that AR applications can reach a wide audience across different devices and operating systems. Their powerful development environments, extensive feature sets, and active communities make Unity and Unreal Engine top choices for developers looking to create innovative and immersive AR experiences across various industries, including gaming, education, healthcare, and retail.

VI. APPLICATION OF AUGMENTED REALITY

Augmented Reality (AR) has revolutionized the entertainment and gaming industries by providing immersive and interactive experiences that blend the digital and physical worlds.

A. Entertainment and Gaming

AR in gaming enhances player engagement by overlaying virtual elements onto real-world environments, creating a more captivating and dynamic gaming experience. Popular games like Pokémon GO and Minecraft Earth have demonstrated the potential of AR to transform traditional gameplay by encouraging players to explore their surroundings and interact with virtual characters and objects in the real world.

In the broader entertainment sector, AR has been used to create innovative experiences in live events, movies, and theme parks. For instance, AR can enhance live concerts and sports events by providing real-time information, interactive overlays, and virtual enhancements that enrich the audience's experience. In film and television, AR applications can bring movie characters to life through interactive marketing campaigns or augmented movie posters, creating deeper engagement with fans. Theme parks and museums are also leveraging AR to create interactive exhibits and attractions that provide educational and entertaining experiences, blending storytelling with cutting-edge technology.

Furthermore, AR is expanding the possibilities for at-home entertainment by integrating with smart devices and home systems. AR applications on mobile devices or AR glasses can transform living rooms into interactive playgrounds, where users can enjoy immersive games, virtual pet experiences, or augmented fitness routines.

B. Education and Training

Augmented Reality (AR) is making significant strides in the fields of education and training by providing interactive, engaging, and immersive learning experiences. In education, AR applications enhance traditional learning methods by overlaying digital content onto physical textbooks, classrooms, and educational tools. This technology allows students to visualize complex concepts, such as anatomy, astronomy, and physics, in three dimensions, making abstract ideas more concrete and easier to understand. For instance, AR can bring historical events to life through interactive timelines and simulations, enabling students to explore and interact with historical figures and environments.

In training, AR offers practical, hands-on learning experiences that can be applied across various industries, from healthcare to manufacturing. Medical professionals can use AR to practice surgical procedures on virtual patients, gaining valuable skills in a risk-free environment. Similarly, engineers and technicians can utilize AR for equipment maintenance and repair training, where AR overlays provide step-by-step instructions and real-time feedback as they work on actual machinery. This approach not only improves the effectiveness of training programs but also reduces the time and cost associated with traditional training methods.

Moreover, AR facilitates remote learning and training by enabling interactive and collaborative experiences regardless of geographical location. Students and trainees can participate in virtual classrooms or training sessions, interacting with digital content and each other in real time. This capability is particularly beneficial in situations where physical presence is impractical or impossible, such as during the COVID-19 pandemic. By enhancing accessibility and engagement, AR is transforming education and training, making learning more effective, interactive, and widely accessible (Wu et al., 2013).

C. Healthcare and Medicine

Augmented Reality (AR) is revolutionizing healthcare and medicine by offering innovative solutions that enhance diagnostics, treatment, and medical education. In diagnostics, AR technology assists healthcare professionals by providing real-time overlays of patient data and medical images onto the patient's body. This allows for more accurate and efficient diagnosis, as clinicians can visualize internal structures, such as bones, blood vessels, and organs, without invasive procedures. For example, AR can be used in ultrasound imaging to guide needle placements or identify abnormalities with greater precision.

In surgical procedures, AR is transforming the operating room by providing surgeons with enhanced visualization and navigation tools. AR headsets and smart glasses can overlay critical information, such as surgical plans, anatomical landmarks, and real-time patient vitals, directly into the surgeon's field of view (Brown & Gupta, 2022). This augmented guidance can lead to more precise and less invasive surgeries, reducing recovery times and improving patient outcomes. Additionally, AR can be used to simulate surgeries and practice complex procedures, allowing surgeons to refine their skills and prepare for challenging cases without the need for live patients.

Medical education and training are also significantly benefiting from AR. Medical students and professionals can use AR applications to study anatomy and physiology through interactive 3D models that offer a deeper understanding of human biology. AR can simulate clinical scenarios, allowing trainees to practice diagnosis and treatment in a controlled, virtual environment. This hands-on approach to learning enhances retention and understanding, preparing future healthcare providers for real-world challenges. Overall, AR in healthcare and medicine is advancing the quality of care, improving surgical precision, and enhancing medical education, ultimately leading to better patient outcomes and more efficient healthcare systems.

D. Retail and Marketing

Augmented Reality (AR) is transforming the retail and marketing landscapes by offering innovative ways to engage customers and enhance their shopping experiences. In retail, AR allows customers to visualize products in their own environment before making a purchase, significantly reducing the uncertainty associated with online shopping. For example, furniture retailers like IKEA use AR applications to let customers see how a piece of furniture would look in their home, helping them make more informed decisions. Similarly, AR-enabled virtual try-on apps for clothing, accessories, and makeup allow customers to see how products will look on them without physically trying them on, enhancing convenience and personalization.

In marketing, AR offers brands unique opportunities to create interactive and immersive campaigns that capture consumer attention. AR can transform traditional print advertisements, packaging, and billboards into interactive experiences that provide additional information, animations, and promotional content when viewed through an AR-enabled device (Park & Yamamoto, 2023). This not only increases engagement but also provides a memorable and interactive way for consumers to connect with the brand. For instance, brands can use AR to create scavenger hunts, interactive games, or virtual showrooms that enhance customer engagement and drive sales.

Moreover, AR is being used to enhance in-store experiences and streamline operations. Retailers can deploy AR navigation systems within stores to guide customers to products, provide real-time promotions, and offer product information through interactive displays. This improves the overall shopping experience by making it more informative and enjoyable. Additionally, AR can assist store associates by providing real-time inventory information and visual merchandising guidelines, ensuring shelves are stocked correctly and efficiently. Overall, AR in retail and marketing not only improves customer satisfaction and engagement but also helps retailers optimize their operations and marketing strategies (Lee & Park, 2022).

E. Industrial and Manufacturing

Augmented Reality (AR) is revolutionizing industrial and manufacturing processes by offering innovative solutions that enhance productivity, efficiency, and safety. In manufacturing, AR enables workers to access real-time information and instructions overlaid onto machinery and equipment, reducing errors and improving workflow efficiency (Archana & Stephen, 2024). For example, AR-guided assembly instructions can provide step-by-step visual guidance directly within the worker's field of view, ensuring accurate assembly and reducing training time for new employees. Similarly, AR can assist maintenance technicians by overlaying diagnostic information and repair instructions onto equipment, enabling faster troubleshooting and minimizing downtime.

In field service and maintenance, AR provides technicians with hands-free access to critical information and remote expert assistance, improving the efficiency and effectiveness of repairs and maintenance tasks. AR-enabled smart glasses or mobile devices can overlay relevant data, such as equipment schematics, service history, and real-time sensor readings, directly onto the technician's field of view. This allows technicians to diagnose problems more quickly and accurately, reducing the need for costly and time-consuming equipment downtime.

Moreover, AR is transforming training and simulation in industrial settings, allowing workers to practice complex procedures and safety protocols in virtual environments. AR simulations can replicate hazardous scenarios or emergency situations, providing workers with realistic training experiences without exposing them to actual risks (Li & Kim, 2023). This improves safety awareness and preparedness while reducing the potential for accidents and injuries in the workplace. Overall, AR in industrial and manufacturing environments enhances operational efficiency, reduces downtime, and improves worker safety, leading to cost savings and increased productivity for organizations across various industries.

VII. CHALLENGES AND LIMITATIONS OF AR

While Augmented Reality (AR) holds immense promise, it also faces several challenges and limitations that need to be addressed for widespread adoption and optimal functionality (Robinson & Evans, 2020).

One significant challenge is related to technical limitations, including hardware constraints and software compatibility issues. AR applications require sophisticated hardware components such as sensors, cameras, and processing units, which may not be available or affordable for all users. Additionally, ensuring seamless compatibility across different devices and platforms can be challenging, leading to fragmentation and interoperability issues within the AR ecosystem.

User experience and interaction issues pose another hurdle for AR adoption. Designing intuitive and user-friendly interfaces for AR applications is complex, as developers must balance providing relevant information with minimizing cognitive overload and ensuring a seamless user experience. Moreover, factors such as latency, tracking accuracy, and occlusion can impact the perceived realism and usability of AR experiences, requiring careful optimization and refinement.

Ethical and privacy concerns also present significant challenges for AR technology. AR applications raise questions about data privacy, as they often collect and process sensitive information about users and their surroundings. Ensuring transparent data collection practices, robust security measures, and user consent mechanisms is crucial to addressing these concerns and building trust among users.

AR technology raises ethical considerations regarding its potential impact on society, including issues related to surveillance, augmented reality addiction, and the blurring of boundaries between the virtual and physical worlds. Addressing these ethical challenges requires thoughtful consideration and collaboration among developers, policymakers, and other stakeholders to ensure that AR technology is used responsibly and ethically.

In conclusion, while AR holds immense potential to transform various industries and improve user experiences, addressing the challenges and limitations outlined above is essential for realizing this potential sustainably and ethically. By tackling these challenges head-on, the AR community can work towards creating more inclusive, intuitive, and responsible AR experiences that benefit individuals and society as a whole.

VIII. FUTURE TRENDS IN AUGMENTED REALITY

The future of Augmented Reality (AR) holds exciting possibilities as the technology continues to evolve and mature. Several key trends are expected to shape the future of AR and drive its widespread adoption across various industries. Advancements in AR technology, particularly in hardware, are expected to play a significant role in enhancing the capabilities and accessibility of AR experiences. The development of more powerful and compact sensors, displays, and processors will enable AR devices to deliver higher fidelity graphics, improved tracking accuracy, and reduced latency, resulting in more immersive and seamless AR experiences. Additionally, the integration of technologies such as 5G connectivity and edge computing will enable AR applications to leverage real-time data processing and cloud-based services, further enhancing their capabilities and performance.

Integration with other emerging technologies, such as Artificial Intelligence (AI) and the Internet of Things (IoT), will also drive innovation in AR applications. AI-powered computer vision algorithms will enable AR systems to better understand and interact with the user's environment, enabling more intelligent and context-aware experiences. Similarly, AR-enabled IoT devices will enable seamless integration with smart homes, connected vehicles, and industrial IoT systems, unlocking new possibilities for enhanced productivity, safety, and convenience.

The proliferation of wearable AR devices, such as AR glasses and smart contact lenses, is expected to drive widespread adoption of AR technology in everyday life. These lightweight and discreet devices will enable users to access AR experiences hands-free, seamlessly integrating digital information and interactions into their daily activities. As AR devices become more comfortable, affordable, and socially acceptable, they will become indispensable tools for communication, productivity, entertainment, and more.

The future of Augmented Reality holds immense potential to transform how we interact with the world around us. By leveraging advancements in technology and embracing emerging trends, AR has the power to revolutionize various industries, improve user experiences, and unlock new opportunities for innovation and creativity. As AR continues to evolve, its impact on society is expected to be profound, shaping the way we work, learn, communicate, and experience the world for years to come.

IX. CONCLUSION

Augmented Reality (AR) stands at the forefront of technological innovation, offering immersive and interactive experiences that bridge the gap between the digital and physical worlds. Through advancements in hardware, software, and integration with other emerging technologies, AR has the potential to transform various industries, including entertainment, education, healthcare, retail, and manufacturing.

AR also faces challenges such as technical limitations, user experience issues, and ethical considerations, which require careful attention and collaboration among stakeholders to address effectively. Looking ahead, the future of AR appears promising, with ongoing developments driving innovation and growth in the field. Wearable AR devices, advancements in AI and IoT integration, and enhancements in AR content creation tools are among the trends shaping the future of AR. By embracing these opportunities and addressing the challenges, we can harness the full potential of AR to create a more immersive, connected, and enriching future for individuals and society as a whole.

Augmented Reality holds immense promise to revolutionize how we interact with information, experience the world, and solve real-world challenges. By leveraging advancements in technology and embracing emerging trends, AR has the power to transform industries, improve user experiences, and unlock new opportunities for innovation and creativity, ultimately shaping the way we live, work, and play in the years to come.

REFERENCES

Archana, T., & Stephen, R. K. (2024). The Future of Artificial Intelligence in Manufacturing Industries. In Industry Applications of Thrust Manufacturing: Convergence with Real-Time Data and AI (pp. 98-117). IGI Global. DOI: 10.4018/979-8-3693-4276-3.ch004

Arena, F., Collotta, M., Pau, G., & Termine, F. (2022). An overview of augmented reality. *Computers*, *11*(2), 28. DOI: 10.3390/computers11020028

Brown, K., & Gupta, M. (2022). The Role of Augmented Reality in Healthcare: A Review of Applications and Challenges. *Journal of Medical Internet Research*, *19*(5), e173.

Carmigniani, J., & Furht, B. (2011). Augmented reality: an overview. Handbook of augmented reality, 3-46.

Edwards-Stewart, A., Hoyt, T., & Reger, G. (2016). Classifying different types of augmented reality technology. *Annual Review of Cybertherapy and Telemedicine*, *14*, 199–202.

Lee, H., & Park, J. (2022). Augmented Reality and Virtual Reality: A Comparative Analysis in Marketing. *Journal of Interactive Marketing*, *58*, 91–104.

Li, X., & Kim, J. (2023). The Impact of Augmented Reality on Manufacturing Processes: A Case Study Approach. *International Journal of Production Economics*, *145*, 312–328.

Linowes, J., & Babilinski, K. (2017). *Augmented reality for developers: Build practical augmented reality applications with unity, ARCore, ARKit, and Vuforia.* Packt Publishing Ltd.

Macedo, M. C., & Apolinario, A. L. (2021). Occlusion handling in augmented reality: Past, present and future. *IEEE Transactions on Visualization and Computer Graphics*, *29*(2), 1590–1609. DOI: 10.1109/TVCG.2021.3117866 PMID: 34613916

Park, C., & Yamamoto, T. (2023). Augmented Reality in Retail: Enhancing Customer Engagement and Sales. *Journal of Retailing*, *98*(2), 215–230.

Robinson, T., & Evans, D. (2020). User Experience Design for Augmented Reality: Challenges and Best Practices. *Human-Computer Interaction Journal*, *36*(4), 567–589.

Vlaminck, M., Luong, H., & Philips, W. (2017, December). A markerless 3D tracking approach for augmented reality applications. In 2017 international conference on 3d immersion (ic3d) (pp. 1-7). IEEE.

Vuta, D. R. (2020). Augmented reality technologies in education-a literature review. *Bulletin of the Transilvania University of Brasov. Series V, Economic Sciences*, *13*(62), 35–46. DOI: 10.31926/but.es.2020.13.62.2.4

Wang, L., & Lee, S. (2021). "Technological Foundations of Augmented Reality: A Comprehensive Overview." *Proceedings of the IEEE International Conference on Computer Vision*, 132-145.

Wu, H. K., Lee, S. W. Y., Chang, H. Y., & Liang, J. C. (2013). Current status, opportunities and challenges of augmented reality in education. *Computers & Education*, *62*, 41–49. DOI: 10.1016/j.compedu.2012.10.024

Chapter 9
Enhancing Automobile Design and Manufacturing Through VR and AR Integration

P Sreenivas
Department of Mechanical Engineering, K.S.R.M. College of Engineering, Kadapa, India

M. Suresh Anand
Department of Computing Technologies, School of Computing, SRM Institute of Science and Technology, Kattankulathur, India

Supriya Bhosale
https://orcid.org/0000-0002-8638-701X
Department of Computer Science and Engineering, Sandip University, Nashik, India

T. V. V. Satyanarayana
https://orcid.org/0000-0002-8764-9982
Department of Electronics and Communication Engineering, School of Engineering, Mohan Babu University, Tirupati, India

Sampath Boopathi
https://orcid.org/0000-0002-2065-6539
Department of Mechanical Engineering, Muthayammal Engineering College, Namakkal, India

DOI: 10.4018/979-8-3693-7287-6.ch009

ABSTRACT

This chapter focuses on the integration of VR and AR technologies with automobile companies, especially through what transformative impacts they bring along for the design and process of automobile manufacturing. Therefore, with the use of these VR and AR technologies, the car manufacturers are facilitated in producing immersive, interactive experiences to streamline the designs thus producing better visions, prototypings, and testing cars as may be needed. These technologies allow designers and engineers, along with many other stakeholders, to provide collaborative working with real-time access to virtual environments that allow exploring and refining of concepts. Apart from this, AR increases efficiency in manufacturing by speeding up on-site assembly as well as reducing errors. This particular chapter includes case studies demonstrating the successful application of both VR and AR to improve workflow productivity, increase precision, as well as fuel innovation in auto-making.

INTRODUCTION

Integration of Virtual Reality and Augmented Reality technologies into the auto-manufacturing industry brought the era of innovation into this segment, revolutionizing dramatically what vehicles could be designed to be produced, manufactured, and delivered into markets. High-speed development rates in such immersed technologies allow automobile-related firms to improve on both creative and technical aspects in the conception of cars towards better productive, accurate, and more innovative designs. This chapter deals with changes that VR and AR bring into design and manufacture, bringing about immersive and interactive experiences, thus streamlining workflows and introducing more precision (Boboc et al., 2020).

The traditional form of automobile manufacturing is based upon the physical prototypes of cars, designing the car in the virtual environment, with manual and labor-intensive adjustments, which always carry a considerable time loss, cost factor, and loss of accuracy. This paradigm has shifted with the advent of VR and AR technologies that offer virtual environments where designers and engineers can simulate real-world scenarios, iterate on designs, and test concepts with more flexibility and precision. Manufacturers are able to create better visions, prototypes, and test vehicles without being limited by physical boundaries. It is more dynamic and adaptive in approach towards the automobile development lifecycle (Aouam et al., 2018).

This basis allows VR technology to transform car designs. Through virtual prototyping, these car designers can develop and interact in full immersion with 3D models of vehicles in the digital space. This has reduced the number of necessary prototypes, as well as lessening the cost and time lead of a product. Fully interactive and real time, engineers and designers will be able to explore in detail all the aspects of the vehicle design: from the structural elements to the aerodynamics, and from interior layouts to user interfaces. The ability to visualize in real-time allows for instant adjustments in a collaborative environment where the stakeholders from the different departments—design, engineering, marketing, and production—can work together efficiently. This way, the final output would also reflect a harmonious combination of functionality, aesthetics, and engineering excellence (Wiegand et al., 2019).

In addition, VR can simulate complex vehicle behaviors, such as crash testing, fluid dynamics, and thermal analysis in virtual space. This accelerates the decision-making process and ensures that the vehicles meet stringent safety and performance standards before physical production. This facilitates a more iterative design process, improving the ability to address design challenges early on, thus reducing costly revisions and minimizing errors during production (Bale et al., 2024).

AR complements VR by adding efficiency to manufacturing processes in assembly and maintenance. AR overposes digital information onto real-world objects in order to provide real-time guidance and contextual support in the performance of manufacturing tasks. For example, in an automobile assembly line, AR can guide workers through the assembly process of complicated procedures by showing them step-by-step instructions on the physical components directly. It thus decreases the likelihood of human error, trains promptly, and offers high accuracy for tasks of assembly that are carried out (Dalle Mura & Dini, 2021).

It further accelerates in-line assembly with access by employees to crucial information that might include schematics, maintenance records, and operation procedures. Real-time and visual cues via AR technology are delivered directly to equipment through which technicians access maintenance while ensuring efficient activity to cut down on repair time or downtime. In a setting of manufacturing involving the use of AR, precision and responsiveness are key demands made for production(Xiong et al., 2018).

This chapter will also focus on analyzing a few case studies about how leading automobile manufacturing companies have been able to incorporate VR and AR technologies in their businesses. For example, companies such as BMW, Ford, and Tesla use VR and AR solutions in designing and producing vehicles for enhanced effectiveness. These companies have exploited VR for virtual testing and iterations of design, while AR gives hands-free guidance in the assembly and maintenance of things. In all these implementations, manufacturers have seen improvements of workflow productivity, accuracy, and innovation toward vehicles which are more responsive to what the customer demands and what technology is advancing(Aslan & Yıldız, 2020).

This apart, the development in VR and AR has driven innovation in the automotive sector, creating new scopes of cooperation and real-time decisions. With this technology, the designing and engineering crew along with all the others, could test innovative concepts through these stages of manufacturing thereby ensuring the final product well manufactured, optimized both on the performance aspect as well as that of user experience(Yıldız et al., 2020).

Hence, the penetration of VR and AR in automobile has opened a relatively streamlined, collaborative, and technologically advanced manufacturing environment. With the help of these immersive technologies, automobile producers are not only improving workflow designs and production but also setting up new benchmarks for innovativeness and efficiency in development. This chapter goes into how VR and AR are changing automobile design and production as means to address an industry in rapid evolution with transformative solutions (Fang & An, 2020).

BACKGROUND VR AND AR TECHNOLOGIES

Definition of VR

Virtual Reality is the technology that creates an artificially generated environment, giving people a fully digital world that they can experience with the help of sensory stimuli. It replaces the real world with a digital one through the simulation of environments and interactions. Users mostly access VR through headsets, controllers, and other devices with sensory capabilities that can monitor movements and give feedbacks. This technology allows one to feel a sense of presence, as if being in a virtual environment in a completely virtual space (Vashisht & Rakshit, 2021).

Applications of VR

a) Games and Entertainment: VR fundamentally changes the way users can engage with digital content - gaming, virtual tourism, and interactive storytelling.
b) Health: VR trains medical professionals and patients as well for therapy and healing. They undergo simulations so that professionals may diagnose it and heal their patients by using virtual reality.

c) Academic and Skill Development Education: In medical, airlines, and other technical activities, students undergo interactive, hands-on experience in what otherwise is quite abstract.
d) Manufacturing and Design: VR is utilized for virtual prototyping to decrease the use of prototypes and also enhance the designs within automobile and aerospace businesses.
e) Real Estate and Architecture: Potential buyers or clients may use VR to feel their houses or designs through virtual media before any building activity takes place.Definition and Applications of AR

Augmented Reality (AR) Technologies

Augmented Reality (AR) is a technology that superimposes digital information-images, sounds or text- onto the physical world through devices such as smartphone, tablet, or a pair of AR glasses. In contrast to VR, which gives a completely virtual environment, AR elevates the real world by putting virtual elements that complement physical settings. AR is defined by the combination of physical and virtual worlds to create interactive experiences that have been seamlessly blended(Yıldız, 2020b).

Applications of AR

a) Retail and E-commerce: AR allows virtual fitting rooms and interactive product displays, thus customers can see the product before buying.
b) Healthcare and Medical Training: AR assists in real-time visualization of medical information during surgical procedures, increasing accuracy and decreasing danger in a clinical environment.
c) Manufacturing and Field Services: AR is widely used in industrial applications to aid maintenance, assembly, and repairs by overlaying instructions or diagnostics directly onto machinery or parts.
d) Education: AR provides a better learning experience to students by visualizing historical events, anatomical structures, or scientific phenomena in real-time.
e) Navigation and Entertainment: AR applications can be interactive in travel, theme parks, and city tours by guiding users to a better understanding of their surroundings.

Hence, both VR and AR are changing several industries as they provide an immersive, interactive, and innovative solution toward design, education, health care, and more.

INTEGRATION OF VR IN AUTOMOBILE DESIGN

Virtual Reality (VR) has emerged as one of the most powerful tools in the automotive design industry, changing the way vehicles are conceptualized, visualized, and refined. The integration of VR into automobile design has enabled manufacturers to streamline the design process, improve accuracy, and enhance the overall user experience. This section looks into the three main aspects of VR integration in automobile design: virtual prototyping, design visualization and iteration, and enhancing user experience(Jing et al., 2020).

Figure 1. Integration of VR in automobile design

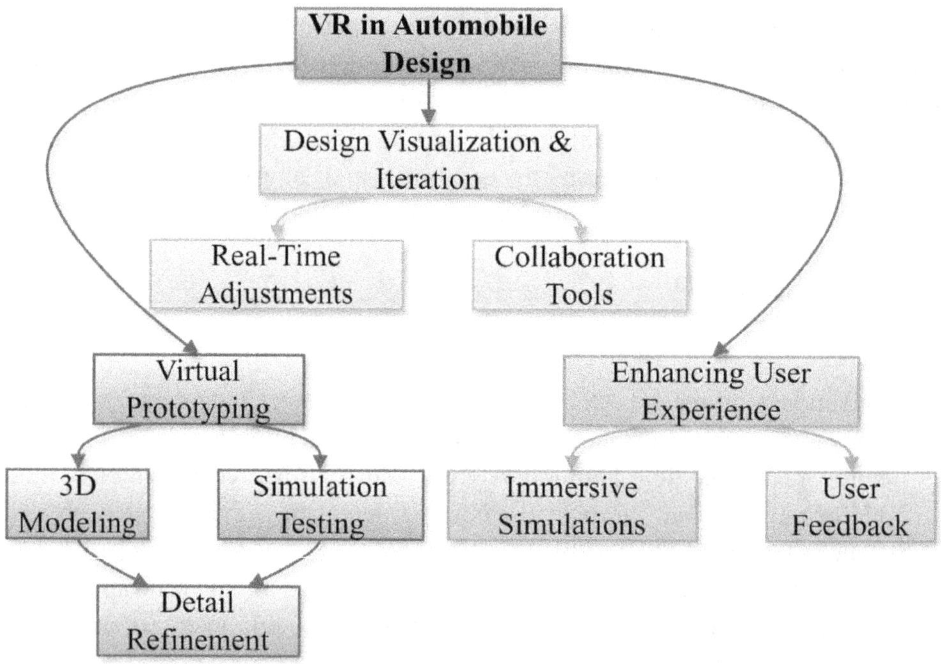

Figure1 illustrates the integration of VR in various stages of automobile design, from prototyping to enhancing user interactions.

Virtual Prototyping

One of the great applications of VR in automobile design is through virtual prototyping. Long before, physical prototypes were used as the basis for testing and perfecting vehicle designs, much time and resources being consumed. A designer can create a detailed 3D model of a vehicle for an immersive experience without its parts. Using VR tools, designers are able to explore different components and systems of a vehicle, like aerodynamics, ergonomics, and interior layouts, in virtual space. This enables the ability to make real-time adjustments and refinements based on simulated performance without needing multiple physical iterations. Additionally, through virtual prototyping that VR enables, engineers and designers can conduct crash tests, material tests, and other analyses to see the final product in detail on how it will be used in real life. This reduces the development cycle and costs of physical prototypes on manufacturers while ensuring a very high level of precision and reliability.

Design Visualization and Iteration

The integration of VR into automobile design improves visualization and iteration. It provides an intuitive way of refining the aesthetic and functionality of a vehicle. Using VR, designers can be immersed in a virtual environment in which they can manipulate various vehicle components to achieve a look and feel that is preferred. This creates quicker feedback loops between design teams, reducing delays often associated with communication and revisions of physical models. With VR simulations, designers can immediately evaluate different design options from exterior styling to color schemes and interior layouts to create a more interactive and collaborative design process. Furthermore, VR fosters cross-disciplinary collaboration where different teams, such as engineers and marketing specialists, can contribute insights and make adjustments simultaneously to ensure that the final design meets all stakeholder expectations. In this iterative approach, the result is a vehicle design that is more refined and cohesive, maximizing aesthetic appeal and functional performance(Yıldız, 2020a).

Enhancing User Experience

Beyond technical design elements, VR is critical to moving the user experience forward in automobile design. Users could simulate driving experiences as well as interactions within the vehicle through virtual environments to learn how they would relate to various features of the automobile. Infotainment, driver-assistance features and more could be visualized in the context of interface integration into the automobile in a more realistic fashion with VR. The comfortability and safety of the driver also get checked with VR simulation with regard to accessibility for successful user experience. VR further allows inferences of emotional response together with the psychological behavior of different vehicles designs on the mindsets of users; meaning it is possible to create great-experience vehicles. This way, designers can anticipate how a driver will really interact with the vehicle in actual conditions, such as negotiating through urban environments, handling road conditions, and interacting with the autonomous systems. In this way, designers are better equipped to optimize performance and user satisfaction(Manimuthu et al., 2022).

The introduction of VR into automobile design has actually changed the way vehicles are designed, from concept to final product. Virtual prototyping streamlines the actual design process that provides precise simulations and least use of physical prototypes. Real-time adjustments and inputs at work enhance the visualization and refinement of design at such a stage so that ultimately, the final product can be met with technical and aesthetic standards. VR actually contributes much to the enhancement of user experience, providing insights into how vehicles are perceived and used by the drivers in real-world situations. These developments place VR at the forefront of innovation within the automotive industry.

INTEGRATION OF AR IN AUTOMOBILE DESIGN

Augmented Reality (AR) has gradually become an unavoidable tool for automobile design and has made the way a vehicle could be visualized, designed, and detailed. Over the physical world, it overlays digital information, so that designers, engineers, and stakeholders immerse, interact, and enhance creativity, precision, and collaboration. This part delves into how AR works with different steps of the automobile design: conceptual visualization, real-time iteration, and refinement(Xiong et al., 2018).

Figure 2. Integration of AR in automobile design

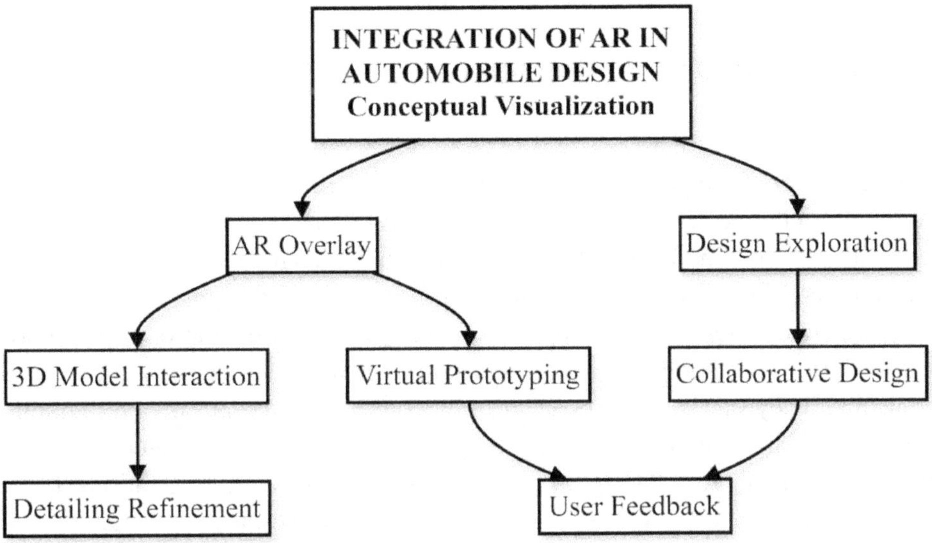

Figure 2 illustrates the integration of AR in automobile design stages, from conceptual visualization to refinement through real-time interaction and feedback.

Conceptual Visualization and Prototyping

Conceptual visualization is the most important application of AR in automobile design. Sketches, clay models, and static 3D renderings are used as traditional methods for exploring ideas within the design process. Often, these methods lack flexibility and realism for deep exploration of design concepts. But AR enables designers to visualize vehicles in real-world environments, with a lifelike experience between virtual and physical design.

By using AR, designers can overlay the digital prototype of vehicle parts over the physical surface, thereby visualizing how things could work in the real world. Designers can overlay a virtual dashboard onto the car interior so that they can test the ergonomics, user interface, and spatial arrangements. Immersion would allow for more precise judgment regarding design concepts, stimulating creativity and innovation at early stages of development.

AR also makes inter-group collaboration easier, whereby possibly engineers, marketing teams, and production managers might collaborate throughout the design process. Indeed, this would be a natural way of increasing the speed of decision making, which will then diminish the need for a final prototype, often leading to repetitive revisions and adjustments.

Real-time Iteration and Refinement

The ability of AR to provide real-time feedback during the design process of an automobile significantly enhances its efficiency in automobile design. With AR, designers can iterate and make improvements based on user feedback, engineering specifications, and environmental constraints quickly. This dynamic approach reduces the need for major manual adjustments and allows design improvements to be integrated seamlessly (Tripathi & Talukder, 2023). For example, in automotive exterior design, AR allows designers to see in real-time a variety of vehicle shapes and forms. They can superimpose virtual modifications such as aerodynamics, body lines, or lighting systems onto the physical car models to determine visual impact and functionality. Designers can experiment with various iterations, testing aesthetics, performance, and user experience in an interactive and immersive way.

In addition, AR improves the accuracy of design elements because it provides accurate spatial information that helps designers assess the proportions, dimensions, and fitment of components with a high degree of accuracy. This eliminates the possibility of misalignments and ensures final designs meet specifications, thereby eliminating costly errors in production.

User-Centered Design and Ergonomics

Another crucial benefit that AR in car design brings is focus on user-centered design, especially in relation to improvement of vehicle ergonomics. Here, AR models how the users interact with various components of the automobile, like controls, displays, seating arrangement, or interior layouts. This means that the designers will achieve a much better understanding of real-life experiences in the vehicle experienced by the drivers and passengers (Gouda & Tiwari, 2024).

For instance, AR enables designers to create prototypes of cockpit interfaces by projecting virtual controls over physical dashboards. This means that the layout is intuitive and comfortable as well as safe. It is also possible for designers to check different seating configurations by visualizing them directly in the vehicle space to enhance driver and passenger experiences. With AR, automobile companies can also conduct usability testing and gather valuable feedback from users. By integrating AR into the design process, companies can refine features according to real-world feedback, ensuring that the final product is tailored to user needs and preferences.

Streamline the Manufacturing and Assembly Process

Besides design, integration of AR involves simplification in the manufacture and assembly. The designers see how components are supposed to be installed, aligned, and assembled to ensure that all design details meet production needs so as to avoid costly errors during manufacturing (Dalle Mura & Dini, 2021).

For instance, through AR, step-by-step assembly instructions can be superimposed onto parts so that workers follow visual cues and annotations on procedures. It improves accuracy and cuts down on the time needed to train workers so that the entire process in production becomes efficient. Moreover, AR enables collaborative work remotely in manufacturing. Engineers and designers can troubleshoot and offer real-time support from anywhere to improve overall efficiency and reduce downtime.

Case Studies and Future Outlook

Already, car giants such as BMW and Audi have been using AR to change the way they design vehicles. The respective companies used AR in improving the way designs are collaborated, reducing the problems associated with prototyping, and enhancing how users experience their automobiles. This technology will continue advancing, thus creating broader applications in automobile designing (Yıldız, 2020a).

In summary, the introduction of AR in car design presents a revolutionary method to visualize, iterate, and refine vehicle concepts. Immersive and interactive experiences allow AR to bridge the gap between virtual and physical design, thus empowering designers to come up with innovative, user-centric, and precise vehicles. This can change the face of automobile development into the future with mature technology, providing unparalleled levels of creativity, efficiency, and precision.

UTILIZATION OF AR IN MANUFACTURING PROCESSES

AR technology has become a revolutionary component in manufacturing processes and transforms efficiency, accuracy, and innovation in various dimensions of production, assembly, maintenance, and quality assurance. The overlay of digital information into the physical world of real life allows for experiential hands-on experiences in support of both operational and training capabilities. This chapter illustrates how AR is being utilized in manufacturing through applications that include assembly and maintenance, real-time interactive training and support, and quality assurance and inspection(Dalle Mura & Dini, 2021). Figure 3 highlights the significant applications of AR in manufacturing, highlighting its enhancement in production, assembly, maintenance, and quality assurance.

Figure 3. Utilization of AR in manufacturing processes

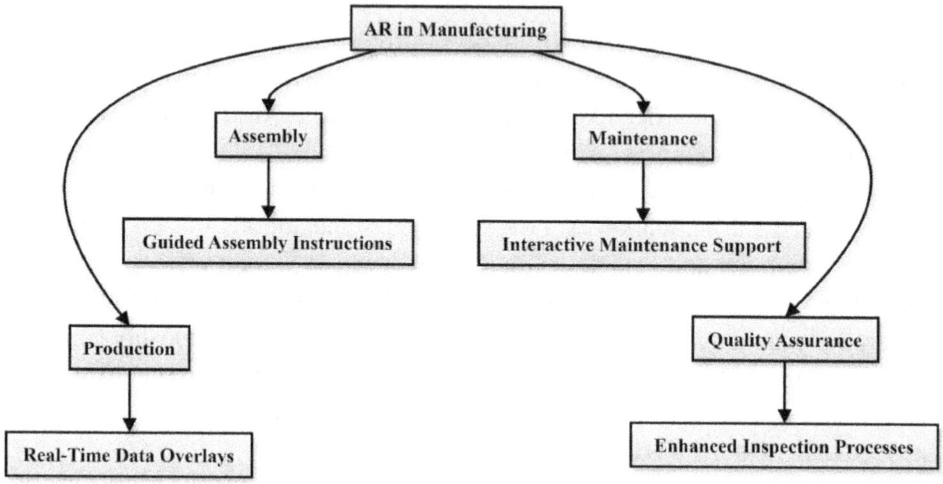

Augmented Reality in Assembly and Maintenance

The two major applications of AR in manufacturing include assembly and maintenance. Complex assembly processes used to be performed through the use of comprehensive manuals and physical documentation, often leading to errors and delays. AR improves this process because it provides real-time visual guidance directly on the physical components or machinery. Using AR-enabled devices, like smart glasses or tablets, enables the worker to see digital overlays of step-by-step instructions that include key assembly steps or maintenance procedures in context with the physical components. It makes for an even more intuitive and streamlined assembly process that reduces mistakes and increases productivity(Aouam et al., 2018; Wiegand et al., 2019). Furthermore, AR offers context relevant for maintenance, replacement parts, performance data, and more, and lets technicians do their work far better. In applications as aerospace, automotive, or even electronics, where precision and accuracy really come into play, with augmented reality, every assembly task or maintenance is conducted in a more precise way than ever before.

Real-time Interactive Training and Support

AR is also changing the face of training and support within manufacturing environments through real-time, interactive experiences. For those in industries where workers are constantly in need of skills development or must learn new processes and technologies, AR presents a hands-on approach to learning. Using AR tools, operators and technicians can access digital overlays that will guide them through complex procedures in real time, allowing them to accomplish tasks with minimal supervision. For example, when setting up machinery or diagnosing equipment, AR can superimpose diagnostic information and step-by-step instructions directly to the machinery. Such real-time, interactive support cuts the time needed for training new employees and reduces downtime related to equipment failure. Moreover, AR-based training ensures that the working staff gain practical insights into complex procedures without being biased toward intellectualized knowledge or class-based education(Boboc et al., 2020).

Quality Assurance and Inspection

Quality assurance and inspection are one of the most important parts of the manufacturing process, especially when accuracy and precision prevail. AR enriches these activities by providing dynamic, real-time inspection capabilities. AR can lay inspection criteria, quality standards, and critical tolerances directly on physical parts or assemblies, allowing workers to inspect with confidence. That way, all components will meet predefined quality standards before being passed on in the production line. AR also detects defects, variations, or non-conforming items against the design standards that may be challenging through naked eyes or by any form of traditional inspection. Automotive, Aerospace, and electronics companies whose minor defects become significant results require a proactive quality control by AR. Adding ML in AR will make automation for increased accuracy and efficiency without mistakes in quality inspection during this process(Dalle Mura & Dini, 2021).

In a nutshell, the utilization of AR in manufacturing is shifting the approach of companies regarding assembly, maintenance, training, and quality assurance. With AR's immersive, interactive, and real-time experiences, the processes are more efficient, accurate, and adaptive to the changing needs of the industries. Whether it be an improvement in assembly workflows, support for maintenance tasks, dynamic training, or the enhancement of quality inspections, AR remains a crucial part of driving innovation and operational excellence in manufacturing environments.

UTILIZATION OF VR IN MANUFACTURING PROCESSES

Virtual reality has transformed the process of manufacturing with the many advantages that it provides, thus boosting efficiency, accuracy, and innovation. It has made the automotive, aerospace, electronics, and heavy machinery industries give virtual, real-time simulations to various aspects of the process of manufacturing, from designing and prototyping to controlling quality and training(Xiong et al., 2018; Yıldız, 2020a). The following section explores how VR is applied in the process of manufacturing and the consequences it brings about in the environment of production. The use of Virtual Reality (VR) in manufacturing processes is depicted in Figure 4.

Figure 4. Utilization of VR in manufacturing processes

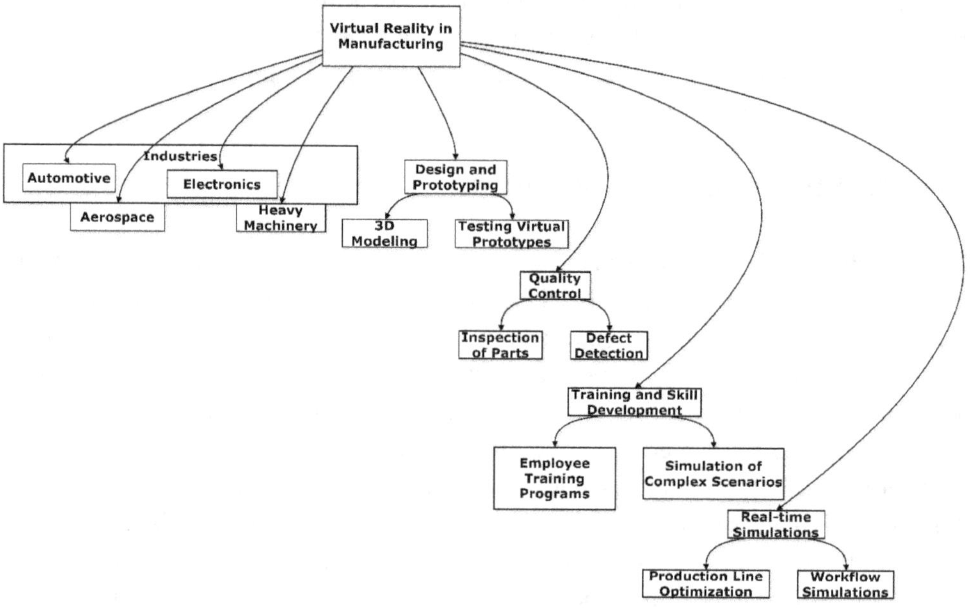

Virtual Prototyping and Design

Perhaps one of the most significant uses for VR in manufacturing is virtual prototyping. A time-consuming, resource-heavy and labor-intensive process where one had to physically make the prototype to test and improve the design. The technology eliminates the need for these physical prototypes by allowing full three-dimensional digital models to be interacted with in real-time by designers and engineers. This helps to see how different designs work, visualize some of the more complex parts, and simulate assembly procedures, stress tests, and working conditions. This is how it would look. For instance, car makers use VR for mock-up purposes regarding the parts in a vehicle; thus, it allows engineers to try out a number of factors related to performance. This will save one time and money spent in physical prototypes but still manage to have detail design adjustments. Also, VR enables cross-departmental collaboration in which design teams, production specialists, and quality assurance teams can easily collaborate in a virtual space that negates communication gaps and innovation.

Training and Workforce Development

VR plays a very important role in training and workforce development within manufacturing settings. In sectors where precision and skill matter, VR-based training equips workers with hands-on experience in a controlled virtual environment before they are assigned to actual tasks. Whether it is operating complex machinery, assembling parts, or conducting quality control, VR offers interactive guided experiences that simulate real-life situations. For example, aerospace manufacturing technicians use VR to practice complex assembly, cutting the time spent in on-the-job training while also increasing the accuracy of the work. The risk of error during critical operations will decrease and improve the general safety of the manufacturing environments. With new technologies and processes discovery, the scalable solution which VR is able to provide for continuous education helps in keeping employees updated with respect to industry advancements and helping them adapt to the new best practices in time.

Quality Control and Inspection

Quality control in the manufacturing process is critical. Simulations enhance quality through the best tools that facilitate the inspection and testing process of products at each point of the production stage. With VR, all issues related to quality control-such as visual inspections, function testing, and the search for defects-can be virtually simulated with high precision in an interactive manner. Manufacturers can, via the use of VR simulations, do virtual quality checks that may reflect flaws or inconsistencies in real time. For example, in precise manufacturing environments where minor defects can result in costly rework or even product recalls, VR ensures that each item will meet requirements before it is sent on to the next step in the production line. Further, with VR, there can be continuous monitoring and real-time feedbacks for maintaining proactivity about upholding quality standards on the products.

Improve Supply Chain Management

Another critical area that VR supports manufacturing processes in is supply chain management. It is used for visualization and optimization of logistics, inventory management, and allocation of resources to ensure a smooth production flow. In these environments, manufacturers are able to simulate scenarios such as demand forecasting, inventory management, and just-in-time production. For instance, aerospace companies use VR to simulate supply chain logistics so that parts and materials move effectively through production facilities. This helps in increasing the transparency of the supply chain operations, and reduces bottlenecks as well as minimizes delay in manufacturing processes.

Product Testing and Simulations

VR is also used in testing products and simulations to help manufacturers test the behavior of products under different conditions. In automotive component manufacturing, for example, engineers can simulate extreme conditions such as high temperatures, vibrations, or heavy loads in VR, thereby making them durable and reliable. It thus saves costly physical tests and gives insight into the actual behavior of products in use. VR also enables simulations of predictive maintenance, wherein manufacturers can test how products will fare for an extended period of use, which helps to predict a failure and address it proactively. Predictive maintenance reduces downtime and maintenance costs, leading to a more efficient manufacturing environment.

The application of VR in manufacturing is making these processes revolutionize the aspect of immersive and realistic simulations that lead the way for streamlined operation, quality improvement, and workforce capability enhancement. From virtual prototyping and training to quality control and supply chain management, VR allows manufacturers to innovate and optimize their production environment in ways that would not have been possible previously. And with the way that this technology is changing every day, its contribution in manufacturing towards the future will continue to increase by solving many of the current challenges faced in the present industrial scenario.

BENEFITS OF VR AND AR INTEGRATION

The integration of VR and AR in various industries, especially in the automotive industry, brings along several advantages that are pushing innovation and efficiency as well as collaboration. Improvement in team collaboration, precision, and accuracy in design and production with these technologies are changing the landscape of business. Below, we discuss the major benefits of including VR and AR in detailed form(Bale et al., 2024; Gouda & Tiwari, 2024; Tripathi & Talukder, 2023). The integration of VR and AR technology offers numerous benefits as shown in Figure 5.

Figure 5. Benefits of VR and AR integration

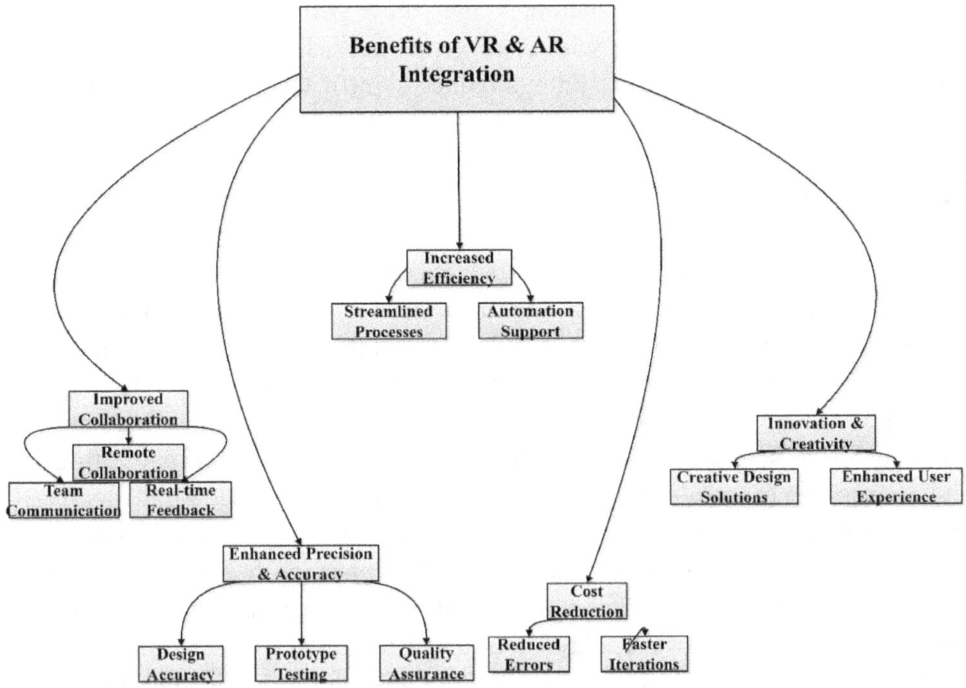

Improved Collaboration Across Teams

One of the primary benefits of integrating VR and AR technologies is to enhance the level of collaboration between teams. Traditionally, workflows consist of siloed departments that work on separate aspects of a project, resulting in communication gaps, misinterpretation, and delayed decision-making. Using VR and AR, however, teams can work in real-time within virtual or augmented environments, which fosters a more unified and efficient approach.

It's possible for cross-functional teams of engineers, designers, marketing professionals, production staff, and suppliers in the automotive design and manufacturing space to collaborate more smoothly. The designers can use VR and create realistic 3D prototypes that engineers and the manufacturing team can interact with and make direct comments on its technical feasibility. Similarly, with AR, on-site teams will overlay virtual information onto physical components. Adjustments and collaboration are therefore achieved in real time and, in some cases, remote environments.

For instance, with vehicle assembly, AR can add digital overlays of assembly instructions right to the physical components; these will guide the workforce step by step through the procedure with clear visual cues. In this real-time approach towards interaction, misunderstandings reduce significantly and the performance accelerates along with productivity levels.

Cost and Time Effective

The integration of VR and AR technologies ensures huge cost and time savings at all stages of design, prototyping, and manufacturing. Physical prototypes developed for testing and revision purposes to implement design change iterations can be costly and time-consuming. VR and AR technologies obviate the need for physical prototypes as they can create fully immersive virtual environments where one can simulate design iterations to test and evaluate performance in a virtual setting.

For example, in automobile design, it enables a designer to view and modify 3D models without any physical prototype; hence, the cost savings on material usage, labor costs, and production machinery. Again, AR enhances the effectiveness of on-site assembly. It gives the worker real-time, hands-free instructions; hence, there is minimal possibility of error and delay from manual interference or miscommunication. Besides this, the AR and VR technologies are known for providing rapid iterations in a cost-effective manner through design changes. The engineers can easily test new concepts in a virtual environment and assess their performance, refining their designs with minimal time consumed. Such efficiency saves time and effectively uses the resources, leading to higher production throughput overall.

Accurate engineering and manufacturing products are critical in design and manufacturing because they ensure functionality, safety, and durability. The features VR and AR provide assure the levels of precision and accuracy for realist simulation models and interactive models necessary in critical analyses and true fine adjustments. For example, in automotive design, AR superimposes digital information onto the physical vehicle. Here, designers can determine the fitment of parts with a precision of millimeters. Thus, parts will be aligned correctly and will not deviate from strict engineering specifications, thus reducing the chance of defects or production errors.

Similarly, in quality control, AR supports the inspection process by allowing real-time defect detection and precise measurement of components. Engineers can use interactive AR overlays to visually inspect surfaces, test tolerances, and identify flaws straight away for significantly improved accuracy and efficiency in quality control processes. It is possible to replicate the actual world in a number of testing environments using VR, thus improving the accuracy in design and production. This allows crash simulations, stress tests, and durability tests to be undertaken on products. In all these applications, there's a lower chance of expensive product recalls or rework.

Applications Across Industries

The benefits of VR and AR integration go beyond automotive manufacturing, into aerospace, healthcare, architecture, and more. For instance, in aerospace, VR is applied in designing and testing aircraft components, while AR enhances maintenance and training by overlaying virtual instructions on physical aircraft parts. Similarly, in healthcare, AR and VR are transforming surgical procedures by offering real-time guidance, visualization, and precision for complex medical interventions. All things combined, the emergence of VR and AR technologies adds to an enormous wealth of benefit toward driving efficiency, accuracy, and collaboration in manufacturing and design processes. Being cost-effective while increasing precision and allowing real-time collaboration makes these immersive technologies shape the future of industries where businesses are innovating in the same direction to help streamline processes in order to comply with all requirements within modern markets.

CHALLENGES AND LIMITATIONS

While VR and AR have tremendous potential to revolutionize businesses such as automobile design and manufacturing, the adoption of these technologies is very challenging and full of limitations in all three technological, cost, and risk-related implementation areas impacting the ease of embedding these future-forward technologies in business processes.

Technological Barriers

One of the biggest hurdles in popularizing VR and AR is technological difficulties. Hardware, software, and infrastructure have not yet been fully developed for mass utilization. For instance, although VR and AR do indeed create immersive experiences, quality renderings of visual interaction, realistic interactivity, and real-time performance are still some steps behind. The most expensive hardware is required to produce high-end VR and AR systems, including advanced GPUs, high-resolution displays, and reliable sensors for precise tracking. Not all companies have the financial resources or technological infrastructure to implement such advanced equipment. Moreover, the constant evolution of hardware and software can result in compatibility issues between different systems and software updates, requiring continuous investment in technology upgrades.

The development of AR and VR applications is also quite complex in their programming, hence time-consuming and resource-intensive. Building customized solutions for the individual needs of a business may require expertise that small companies will find difficult to deploy.

Cost Inclusions

The second major issue in integrating VR and AR into automobile design and production is their cost. With these long-term benefits, though, high initial expenses might be nothing but a stumbling block for SMEs at least.

This involves significant investments in high-quality VR/AR hardware and software, as well as training for employees. Creating and maintaining AR/VR content such as 3D models and simulations is also expensive. Hence, the access to these solutions will be limited to only big organizations, with deeper pockets. Many organizations are however, challenged by parity in cost for traditional versus new technologies. Although the initial investment for VR and AR technologies will be expensive, the payback down the road can be through time savings, reduced errors, and improved productivity. However, smaller organizations are challenged in presenting these costs since the full effects of the technology cannot be demonstrated without a complete-scale adoption and integration across fields or industries.

Risks of Implementation

The implementation of VR and AR solutions in the automobile industry is associated with certain risks primarily related to organizational change, adaptability, and the risk of disrupting established workflows. Companies that have always relied on conventional design and manufacturing processes face problems when adopting immersive technologies. Some of the main risks involve employees' resistance to change who might not be accustomed to using new technologies. Training the workforce on the proper use of VR and AR tools requires extensive time and effort, which, in turn, limits the adoption process in terms of a lack of knowledge of its use. The addition of VR and AR may require massive adjustments to existing processes and thus might disrupt during the implementation process.

Data safety and privacy is another serious threat. As such, most of the technologies in VR and AR often require real-time access to sensitive information regarding designs and production; companies need robust cyber security measures against unauthorized access and data breaches. This means that reputational damage as well as possible legal suits may be on the way of failure to sufficiently deal with this. Lastly, scalability across different manufacturing plants and global teams can be challenging. Differences in infrastructure, hardware, and software capabilities will create inconsistency in how VR and AR applications are deployed, thereby reducing the overall effectiveness of these solutions.

Hence, there are lots of benefits brought about by virtual reality and augmented reality to automobile design and manufacturing, but the challenges lie in their technological limitations, cost-related issues, as well as risks involved in implementing these concepts. Strategically planned and continuously innovated, companies can unlock their full potential through these revolutionary technologies.

FUTURE TRENDS AND INNOVATIONS

The automobile industry is not static; there are always new trends emerging to help bring innovation in designing and manufacturing the product and better experience for customers with the advent of Virtual Reality (VR) and Augmented Reality (AR). The more advancement these technologies undergo, the higher is the scope for revolution in the automobile sector. Following are some of the emerging trends, advancements in automotive simulation, and how AI could better VR/AR applications.

Emerging VR and AR Technologies

Future advancements of VR and AR in the automotive are supported by innovations that catapult immersive experiences and interactive solutions. Refined wearable devices as well as mixed reality solutions present exciting progressions toward creating smoother, more portable, and fully integrated AR/VR experiences. Companies are working with technologies to create lightweight hands-free devices that offer superior resolutions, enhanced tracking capacity, and lower latency toward creating real-time interaction experiences. Another revolution that 5G technology promises to bring into AR and VR applications in car design and manufacture is much faster, and more reliable networks to transfer high-quality data in real time, making collaboration easier and working processes much simpler. Edge computing, with AI algorithms, will make it easier to handle greater complexity in simulation and higher-fidelity environments and ensure smooth flow even with more extensive datasets.

Advances in haptic feedback systems will also make VR and AR more tactile. This would provide the user with greater control and interaction with virtual models. They will be very useful for complex vehicle component design and simulation, letting designers and engineers interact with their designs using a combination of both visual and auditory and tactile feedback.

Advances in Automotive Simulation

Simulation technology in the automotive sector is advancing at breakneck speed, with VR and AR playing a major role in enabling more realistic and comprehensive virtual experiences. Automotive simulations are becoming increasingly sophisticated, allowing engineers and designers to test various scenarios—such as crash simulations, thermal management, and aerodynamics—within virtual environments. Physics-based simulations, one of the important innovations, have been able to give high-level accuracy for real-world conditions. Manufacturers can test vehicles under extreme weather conditions, on different road conditions, and wear and tear without having to physically prototype. As a result, virtual simulations will promote faster decision-making, reduce testing time, and cost savings related to traditional physical prototyping (Dalle Mura & Dini, 2021; Yıldız et al., 2020).

In addition, AI-driven analytics in simulations improve predictive capabilities, giving insights on vehicle performance and safety. The integration of machine learning models into the simulation makes the simulation dynamic because it can adapt to evolving scenarios while offering deeper understanding of vehicle behavior in real-time.

Role of AI in Enhancing VR/AR Applications

Artificial Intelligence (AI) is a critical booster of VR and AR applications in the automotive sector. AI algorithms provide accurate data processing, object recognition, and pattern analysis so that users have a smart and personalized experience. Based on AI insights, the simulation through VR and AR can be provided to a hyper-personalized experience meeting the exact requirements of the designers', engineers', and manufacturers' (Aftab et al., 2020; Aslan & Yıldız, 2020; Wiegand et al., 2019).

In automotive design, AI allows more automation of repetitive tasks like material analysis, component testing, and performance optimization. For example, AI-powered VR and AR systems can make suggestions to designers based on predictive analytics and historical data that gives real-time feedback to constantly improve. In like fashion, AI-driven AR applications aid real-time decisions by overlaying contextual information, giving engineers and technicians actionable insights in the assembly and maintenance process. In addition to that, complex combinations between VR/AR and IoT allow vehicles to be seen with perpetual monitoring and virtual simulators that would be very responsive with such interlinks. Their union supports more efficient predictive maintenance with advanced quality control plus assurance towards enhanced vehicle performance.

The light it shines will be bright about the future of VR and AR in the automotive industry; emerging technologies, enhanced simulation techniques, and AI are innovations. These developments will redefine automotive designing and manufacturing in greater efficiency, precision, and creativity-a new change altogether for good in the auto market. As these trends start to emerge, opportunities for immersive, interactive, and data-driven experiences in automotive applications are likely to grow and form the new era of innovation.

CONCLUSION

Integration of Virtual Reality and Augmented Reality in the Automotive Industry is transforming how a car is designed, produced, and experienced. That includes designing, prototyping, visualization, and even smoothing out the production processes and collaboration among various teams when immersed with technologies that have shaped this new wave of innovation. As discussed in detail in this chapter, VR and AR come with several benefits, such as precision, reduced time and cost inefficiencies, and the ability to create more interactive and intuitive user experiences.

They are destined to revolutionize the very landscape of what may be achieved in the process of developing automobiles. As these come with VR and AR advancements, these would consist of upcoming trends such as 5G integration, AI integration, and most sophisticated simulations. This will give real-time access to data, improve the simulation process, and even make collaboration seamless, ensuring that it has an efficient and dynamic workflow. AI-driven insights in predictive analytics also provide useful feedback for continuous improvement in design and manufacturing processes.

However, these advancements bring along with them challenges that include technological barriers, cost implications, and risks of implementation. Companies will have to invest in strong infrastructure, training, and cybersecurity measures to fully leverage the capabilities of VR and AR. Innovation must be balanced against scalability and adaptability as more businesses adopt these transformative tools.

Thus, the world of cars can be changed in order to become an immersive experience where it drives efficiency, accuracy, and innovation using VR and AR. And for this is the future trend, not only will improvements happen over design and manufacture, but driving experience should head in the direction of an extensively connected, intelligent, and sustainable automobile ecosystem.

REFERENCES

Aftab, A. R., von der Beeck, M., & Feld, M. (2020). You have a point there: Object selection inside an automobile using gaze, head pose and finger pointing. *Proceedings of the 2020 International Conference on Multimodal Interaction*, 595–603. DOI: 10.1145/3382507.3418836

Aouam, D., Benbelkacem, S., Zenati, N., Zakaria, S., & Meftah, Z. (2018). Voice-based augmented reality interactive system for car's components assembly. *2018 3rd International Conference on Pattern Analysis and Intelligent Systems (PAIS)*, 1–5.

Aslan, B., & Yıldız, A. R. (2020). Optimum design of automobile components using lattice structures for additive manufacturing. *Materials Testing*, *62*(6), 633–639. DOI: 10.3139/120.111527

Bale, A. S., Patil, M., Prathima, M., Telgane, S. V., Ghorpade, N., & Anish, B. (2024). Augmenting Knowledge Management with Immersive Technologies: Exploring Transformative Use Cases for AR and VR. *2024 International Conference on Advances in Computing, Communication and Applied Informatics (ACCAI)*, 1–6. DOI: 10.1109/ACCAI61061.2024.10602348

Boboc, R. G., Gîrbacia, F., & Butilă, E. V. (2020). The application of augmented reality in the automotive industry: A systematic literature review. *Applied Sciences (Basel, Switzerland)*, *10*(12), 4259. DOI: 10.3390/app10124259

Dalle Mura, M., & Dini, G. (2021). An augmented reality approach for supporting panel alignment in car body assembly. *Journal of Manufacturing Systems*, *59*, 251–260. DOI: 10.1016/j.jmsy.2021.03.004

Fang, W., & An, Z. (2020). A scalable wearable AR system for manual order picking based on warehouse floor-related navigation. *International Journal of Advanced Manufacturing Technology*, *109*(7-8), 2023–2037. DOI: 10.1007/s00170-020-05771-3

Gouda, G. K., & Tiwari, B. (2024). Unravelling the enablers of Industry 4.0 in Indian automobile industry amid COVID-19: An integrated TISM and fuzzy MICMAC approach. *Journal of Business and Industrial Marketing*, *39*(2), 227–243. DOI: 10.1108/JBIM-02-2022-0067

Jing, R., Yuan, C., Rezaei, H., Qian, J., & Zhang, Z. (2020). Assessments on emergy and greenhouse gas emissions of internal combustion engine automobiles and electric automobiles in the USA. *Journal of Environmental Sciences (China)*, *90*, 297–309. DOI: 10.1016/j.jes.2019.11.017 PMID: 32081326

Manimuthu, A., Venkatesh, V., Shi, Y., Sreedharan, V. R., & Koh, S. L. (2022). Design and development of automobile assembly model using federated artificial intelligence with smart contract. *International Journal of Production Research*, *60*(1), 111–135. DOI: 10.1080/00207543.2021.1988750

Tripathi, S., & Talukder, B. (2023). Supply chain performance and profitability in Indian automobile industry: Evidence of segmental difference. *Global Business Review*, *24*(2), 371–392. DOI: 10.1177/0972150919898302

Vashisht, S., & Rakshit, D. (2021). Recent advances and sustainable solutions in automobile air conditioning systems. *Journal of Cleaner Production*, *329*, 129754. DOI: 10.1016/j.jclepro.2021.129754

Wiegand, G., Mai, C., Holländer, K., & Hussmann, H. (2019). Incarar: A design space towards 3d augmented reality applications in vehicles. *Proceedings of the 11th International Conference on Automotive User Interfaces and Interactive Vehicular Applications*, 1–13. DOI: 10.1145/3342197.3344539

Xiong, F., Wang, D., Ma, Z., Chen, S., Lv, T., & Lu, F. (2018). Structure-material integrated multi-objective lightweight design of the front end structure of automobile body. *Structural and Multidisciplinary Optimization*, *57*(2), 829–847. DOI: 10.1007/s00158-017-1778-1

Yıldız, B. S. (2020a). Optimal design of automobile structures using moth-flame optimization algorithm and response surface methodology. *Materials Testing*, *62*(4), 371–377. DOI: 10.3139/120.111494

Yıldız, B. S. (2020b). The spotted hyena optimization algorithm for weight-reduction of automobile brake components. *Materials Testing*, *62*(4), 383–388. DOI: 10.3139/120.111495

Yıldız, B. S., Pholdee, N., Bureerat, S., Yıldız, A. R., & Sait, S. M. (2020). Sine-cosine optimization algorithm for the conceptual design of automobile components. *Materials Testing*, *62*(7), 744–748. DOI: 10.3139/120.111541

Chapter 10

Augmented Reality-Driven Heads-Up Displays for Revolutionizing Vehicle Driver Assistance

M. K. Srilekha
https://orcid.org/0000-0002-7191-7620
Department of Electronics and Communication Engineering, SRM Institute of Science and Technology, Kattankulathur, India

A. Maria Jossy
https://orcid.org/0000-0001-8522-7025
Department of Electronics and Communication Engineering, SRM Institute of Science and Technology, Kattankulathur, India

S. Hema Priyadarshini
https://orcid.org/0000-0003-1108-6252
Department of Medical Electronics Engineering, Dayananda Sagar College of Engineering, Bangalore, India

G. V. Krishna Pradeep
Department of Mechanical Engineering, Aditya University, Surampalem, India

S. Devi
Science and Humanities, Prince Shri Venkateshwara Padmavathy Engineering College, Chennai, India

Arun Chokkalingam
https://orcid.org/0009-0009-4897-801X
Sona College of Technology, India

ABSTRACT

Augmented Reality (AR)-based HUD (Heads-Up Display) systems have changed the

DOI: 10.4018/979-8-3693-7287-6.ch010

face of driver assistance in vehicles, which significantly improves safety, situational awareness, and driving efficiency. These systems project key information directly onto the windshield, superimposing navigation directions, speed, and collision avoidance alerts in a way that merges the real and digital worlds. AR-HUDs have solved some key problems with a society of current transportation by taking away driver distraction and making reaction times better. Utilizing advanced sensors, machine learning, and real-time data processing, these displays can dynamically adapt to changing conditions and ultimately provide a driver-specific experience. This chapter discusses the mechanism, the design principle, and the technology behind the AR-HUDs with a better effect on accident reduction, increasing the safety on the road overall. It includes current limitations, advancements on the way, and how these are being examined for autonomous and connected vehicles.

INTRODUCTION

Augmented reality (AR) in vehicle heads-up displays (HUDs) is an exciting new feature for the vehicles of today—a technology that will forever augment the interaction between driver, vehicle, and the environment. AR-powered heads-up display (HUD) systems are an impressive example of how AR could provide a gradual blending of the virtual world with the real world by superimposing digital information over your car's view of the road through the windscreen. This new age of cutting-edge innovation, robotiqz, are brilliant all the time and conquer major challenges in driver help with advancement to security, instinct to surrounding knowledge, and diminishing cognitive burden. This is an enormous push towards the future for a savvy transportation framework(Murugan et al., 2022a).

Why not take it a step further and apply the idea of HUDs to other forms of mobility? Over time, the automotive world started to embrace this technology and even to repurpose it, moving on from basic projections of your speed and fuel levels to systems that use augmented reality to provide you with live navigation, obstacle detection, and advanced driver assistance systems. This technologically advanced measure is widely recognized and adopted for the convergence of AR and HUD technologies, where the object's so-called virtual elements allow it to be adjusted in accordance with the driver fitting into the cognitive processes by presenting the vital information for the driver directly in the line of vision, thus lowering the distraction or drastically accelerating the decision-making time in terms of a good performance compared to the conventional way of doing the tasks on the road and hence reducing the considerable number of accidents(Currano et al., 2021).

Augmented Reality (AR)-based HUDs are no more simple systems, barely hardware or software, but merged to work in sync. The hardware typically features high-resolution projectors, optical systems, and also transparent displays that can overlay the digital visuals on the windshield. Those components are paired with a myriad of sensors—LiDAR, radar, and cameras—that constantly gather and process data about what surrounds the vehicle. The software has been trained using machine learning algorithms, in combination with computer vision and real-time processing of environmental parameters, so that the projected information can adapt in real-time. The result is a company with an augmented, contextually aware process providing its drivers with information that is actionable without causing unnecessary disruption in their focus on the road(Winkler & Soleimani, 2025a).

One of the most impressive aspects about AR head-up displays is the route guidance that can be projected in real-time as "direction arrows, lane markings, and destination indicators" on the road ahead of the driver. Conventional navigation systems compel a driver to take their eyes off the road and look at another display, while AR-HUDs tuck these cues right in the driver's line of sight, contributing to safer and better navigation. In addition to this, the systems can also display visual markers—such as alerts color-coded with an overlay that will animate—showing potential hazards such as pedestrians, bicycles, or cars in blind spots of a moving vehicle. In addition to collecting and relaying flight data and notifying the driver of changes to conditions, the system takes a proactive approach to safety, allowing a driver to react quickly to the dynamic environment to reduce the risk of a crash(Doshi et al., 2008a).

The second significant augmenting application of AR-HUDs is adaptive traffic managing and scanning speed. Using real-time traffic data and vehicle-to-road (V2I) communication, these systems can communicate information from increasing traffic congestion and road closures to speed limit changes to help drivers. An AR-HUD, for example, could display a live speedometer that adjusts to the different speed limits or suggest possible alternate routes to circumvent traffic. This does not only make the driving more comfortable but also helps to reduce traffic and reduce fuel consumption(Doshi et al., 2008b).

Beyond their functional benefits, AR-HUDs represent a future battleground in user experience in connected and autonomous vehicles. With automotive brands placing more focus on smart, connected vehicles, AR-HUDs represent an important next step in providing intuitive, contextual information to drivers and passengers. For example, in semi-autonomous driving modes, an AR-HUD can help maintain driver initialization and awareness of the driving scene through which the car is traveling—such as when lane changes and adaptive cruise control are active, etc. Moreover, these systems can enhance that experience using augmented visuals to entertain passengers on the ride(Liu & Wen, 2004).

While they could revolutionize driving, AR-HUDs aren't problem-free. One reason the technology is not widely adopted is the high development and manufacturing expense. This is no easy mission, requiring the seamless integration of hardware and software elements, in which not just must the data be transferred without degradation of quality and integrity whilst the conditions of bright sunlight and torrential rain, amongst other things, change. Also, the enhancement and reduction of distracting details from AR-HUD is a challenge to address for AR-HUD user adoption(Chouksey & Sirsikar, 2016).

The future of AR-HUDs is full of possibilities, accelerated by advances in artificial intelligence (AI), cloud mash-ups, and sensing technologies. These systems are used to improve vehicles through advances in training AI (artificial intelligence) for the transition from human drivers to fully autonomous vehicles and serving as an intermediary between human drivers and machine intelligence. In the case of AR-HUDs, they can support self-driving car adaptation as they can offer real-time explanations of the decision-making of the car, so users become comfortable and trust these technologies. Also powered by 5G connectivity as well as edge computer technologies, AR-HUDs would also be capable of obtaining and analyzing large volumes of data from external sources, elevating their flexibility in function and performance to a whole new stratosphere(Murugan et al., 2022b).

Hence, it also marks a paradigm shift in vehicle driver assistance, an incredible metamorphosis that perfectly exemplifies how technology will continue to combine with things that will matter more and help make the use of the vehicle as a whole have a more positive effect on our day-to-day life, adopting safer, easier, and more entertaining life on the road. They not only address his longstanding problems of distraction and situational awareness, but they also give him vital information in his line of sights, ushering in a new automotive age of intelligence(Bark et al., 2014). As the automotive industry is shifting towards innovative approaches, AR-HUDs will find their realm in connected & autonomous vehicles, and they have the potential to revolutionize how we perceive mobility and our relationship with technology.

Objectives

To explore AR-based HUDs role in the evolution of vehicle driver assistance. Here we use this data to examine the state of hardware and software adoption that AR-HUDs are capable of in order to deliver real-time data, seamlessly fused. This study aims to analyze the impact of AR-HUDs on the safety of the road, awareness of the driver, and prevention of accidents. It also highlights current obstacles such as technical constraints and hurdles that consumers must conquer before looking ahead to emerging trends that could make the correlation between AR-HUDs and autonomous and connected vehicles more concrete. Feed to promote future developments in AR-HUD technologies and innovative trends for smarter, safer, and more sustainable transportation.

TECHNOLOGICAL COMPONENTS OF AR-HUDS

Figure 1. Technological components of AR-HUDs, focusing on hardware and software integration

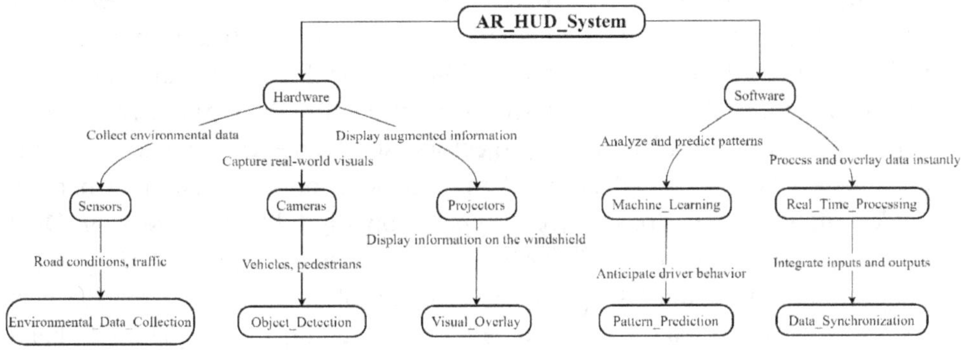

Figure 1 depicts the technological components of AR-HUDs, focusing on hardware and software integration. It explores the structure of AR-HUD systems, focusing on their hardware components like sensors, cameras, and projectors, and their interaction with software like machine learning and real-time processing(Ma et al., 2021).

AR Hardware: Sensors, Cameras, and Projectors

AR allows for these types of heads-up displays, but the hardware aspect of those systems is where the function comes from. They depend on the use of multiple components—sensors, "cameras, and projectors" —that each serve a distinct and specific purpose yet work together to create a responsive, intuitive experience. AR-HUD systems are powered by sensors that collect real-time data from the environment around the car. These sensors are based on LiDAR (Light Detection and Ranging) for accurate distance measuring (some of these), radar for detecting objects in different weather conditions (other types of the sensors), and GPS modules for accurate positioning observation (the last category of these). Automated vehicles observe vehicle dynamics, the distance to obstacles, and the road surface using ultrasonic sensors, inertial measurement units (IMUs), and other sensors in their effort to improve environmental awareness as well. When used together, these sensors deliver a comprehensive view of driving surroundings, which allows the system to react to real-time changes(Charissis et al., 2021).

Cameras—these are the eyes in this augmented-reality reality—the eyes in place to assemble the place data and feed the pictures into the AR-HUD engine. Forward-facing cameras watch the road ahead, spotting lane markings, traffic signs, and vehicles. This feature gives you a view above the car so you can see how close you are to another car during parking or a potential blind spot. Infrared cameras are additionally used for night vision to help spot pedestrians, animals, and other objects in low-visibility conditions. These cameras are combined with machine learning algorithms that use the data to generate actionable insights that the driver can then see on a display. The AR experience works by projecting the digital overlays onto the windshield through projectors. Its projectors use advanced optics to display even the most intricate details clearly, regardless of the level of illumination. Depending on the design, they use various projector types, from digital light processing (DLP) to liquid crystal on silicon (LCoS) or laser-based systems. The projection system projects the digital content along the driver's line of sight, minimizing the need to look away from the road. Advanced calibration techniques, coupled with head-tracking systems that monitor the driver's position, allow these displays to remain aligned so that the driver's constantly changing line of sight is in line with the AR imagery as the vehicle proceeds through the real world.

All these hardware elements come together to guarantee that the AR-HUDs provide drivers with accurate, real-time, and contextually relevant data. Not only does the integration of sensors, cameras, and projectors improve the sense of context, but it also reduces distractions, leading to safer and smoother navigations. The continuous development of these components will lead to better and improved AR-HUD technology.

Software Integration: Machine Learning and Real-Time Processing

The emergence of soft technologies such as machine learning (ML), real time data processing, and integration of these soft technologies is creating opportunities to significantly enhance the usability and performance of augmented reality heads-up displays (AR-HUDs). These systems are software intensive, absorbing data from myriad sensors, cameras, foreign networks, and other inputs to manage thousands of pieces of information in a way that generates meaningful output. The algorithms that are presented in this field are the simplest forms that are the basis to build modern-day artificial intelligence, where data is received and processed based on the driver behavior, the road networking, and basics of the environmental interactions. Such intelligent features develop by ML models that analyze historical data and real-time inputs, allowing AR-HUDs to guide drivers in a highly precise and contextual way, such as by using adaptive navigation, dynamic lane assistance, and collision notifications(Blankenbach, 2023).

Real-time processing is another integral aspect of AR-HUD software embedding. For instance, distance to obstacles and speed of vehicles must be reactively processed by the system in a matter of milliseconds, and visual representations need to be built in many cases to mitigate situations such as a crazy ride due to a laggy system. AR-HUDs rely on high-performance computing platforms as well as optimized data pipelines that can ultra-fast transmit and process data. Real-time sensor fusion would fuse the output of different providers (LiDAR, radar, GPS, vehicle sensors, etc.) into a single, unified, accurate representation of the context. The driver explores this information while it's being projected onto the driver's field of vision so that the information can be brought into direct availability to real-life environments, opportunities, risks, etc. Other than basic heads-up displays, augmented-reality heads-up displays can provide adaptive configuration, meaning that the software can modify the system based on driver preferences and the driving conditions. For instance, the machine learning models can adapt the display based on lighting, weather, or even the driver's cognitive load to show that content in a way that it will be at a maximum visibility and usability point. AR-HUDs are also increasingly adopting V2X communication thanks to an ADAS (advanced driver-assistance systems) acceleration of the V2X, which enables them to easily process external data (traffic signals, surrounding cars, roadwork updates, etc.) seamlessly in real time.

Software-powered features will not only improve the safety and comfort of AR-HUDs but also prepare their adoption in sophisticated ecosystems like connected and autonomous vehicles. But with these innovations also come challenges, including the necessity for cybersecurity systems to help safeguard against threats and pseudonymization across a number of automotive platforms. However, the progress in machine learning and real-time processing continues to evolve to provide a final consistent advancement to the AR-HUD, which will be an indispensable component of modern and future automotive systems.

Display Technologies: Optical and Digital Integration

Navigating AR-powered Heads-Up Displays (HUDs) also leans on display technologies that augment virtual data overlaid on the real-world ambiance. Today's AR-HUD systems depend on the marriage of optical and digital information to supply drivers with accuracy and context to the images they're seeing. Optical integration is relatively sparse, using a combiner screen and waveguide optics and (when the display is external) a holographic display to project images on the windshield or on a dedicated transparent display. In these systems, light-guiding techniques are used to display computer graphics content with minimal distortion and high data and environmental brightness. Holography optical components can reduce each optic size with no image quality trade-off(Winkler & Soleimani, 2025b).

Digital Integration, on the other hand, is apt for managing real-time data flows from many sensors, such as cameras, LiDAR, GPS, and moving vehicle sensors, and also for generating dynamic content. Advanced algorithms process this information to overlay navigation directions, speed limits and hazard warnings in the driver's line of sight. Technologies such as DLP (Digital Light Processing) and TFT (Thin Film Transistor) LCD are used to create the high-resolution projections with brilliant colors and sharply contrasting images. A display system also employs adaptive brightness adjustment controls and anti-reflective coatings to consistently provide great readability across environmental use cases, such as bright sunlight on the road or driving at night.

It's the optical and digital technology hybrid that allows AR-HUDs to display precise, contextual information while ensuring the driver is not overly distracted. The evolution of AR-HUDs will certainly be augmented by gaps in microminiaturization and waveguide technology (e.g., micro-LED displays, quantum dot technologies, better waveguides) in the coming decades. Not only does this enhance the driver's situational awareness, but it also takes a greater role in making its way into autonomous and connected vehicles, using visualization of data in use instantaneously to ensure safe and ultimately productive operations.

Figure 2. Functional features of AR-HUDs

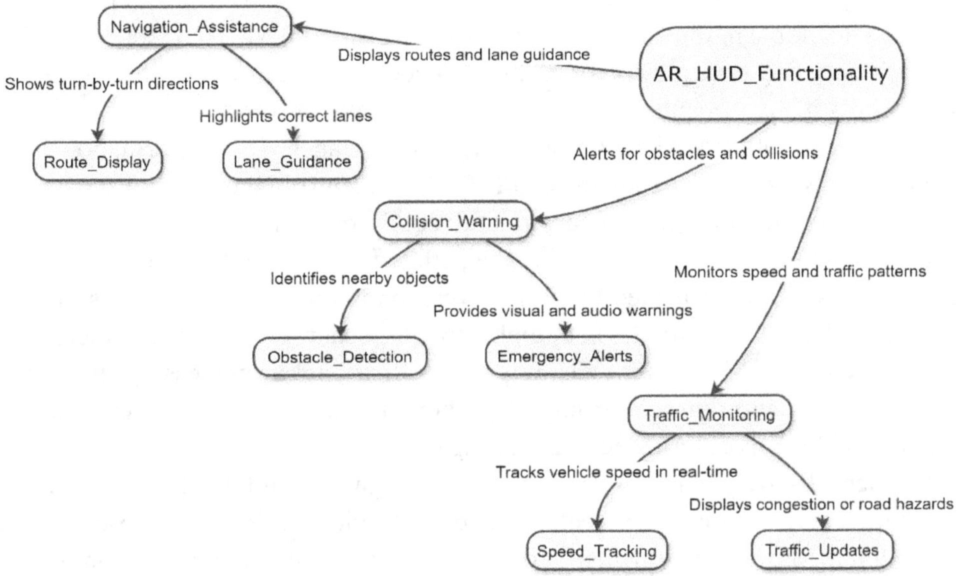

FUNCTIONAL FEATURES OF AR-HUDS

Figure 2 showcases the functional features of AR-HUDs, highlighting how each component enhances the driving experience.

Navigation Assistance and Lane Guidance

With indispensable functional features like navigation assistance and lane guidance, augmented reality (AR)-based head-up displays (HUDs) play a vital role in driver comfort and road safety. These systems use AR technology to project items like arrows for navigation and lane information directly onto the windshield in the driver's field of vision in real time. AR-HUDs can be installed in the driver's view line, allowing digital content to be superimposed onto the physical view of the road, thus not requiring the driver to look away to another screen or another device, reducing distractions and enhancing situational awareness(Jiang & Guo, 2023). Usually, AR-HUDs provide additional information to enhance navigation, such as directional arrows, zone or line maps, and distance displayed over the road ahead. This dynamic direction-finding employs a synergy of GPS systems, digital cartography, and vehicle sensors to ensure precision and immediacy. The position, heading, and speed of the vehicle are constantly fed into the map in real-time, enabling the system to generate dynamic representations and give real-time turn-by-turn navigation, which can adjust to the changing on-road conditions. For example, AR-HUDs can visualize the correct routing on highways or highlight where the next turn is so that the driver doesn't miss it, significantly lowering the risk of driver errors.

Another essential AR-HUD function is lane guidance. Uses information from advanced sensors such as cameras and LiDAR to identify lane markings and road borders, including low-light or heavy rainfall conditions. Visual prompts tailored to each lane — for example, highlighting the path to follow · or adding boundaries on the lane lines — are projected to assist drivers to maintain appropriate positioning inside their lane. For example, in complex traffic scenarios such as lane merges or an intersection, AR-HUD provides improved situational awareness by indicating the best lane in addition to warning of possible hazards, such as the presence of vehicles in blind spots or sudden lane changes.

Further AR-HUDs can utilize traffic sign recognition and display speed limits, stop signs, caution signs, etc. in real time to comply with traffic rules. These systems also might incorporate augmented prompts, such as illuminating pedestrians or cyclists, making the roads even safer. Navigation assistance and lane checking for the driver not only simplify the driving experience but also act as a preventive measure against accidents by easing driver route action and improving decision-making. Through these improvements, AR-HUDs will evolve, creating better-integrated and safer transportation systems of the future, including predictive navigation, integrated with autonomous driving guidance.

Collision and Obstacle Warning Systems

Augmented reality technology, better known as AR technology, allows for enhanced perception from the driver's side in the areas of collision avoidance systems and obstacle warning systems; hence it is one of the vital steps towards vehicle driver assistance. This is where AR-HUDs are relevant, as these systems feed real-time, context-sensitive warnings straight into the driver’s line of sight, so response times to warnings can be reduced, or in some cases the driver can avoid them entirely. These systems make use of data from sensors, cameras, LiDAR, and radar for the detection of potential road hazards, which may include other vehicles on the road, pedestrians, cyclists, and obstacles such as road debris or construction barriers. Using an AR-HUD overlay, the apparatus overlays caution messages on top of hazards a driver would recognize via the windshield, so a driver can react in real time, as all of those visual data elements collapse into one signal/real-world component(Xu et al., 2023).

In cases where the sensor system identifies a potential collision, the AR-HUD can overlay navigation opportunities within the existing infrastructure over the actual environment to create a visual display of the hazard, such as a warning icon, arrow, or color-coded system (where red signals an immediate threat with yellow indicating potential danger). These notifications, displayed both in real time and synchronized to the driver's vantage point in manners easily digestible at a glance without taking the driver's eye off the road. In extremely advanced systems, HUD may even attempt to predict dangerous trajectories (i.e., if a side-on collision is plausible) and recommend maneuvers that the driver can perform, such as turning away from the obstacle or decelerating. The depth perception and spatial awareness are two of the most important features of AR infotainment systems when integrated with warning systems. By using both visual data from the sensor in the vehicle and from the AR display, the system can determine the distance between the vehicle and any obstacles and then visually display the real threat on the AR display itself. This spatial feedback serves to mitigate superordinate panic while simultaneously showing the driver how serious the event is. These systems may also adapt to different driving conditions, such as tailoring their warnings for the speed, conditions, and type of road.

Adaptive Traffic and Speed Monitoring

Key Functional Feature of AR-Powered HUDs Adaptive traffic and speed monitoring is probably one of the most critical functional features of AR-powered heads-up displays that can make driving much safer and more enjoyable. This feature uses sensors to ascertain real-time information of road traffic and the speed of other vehicles and sends visual messages that are more relevant to the current situation with increased vigilance to guide the driver through various situations. This monitoring process forms a closed control loop where the sensor information about the traffic environment is obtained from the sensors in the vehicle, cameras watching the vehicle's surroundings, and external sources of information such as GPS, cloud-based traffic feeds, among others. The driver then views the more pertinent information displayed onto the windshield with an active view of the current traffic and speed limits. Adaptive speed monitoring is also highly essential, making sure the drivers are at the legal speed limits while also adjusting for the varied terrains and weather conditions(Ding & Li, 2024). In practice, AR-HUDs can proactively highlight or flag objects in your view based on the real-time information for the driver. For example, as you approach a reach of lower speed, such as a construction area or the front of a junction, the new speed limit information is automatically highlighted in your view, giving you a visual of your surroundings. Moreover, the HUD can also flash visual warnings when one exceeds the legal speed limit. This aspect is also critical not only for urban roads but also for high-speed highway driving, where a reaction time of less than one second may be the difference between an incident and safe driving.

The system's traffic flow analysis helps drivers make better decisions in advance, allowing them to anticipate bottlenecks, slow points, or potential hazards ahead. For example, these visual cues contribute to making better and more timely decisions regarding what maneuvers to perform, as well as how to adjust your driving in relation to traffic density, obstacles ahead, or sudden lane changes. For example If the vehicle in front of the car suddenly slows down or shifts lanes, the HUD may bring attention to that to provide an early warning to the driver to react. As this technology matures, its amalgamation with vehicle-to-vehicle (V2V) and vehicle-to-infrastructure (V2I) communication networks will further improve the accuracy and the promptness of the adaptive traffic and speed monitoring, as this feature will emerge as a cornerstone in the pursuit of safer, smarter transportation systems.

Figure 3. Impact on driver safety and experience

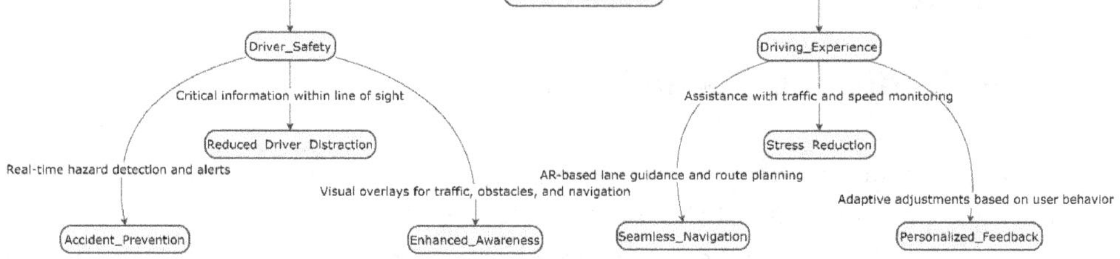

IMPACT ON DRIVER SAFETY AND EXPERIENCE

Figure 3 demonstrates how AR-powered HUDs enhance safety and driving experience by incorporating real-time information, intuitive navigation, and personalized features.

Reduction in Driver Distraction

Perhaps the most important way that an AR-powered Heads-Up Display (HUD) affects driver safety is the elimination of driver distraction. Conventional dashboard displays and navigation systems frequently demand drivers turn their heads away from the road to look at the infotainment screen or the instrument cluster, which can slow their reaction time and result in accidents. Unlike such traditional HUDs, AR-HUDs overcome this with hardware that shows important info on the windshield itself, keeping the driver looking ahead while having all the needed info right in front of their line of sight. AR-HUDs provide information with less cognitive burden on drivers through not requiring them to divert attention from the road to separate displays, allowing them to remain aware of their surroundings without being distracted by irrelevant data(S. Wang et al., 2017; W. Wang et al., 2018).

These are able to present the driver with real-time navigation directions, speed limits, upcoming turns, and hazard warnings in a way that is directly along their prescriptive line of sight across the road in front of them, granting variable access to the most important information at the most relevant time. When the vehicle is approaching an intersection or complex turn, the embedded HUD is capable of projecting the intent of the upcoming route on the windshield in real time, enabling drivers to follow where they need instead of searching for directions on the separate second screen. This makes the driving experience easier as well as reduces the tendency to second-guess themselves based on last-minute or vague information. AR-HUDs offer other advantages, such as the ability to prioritize the data presented based on what is most relevant to the current context, with more salient information, for example, a looming hazard, being given more weight than basic changes in the driving environment. Whereas the existing systems can only indirectly prompt by auditory signals, if an object is detected ahead, the system can immediately project a visual warning on the windshield. Elements that are data-driven spring to life just to represent what's needed in that moment and to minimize any visual clutter so we can focus on the issues that really matter.

Enhanced Situational Awareness

AR-Weather in Augmented Reality (AR) significantly improves situational awareness, providing context-aware data for drivers through heads-up displays (HUDs), which have become ubiquitous in everyday life. This change lets the drivers focus attention on the road and glance at a display on a dashboard, navigation system, or other instrument. AR-HUDs project key information—about upcoming turns, speed limits, traffic signs, and hazards, among other things—directly onto the windshield to help drivers stay focused on their surroundings while imparting the information needed to make better decisions. By linking digital content to the real world, drivers can remain connected to near-distance and far-distance environments alike, and in complex driving situations such as merging on highways, passing through an intersection, or traversing an area dense with obstacles(Langlois & Soualmi, 2016; S. Wang et al., 2017).

Reduced Response Time in Emergencies One of the key advantages of improved situational awareness is the reduction in response time in an emergency. AR-HUDs can build on those that would otherwise not be found in the expensive information bulk that is making the driver easily shy away from a footman trying to cross the road with the aid of a car blocking its way and where the blind spots are. How can it be most reappearing, and when would that dreaded—glued-to-the-road-eyes moment—all AR-HUDs helping the driver keep their eyes on the road at more speed. That kind of ahead-of-the-curve hazard detection is as important for preventing crashes and decreasing injuries as collision-avoidance technology. And through constantly rehashing and relevantly displaying, in situational context, the most useful and relevant information regarding, and filtered by, the driver's position and other local traffic conditions, AR-HUD is, for the moment, a situationally aware, tailor-made experience for the driver. Example such as with a twisting or sudden alteration of the road topology in front, the AR system will adapt how the information is shown so that the driver has sufficient latency to psychologically prepare for the environment decomposition.

In addition to preventing crashes, AR-HUDs offer a more relaxed way of driving. Drivers can obtain a feeling for control and situational awareness even in late-breaking circumstances similar to a dense fog or dense visitors. Arkitech says the AR-HUDs contribute to an experience that is safer and less stressful by providing an intuitive and streamlined interface that reflects the driver's natural behavior. These systems are being improved further and, as a result, will become the drivers of road safety (pun intended) around awareness and responsive drivers.

Contribution to Accident Prevention

AR-HUDs are an easier way to drive, on top of preventing accidents. You can understand the sense of control and situational awareness even in conditions that require late braking, such as a dense fog or heavy traffic. As per Arkitech, AR-HUDs keep the driver in their natural behavior through a safe and less stressful experience with an intuitive yet streamlined interface. As these systems develop, there will be synergistic mechanisms put in place where they've become an influence for driving on the road safely and a help around perception and agile drivers(Langlois & Soualmi, 2016; Liang et al., 2023).

AR-HUDs have proven to minimize road accidents quite effectively. Routed through high-definition sensors, cameras, and radar systems, the HUD can detect surrounding cars, pedestrians, or obstacles, and, when any of these represent a potential threat of collision, a visual cue is immediately provided on the windshield. There are warnings displayed on the vehicle when it approaches another vehicle too closely or when they are near the edge of the road, which need to be corrected by either braking, adjustment of steering, or its speed. These early warnings significantly reduce the chances of rear-end collisions, side-impact collisions, and lane-exit crashes. Apart from collision avoidance, AR-HUDs also make travelling safer through lane-keeping assistance. This comprises visualizing lane boundaries and providing warnings to the driver to help prevent drifting into another lane, especially useful in such conditions as poor visibility, sharp turns, and driving at high speeds in a highway environment. This feature is useful for driving at night or during bad weather when it's harder to see road markings.

Potentially, AR-HUDs will also help prevent accidents by allowing them to relay real-time alerts about road hazards (traffic jams, road construction, icy patches). The advanced nature of AR-HUDs allows drivers to adapt their driving behavior by receiving information about a potential hazard, even long before its approach, minimizing the potential threat of inefficient traffic flow and potential collision with sudden obstructions on the road. That potential will only increase as the technology evolves, as we're sure will be the case for autonomous systems and vehicle-to-vehicle (V2V) communications.

Figure 4. Real-world applications of AR-HUDs

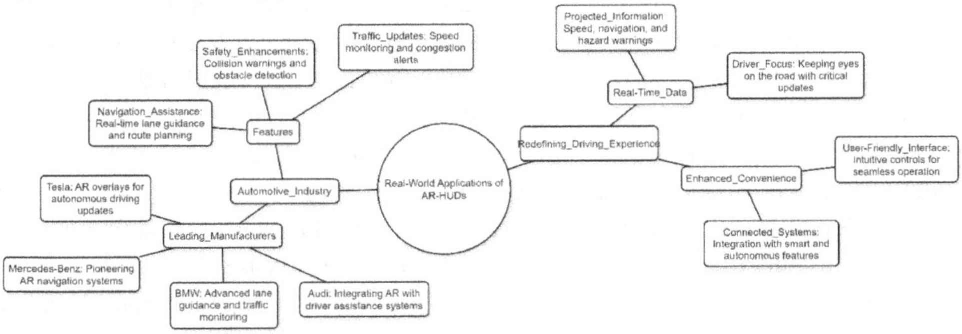

REAL-WORLD APPLICATIONS

Many leading automotive brands are pioneers in the use of Augmented Reality (AR) Heads-Up Display (HUD) in their vehicles, and they do so for practical utility to enhance driving safety, navigation, and comfort. Take the Honda and Jaguar companies, which have redirected AR-HUD technology and made driving a unique experience by grouping these features, projecting them into the driver's view so that the driver keeps their focus on the road, bringing real-time updates. Figure 4 shows AR-HUDs in various real-world applications along with specific manufacturers and their integrations/features, which provide different anomalies to the driving experience(Ding & Li, 2024; Winkler & Soleimani, 2025b; You et al., 2024).

For instance, the BMW 7 Series was the first car from a carmaker to have AR-HUDs. The next level would be BMW's AR-HUD system that lays navigation directions, warnings, and other important information right on the windshield, seamlessly creating a layer of virtual data right there in the real world. This places features such as traffic, vehicle speed, or distance from other vehicles on such a large screen that, in effect, it guides the driver with minimal distraction. This is not to mix it with other heads-up display systems; it is also based on such info that the AR-HUD on the BMW 7 Series is a more precise alternative that offers augmented reality, reaching a larger surface and providing a simply larger and immersive area compared to classical HUD. Similar to in-car systems for other driver-assistance systems like adaptive cruise control and lane-keeping assistance, BMW's augmented reality technology is intended to assist in keeping the driver aware of their environment.

Audi has also adapted this AR-HUD technology in their automobiles, such as the Audi Q7 and Audi A8, which would make the drivers relate towards their better guidance. Audi debuted AR-HUD in 2019 as part of the AI:ME concept and brought the tech to market here in the production-ready Q4 e-tron, which has basic functions in terms of showing speed, navigation, and safety prompts with fancy graphics to match, but what brought this tech to life in both the show car world and the production car world was its rules for what to show, how to show it, and how to overlie the real environment in everyday scenarios. Audi's dynamic path-planning system, for example, projects navigation instructions while merging the vehicle's position on a road with real-time traffic updates. This visual input helps users anticipate turns and upcoming junctions, leading to smoother and safer driving.

Another heavy hitter, Mercedes-Benz seamlessly incorporates augmented reality functionality into their Mercedes-Benz S-Class series AR-HUDs for a very tailored and sophisticated experience. For an additional layer of assistance, the navigation system overlays arrows to guide drivers on the right path, allowing drivers to make their way to destinations that may be confusing in other vehicles. Additionally, it enhances driver situational awareness through critical safety alerts presented directly as heads-up on the windshield by warning the driver of surrounding pedestrians and other vehicles to promote quicker, more efficient responses to dangerous situations.

These technologies provide a varying degrees of sophistication, integration, and experience where manufacturers are comparing themselves against technologies offered by third parties. All the while, BMW and Audi zero in on the ability to render higher fidelity, more tempting displays, whereas Mercedes-Benz might favor a more delicate and elegant experience, incorporating aspects from the virtual and physical realms for dialed-down, effortlessly easy cognitive processing. Augmented-Reality Head-Up Display: The AR-HUD system from BMW features a wide, high-resolution area display that occupies a more extensive part of the windshield, providing assistance during long-distance driving, where the information is required malevolently. Whereas Audi's systems for AR-HUDs are more about accurate alignment, so the data appears to be part of the road.

The most prominent difference is the level of system integration. BMW and Audi have paired their AR-HUDs to advanced driver-assistance systems (ADAS)—(adaptive cruise control lane-keeping assistance automated braking)for another complete take on vehicular safety. The company, however, says the system is intuitive and will adjust based on any alterations in driving conditions, providing the right information to suit an individual environment.

As the technology of the holographic display module evolves, we are likely to see more advanced display quality, seamless integration with autonomous vehicle systems, and new features like vehicle-to-vehicle communication, enhancing the driving experience by making it safer and more convenient for drivers. While the systems' exact capabilities and displays differ by manufacturer, they all are ultimately heading toward the same destination: greater safety, navigation accuracy, and a connected, intuitive driving experience.

CHALLENGES AND LIMITATIONS

The same is true for Augmented Reality (AR) Heads-Up Displays (HUDs) in the automotive segment; much great development and opportunities have been created with their use, but there are so many obstacles and limitations that are not yet allowing them to reach a broader audience and get them to work perfectly. This is because there are various barriers you will face, ranging from technical barriers, costs, environmental and lighting conditions, to user adoption and learning curve(Murugan et al., 2022a).

Technical Barriers and Costs

The cost and complexity of systems is perhaps the most difficult challenge to overcome for AR-HUD systems. Real-time data processing of advanced glass components, high-resolution liquid crystal displays, and sensors from liquid crystal displays and sensors requires complex hardware and software. That's a lot of engineering under the hood to deliver a compact and lightweight AR-HUD that can seamlessly and accurately project images wherever it is desired. Ditto for active safety technology: Even if one car model embraces it, automakers can recover costs through add-on prices for buyers, and therefore consumers still will have to pay for the systems because that price premium should have already been factored into the prices for high-end cars. For mass-market adoption, production cost is a technical hurdle that must be lowered. This would require massive reductions on the part of automotive manufacturers for it to be ubiquitous.

Another technical challenge of AR-HUDs is projecting virtual information onto real-world objects with live accuracy, without distortion. VocentoGlobe has been an analyst, and in this space you need the best sensors and processors to be able to process millions of data every second. As lovely as this sounds, the translation is highly involved; synthesizing information from many data streams – cameras, radar, GPS, etc– with the required synchronism to prevent projection errors is still a huge problem. Hitting just the right amounts of clarity, brightness, and contrast is particularly challenging on the display because it has to be legible under differing driving conditions, as well as a range of lighting conditions.

Environmental and Lighting Constraints

Environmental factors such as lighting conditions, road surface, and weather lead to formidable challenges in the effectiveness of AR-HUD systems. In bright sunlight or in case of glare, for example, the driver will not be able to properly see the projected information on the windshield if the specific display lacks adaptive brightness capability. But while some AR-HUDs include sensors that can adjust the display based on the instance, displaying perfectly in all situations is somewhat of a technical conundrum(Chu et al., 2008; Okabayashi et al., 1989).

The AR-HUD systems also face an extra challenge from weather conditions (rain, fog, snow, etc.). These considerations can cause the data to be misconstrued or concealed, making it susceptible to incorrect logic, which can be dangerous to follow. But if the windshield of the vehicle is soiled or covered with snow or frost, that would present a challenge to the optical clarity required for AR projections to function. In these conditions, it can rain, and the AR-HUD becomes less effective, possibly leading to the driver missing out on critical alerts or navigation index, which come in too late. Another environmental constraint is due to the physical properties of the windscreen of the car. The curvature, reflection, and coating of the windshield affect the perception of the image that is projected onto it. Even a slight misalignment or distortion of the display can be distracting and confusing for the driver, thereby undermining the safety advantages of heads-up devices.

User Adoption and Learning Curve

User adoption and the corresponding learning curve are also challenges faced by AR-HUDs. With the safety and convenience gains the new technology aims to offer, some drivers may be challenged by the concept of this new mode of interaction. AR-HUDs require drivers to see and react to overlays visually in real time, which can lead to cognitive overload for some users—especially in high-stress moments or among users less exposed to the technology(Doshi et al., 2008b; Maroto et al., 2018).

And certainly, the idea of a display shining virtual information onto a driver's line of sight is not yet universally accepted, as it can all too easily be viewed as a distraction or an added complication. AR-HUDs bypass the frustration of the traditional straight-line HUD that was nothing but a voiceless group of data, such as speed or fuel levels, but this kind of information became intuitive enough not to require integrated visual aids, and yet the drivers should prepare for the upcoming information set they'll have to process; augmented navigation routes or danger warnings may end up being overwhelming or even contradictory, so a well-designed AR-HUD interface would remain a major part of the task. It will take a little bit of a learning curve to adopt this technology, and good UI design will help to make this as smooth as possible. If the system is not intuitive and user-friendly, it can lead to frustration or even misuse of the advanced features by drivers. Hence, manufacturers need to provide training material and an onboard short introduction to the system to users illustrated in the car. But even with that assistance, the learning curve for AR-HUDs can be so steep that some will never fully adopt and use them.

In short, AR-HUDs are not science fiction because they can improve driving safety and comfort, but technical, environmental, and broad-based user challenges remain to be overcome. Overcoming these hurdles will require continual iteration of software and hardware and lowering costs, as well as a human-centered automotive experience so that AR-HUDs become ubiquitous in any class of vehicle; this is a road to maximizing its potential for drivers of all walks of life.

FUTURE DIRECTIONS AND INNOVATIONS

Future trends concerning Augmented Reality (AR)-powered Heads-Up Displays (HUDs): transformative innovations, enhancing connectivity with autonomous driving, AR integrated with advanced cloud and AI technologies, and connected and smart vehicle evolution. The hope is that these improvements will not only make AR-HUD systems far more effective but, at the same time, change the entire experience of what being in a car is about(Jing et al., 2022; Park et al., 2013; Winkler & Soleimani, 2025b). The future of Augmented Reality (AR)-powered Heads-Up Displays (HUDs) is depicted in Figure 5.

Figure 5. Future of Augmented Reality (AR)-powered Heads-Up Displays (HUDs)

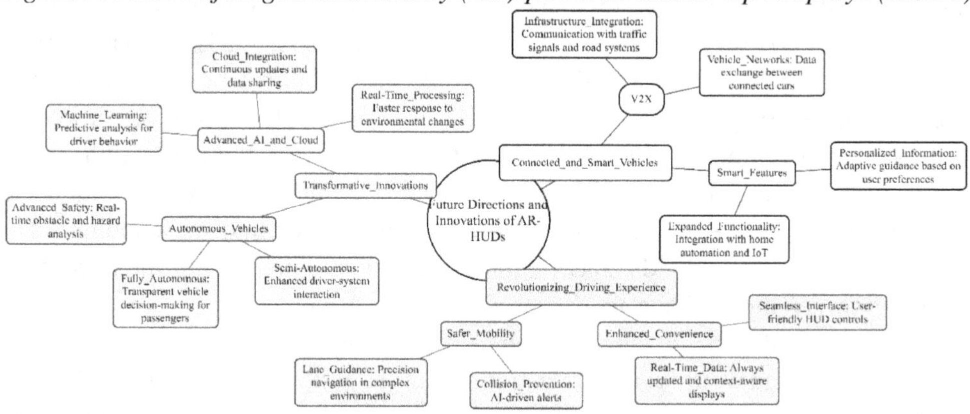

Integration with Autonomous Vehicles

As we head toward the ultimate goal for the automotive industry of fully autonomous vehicles, AR-HUDs will be a vital part of that journey. With autonomous vehicles, the driving task changes from controlling a vehicle to interacting with the environment around it. By 2030, AR-HUD systems will be high-level assistive technologies that will all be used as a means of giving feedback and information to passengers and operators in semi-autonomously driven vehicles(Doshi et al., 2008b; Okabayashi et al., 1989).

The evolution of AR-HUDs in fully autonomous vehicles will likely see them move from a driver-focused role to one focused on the interaction between the passengers and the vehicle. Such systems could display key information about the vehicle's environment, upcoming maneuvers, or automated driving decisions, enabling passengers to grasp and trust the vehicle's behavior. For example, the AR-HUD could show the vehicle's awareness of obstacles, the route planning, or upcoming action decisions, such as lane changes and stops. This transparency will boost passenger confidence in the vehicle's capabilities and contribute to a smoother and more comfortable autonomous travel experience. Even in semi-autonomous vehicles, where human drivers retain the capacity to intervene, AR-HUDs can provide benefits by developing the situational awareness necessary to make sense of the autonomous system's decision-making processes. For example, if the situation is abnormal (e.g., road conditions) or the system needs human intervention, the HUD can highlight the situation and guide the driver regarding how to respond.

Use of Advanced AI and Cloud Technologies

Cloud technology integration with new-age AR-HUDS with advanced artificial intelligence (A.I.) will further help in the scalability and efficiency of the grid system. AI can analyze that vast storehouse of data collected through the array of sensors, cameras, and radar built into a vehicle in real time so that the HUD can deliver highly relevant, context-aware information tailored specifically to the driver. Using environmental data, AI can optimize the display—ensuring that relevant data is shown to the driver or passengers at any given time(Chu et al., 2008; Winkler & Soleimani, 2025a).

It can also use traffic, historical behavior, and road conditions to predict driving patterns. This ability to predict the near-future environment would allow AR-HUDs to deliver proactive advisories such as upcoming traffic jams, pending hazards, or blockages. Such a system could also possibly preemptively alter directional advice based on traffic patterns, so as to prevent motorists from hitting hot spots of congestion or delay. With the advent of cloud technology, it is now possible to update AR-HUD systems instantly. Believers say that AR-HUD systems will receive real-time updates and synchronize data with cloud servers. By pairing with the cloud, AR-HUDs can be fed with up-to-date data—traffic or weather updates or warnings of road hazards—displayed in the AR realm, helping to keep content on the display up to the nanosecond. Cloud connectivity can also harness vehicles' ability to communicate with one another (V2V) and with the infrastructure (V2I), enabling more communication and situational awareness across a fleet of vehicles.

Expanding Applications in Connected and Smart Vehicles

As connected and smart transportation ecosystems continue to mature, AR-HUDs will further be adapted into more use cases. In chatting networked cars, the HUD will also be able to use information from infrastructure surrounding it (traffic lights, road sensors, parking management), for instance. Making your drive more productive and less dangerous—advice on optimized parking, in-route optimization taking into account traffic as well as the availability of chargers for EVs(Winkler & Soleimani, 2025a; You et al., 2024).

In smart cities, AR-HUDs would allow drivers to know about the availability of public transit, points of interest, or surrounding amenities that would come to be intelligently processed before the driver in real time. It seems as if the system could show information about what the driver is seeing, what cyclists and pedestrians look like, how far apart destinations are, and maybe integrate with public transport schedules, instead of clumps of discrete cars positioned as though on neighboring pasturelands. As cars evolve into more autonomous vehicles and make increasing use of connected global systems, AR-HUDs could serve as a bridge between the driver and the wider world. For example, the AR-HUD of the future could potentially connect with home automation systems and keep track of the security status of the house, energy consumption, and operational status of appliances while driving. It might even hook into the driver's calendar, displaying appointments or reminders and guiding the driver to the next destination based on live traffic or appointment schedules.

Hence, the AR-HUDs technology is creating an era of opportunities with advancements in the urinary vehicles with the integration of AI and cloud technologies, thus realizing connected and smart vehicles. This means AR-HUDs will not only enhance the driver experience but also comprise an integral part of the innovations that will define the future of mobility—evolving into safer, smarter, and more personalized transport solutions.

CONCLUSION AND OUTLOOK

AR-based HUDs (heads-up displays) are a disruptive technology for automobile applications that are establishing a new benchmark for driver assistance, safety, and driving experience. AR-HUDs project information about the real world to help the drivers that even you possibly saw on Arkash's video for better thinking about AR-HUDs render life-changing info straight into the drivers field of vision, minimizing the distractions as well as the situation awareness. The benefits of things like adaptive traffic, adaptive speed monitoring, collision warnings, and lane-keeping assistance in the prevention of accidents have long been established, and an AR-HUD proves itself to be a trustworthy co-pilot that guides the way to safer travelling.

According to the report, even with their excellent potential, the implementation of AR-HUD systems is still being held back by a number of hurdles, such as high engineering costs, environmental concerns, and needs to improve user acceptance. The next chapter of the AR-HUD story is their rollout in autonomous cars, where they will enable a more fluid and transparent relationship between passengers and vehicles. Additionally, the synergy of powerful generative AI utilities and cloud-based environments will enable unprecedented levels of personalized interaction and real-time adjustment, helping the system better meet evolving needs and unexpected events.

As AR-HUDs grow more prominent in connected and smart vehicles, they will be more than just basic navigation or safety features, aiding smarter transport networks. Ultimately, AR-HUD technology represents a paradigm shift in how we interact with our vehicles, paving the way for a new era of safer and more connected driving that is likely to be increasingly integrated with the development of autonomous and intelligent mobility solutions.

REFERENCES

Bark, K., Tran, C., Fujimura, K., & Ng-Thow-Hing, V. (2014). Personal navi: Benefits of an augmented reality navigational aid using a see-thru 3d volumetric hud. *Proceedings of the 6th International Conference on Automotive User Interfaces and Interactive Vehicular Applications*, 1–8. DOI: 10.1145/2667317.2667329

Blankenbach, K. (2023). Augmented Reality HUDs: Challenges, Solutions and Competitors. *2023 30th International Workshop on Active-Matrix Flatpanel Displays and Devices (AM-FPD)*, 28–31.

Charissis, V., Falah, J., Lagoo, R., Alfalah, S. F., Khan, S., Wang, S., Altarteer, S., Larbi, K. B., & Drikakis, D. (2021). Employing emerging technologies to develop and evaluate in-vehicle intelligent systems for driver support: Infotainment AR HUD case study. *Applied Sciences (Basel, Switzerland)*, *11*(4), 1397. DOI: 10.3390/app11041397

Chouksey, S., & Sirsikar, S. (2016). A prototype of low cost heads up display for automobiles navigation system. *2016 International Conference on Computing, Analytics and Security Trends (CAST)*, 205–210. DOI: 10.1109/CAST.2016.7914967

Chu, K., Brewer, R., & Joseph, S. (2008). *Traffic and navigation support through an automobile heads up display (a-HUD)*.

Currano, R., Park, S. Y., Moore, D. J., Lyons, K., & Sirkin, D. (2021). Little road driving hud: Heads-up display complexity influences drivers' perceptions of automated vehicles. *Proceedings of the 2021 CHI Conference on Human Factors in Computing Systems*, 1–15. DOI: 10.1145/3411764.3445575

Ding, Y., & Li, J. (2024). Understanding User Intention to Adopt Augmented Reality Head-Up Display: A Study Based on Design Factors. *International Journal of Human-Computer Interaction*, •••, 1–19. DOI: 10.1080/10447318.2024.2425884

Doshi, A., Cheng, S. Y., & Trivedi, M. M. (2008). A novel active heads-up display for driver assistance. *IEEE Transactions on Systems, Man, and Cybernetics. Part B, Cybernetics*, *39*(1), 85–93. DOI: 10.1109/TSMCB.2008.923527 PMID: 19068432

Jiang, Q., & Guo, Z. (2023). AR-HUD optical system design and its multiple configurations analysis. *Photonics*, *10*(9), 954. DOI: 10.3390/photonics10090954

Jing, C., Shang, C., Yu, D., Chen, Y., & Zhi, J. (2022). The impact of different AR-HUD virtual warning interfaces on the takeover performance and visual characteristics of autonomous vehicles. *Traffic Injury Prevention*, *23*(5), 277–282. DOI: 10.1080/15389588.2022.2055752 PMID: 35442130

Langlois, S., & Soualmi, B. (2016). Augmented reality versus classical HUD to take over from automated driving: An aid to smooth reactions and to anticipate maneuvers. *2016 IEEE 19th International Conference on Intelligent Transportation Systems (ITSC)*, 1571–1578.

Liang, Y., Zheng, P., & Xia, L. (2023). A visual reasoning-based approach for driving experience improvement in the AR-assisted head-up displays. *Advanced Engineering Informatics*, *55*, 101888. DOI: 10.1016/j.aei.2023.101888

Liu, Y.-C., & Wen, M.-H. (2004). Comparison of head-up display (HUD) vs. Head-down display (HDD): Driving performance of commercial vehicle operators in Taiwan. *International Journal of Human-Computer Studies*, *61*(5), 679–697. DOI: 10.1016/j.ijhcs.2004.06.002

Ma, X., Jia, M., Hong, Z., Kwok, A. P. K., & Yan, M. (2021). Does augmented-reality head-up display help? A preliminary study on driving performance through a VR-simulated eye movement analysis. *IEEE Access : Practical Innovations, Open Solutions*, *9*, 129951–129964. DOI: 10.1109/ACCESS.2021.3112240

Maroto, M., Caño, E., González, P., & Villegas, D. (2018). Head-up Displays (HUD) in driving. *arXiv Preprint arXiv:1803.08383*.

Murugan, S., Sampathkumar, A., Kanaga Suba Raja, S., Ramesh, S., Manikandan, R., & Gupta, D. (2022). Autonomous vehicle assisted by heads up display (HUD) with augmented reality based on machine learning techniques. In *Virtual and augmented reality for automobile industry: Innovation vision and applications* (pp. 45–64). Springer.

Okabayashi, S., Sakata, M., Fukano, J., Daidoji, S., Hashimoto, C., & Ishikawa, T. (1989). Development of practical heads-up display for production vehicle application. *SAE Transactions*, •••, 638–647. DOI: 10.4271/890559

Park, H. S., Park, M. W., Won, K. H., Kim, K.-H., & Jung, S. K. (2013). In-vehicle AR-HUD system to provide driving-safety information. *ETRI Journal*, *35*(6), 1038–1047. DOI: 10.4218/etrij.13.2013.0041

Wang, S., Charissis, V., & Harrison, D. K. (2017). Augmented reality prototype HUD for passenger infotainment in a vehicular environment. *Advances in Science, Technology and Engineering Systems Journal*, *2*(3), 634–641. DOI: 10.25046/aj020381

Wang, W., Zhu, X., Chan, K., & Tsang, P. (2018). Digital holographic system for automotive augmented reality head-up-display. *2018 IEEE 27th International Symposium on Industrial Electronics (ISIE)*, 1327–1330.

Winkler, M., & Soleimani, M. (2025a). A Review of Augmented Reality Heads Up Display in Vehicles: Effectiveness, Application, and Safety. *International Journal of Human-Computer Interaction*, •••, 1–16. DOI: 10.1080/10447318.2024.2443252

Winkler, M., & Soleimani, M. (2025b). A Review of Augmented Reality Heads Up Display in Vehicles: Effectiveness, Application, and Safety. *International Journal of Human-Computer Interaction*, •••, 1–16. DOI: 10.1080/10447318.2024.2443252

Xu, N., Yu, F., Xu, J., Jin, D., Feng, H., Xu, Z., Li, Q., & Chen, Y. (2023). HUDNet: A dynamic calibration method for automotive augmented reality head-up-displays. *Displays*, *78*, 102453. DOI: 10.1016/j.displa.2023.102453

You, F., Fu, Q., Yang, J., Chen, H., Cui, W., & Wang, J. (2024). Team situation awareness-based augmented reality head-up display design for security requirements. *International Journal of Human-Computer Interaction*, *40*(18), 5119–5138. DOI: 10.1080/10447318.2023.2231281

Chapter 11
Augmented Reality: Transforming Manufacturing in the Industry 4.0 Era

Sunil Sharma
https://orcid.org/0000-0001-9936-5103
Lovely Professional University, India

ABSTRACT

This chapter provides a comprehensive examination of Augmented Reality (AR) in manufacturing, covering its introduction, types, implementation strategies, advantages, challenges, applications, and methodologies for implementation. We present a brief overview of AR technology in the context of Industry 4.0 and its significance in manufacturing. It highlights the potential of AR to bridge the gap between the physical and digital worlds. We then discuss the various types of AR technologies relevant to manufacturing along with the working of AR technology. The numerous benefits of adopting AR in manufacturing and the systematic methodology for implementing AR in manufacturing environments is presented. Five applications of AR in manufacturing from real industries have been presented. The advantages, common challenges and barriers to AR implementation in manufacturing are discussed. This research provides a comprehensive roadmap for leveraging AR technology in manufacturing, offering insights into its potential, challenges, and practical implementation strategies.

1. INTRODUCTION

Augmented reality (AR) constitutes an interactive encounter that enriches the physical world by integrating computer-generated sensory data. Through the utilization of software, applications, and devices like AR glasses, this technology superimposes

DOI: 10.4018/979-8-3693-7287-6.ch011

digital elements onto real-world settings and objects. Such augmentation enhances user engagement, transforming everyday surroundings into interactive learning spaces, which is especially beneficial in fields like manufacturing and Industry 4.0. It enables industrial practitioners to seamlessly integrate with machinery and systems, optimizing technology and Internet of Things (IoT) networks through human insight, observation, and inventive thinking. AR applications are widely available on smartphones and portable devices.

1.1 Understanding Augmented Reality

AR devices are hardware tools that improve our senses of sight, sound, and touch by superimposing virtual information over the real world. Table 1 provides an explanation of augmented reality from various perspectives. These devices can be divided into a number of categories. Headsets and smart glasses, handheld devices, projection-based systems, spatial augmented reality (SAR), head-up displays (HUD), and augmented reality contact lenses are the first six groups of this categorization (Azuma et al., 2001; Mendoza-Ramírez et al., 2023; Milgram et al., 1995). This categorization is also represented in Figure 1.

Table 1. Augmented reality is represented using various perspectives

Aspect	Description
Definition	Augmented reality superimposes digital information over the actual environment, improving the user's experience of reality. (Chatzopoulos et al., 2017).
Implementation	Operates mainly through mobile devices such as smartphones or tablets. In industries where hands-free operation is crucial, AR experiences are facilitated through glasses or headsets (Kipper and Rampolla, 2013).
Applications	Applied in various fields, including gaming, retail, navigation, and manufacturing, enhancing tasks with digital overlays and information (Dargan et al., 2023).
Interaction	Enables users to interact with digital information overlaid onto the physical world, often through gestures or voice commands.
Use of Devices	Primarily accessed through smartphones or tablets; in specific industries, glasses or headsets are preferred for hands-free operation.
Example Scenario	In manufacturing, AR glasses overlay assembly instructions onto machinery, allowing technicians to perform tasks with real-time guidance.

Augmented Reality operates by overlaying digital content onto real-world objects, enabling users to engage with a blended 3D environment that merges physical and digital elements. However, AR's effectiveness lies in its integration within a connected Industry 4.0 framework, incorporating various technologies like big data and automated robotics. Table 1 provides an explanation of augmented reality from various perspectives.

This study looks at augmented reality for manufacturing applications in the framework of Industry 4.0. We first understand AR, provide a literature review, understand its importance in the context of Industry 4.0, and then discuss augmented reality in manufacturing. It is significant because this study contributes to the knowledge of AR's role in manufacturing within the context of Industry 4.0, shedding light on its significance, challenges, and opportunities for driving technological advancements and operational excellence in the manufacturing sector.

Figure 1. Types of AR devices

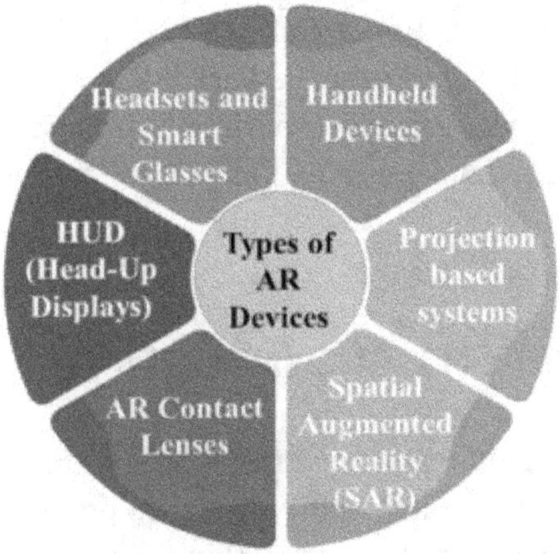

2. BACKGROUND

The applications of AR range from sports, video games, marketing, sales, education, travel, and manufacturing (Nee et al., 2012). Workers are using immersive technologies, like AR and VR, to help them overcome these difficulties by offering support, modeling, and improving industrial processes before implementing them in a production setting (Eswaran and Bahubalendruni, 2022). AR has been used in manual assembly station planning (Reinhart and Patron, 2003), for products assembly guidance in product assembly (Sääski et al., 2008; Yuan et al., 2008), for maintenance (Deac et al., 2017), as a virtual Lean tool using AR apps, in metrology and for solutions of in-line quality control (Ho et al., 2022). Using AR, complex technical systems can be represented digitally in a realistic virtual environment (Bellalouna, 2021).

Numerous case studies have been conducted, industrially and academically, including one on the integration of AR-smart glasses into warehouse logistics procedures (Plakas et al., 2020), augmenting digital virtual prototypes with physical products (Di Gesù et al., 2000), implementation of an AR-based tool that can be used in the manufacturing industry to support operators during assembly operations (Botto et al., 2020), data glove-based virtual assembly (Valentini, 2009) and replacing and physical manuals with augmented virtual contents (Wiedenmaier et al., 2003). However, AR technology has been asked to be user-centric for application development, requires user acceptance, and needs to overcome technological and organizational challenges (Egger and Masood, 2020). The various Industrial AR tools exhibit remarkable performance variances. The vast bulk of existing Industrial AR hardware devices are still experimental. Their integration with current Industrial AR frameworks and the creation of new functionality is complicated (Blanco-Novoa et al., 2018)

3. AUGMENTED REALITY AND INDUSTRY 4.0

In this section, we first present the various technologies in Industry 4.0 to emphasize the standing of Augmented reality. We then present various objectives that Augmented Reality achieves for Industry 4.0 in the context of manufacturing.

3.1 Classification of Cutting-edge Technologies in Industry 4.0

Industry 4.0 introduces innovations that surpass what was previously conceivable through four fundamental categories of revolutionary technologies (Figure 2) that are applicable across the entire value chain (Alcácer and Cruz-Machado, 2019; Bajic et al., 2021, p. 4; Xu et al., 2018; Zheng et al., 2018). This classification for cutting-edge technologies in Industry 4.0 is as follows.

Connectivity, data, and computational prowess: Technologies like cloud computing, the Internet, blockchain, and sensors enhance connectivity and computational capabilities.

Analytics and intelligence: Advanced analytics, machine learning, and artificial intelligence provide deeper insights and intelligence.

Human-machine interaction: Virtual reality (VR), augmented reality (AR), robotics, automation, and autonomous guided vehicles facilitate seamless interaction between humans and machines.

Advanced engineering: Additive manufacturing (e.g., 3-D printing), renewable energy solutions, and nanoparticles constitute the forefront of engineering advancements.

Figure 2. Categorization of cutting-edge technologies in Industry 4.0

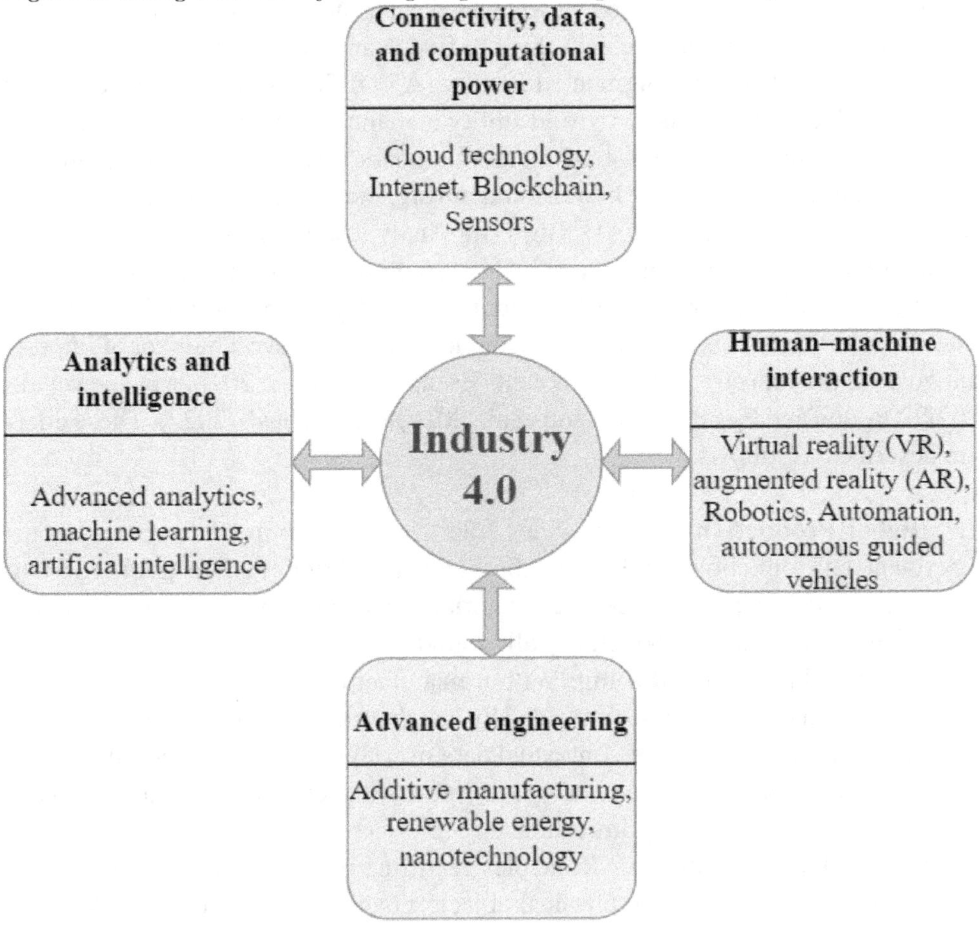

3.2 Importance of AR in Industry 4.0

According to Analytics Insight, the industrial AR and VR market is expected to grow to a US$2.8 billion value by 2023. Virtual reality (VR) and augmented reality (AR) have changed over the past ten years from being mainly connected with gaming to being useful technologies that improve frontline workers' responsibilities. The global AR and VR industry is expected to increase significantly, from US$3 billion in 2020 to US$28 billion by 2025, according to financial analysts. This growth is expected to be driven by breakthroughs in technology and a decline in associated costs. PWC estimates that by 2030, around 23 million occupations will use AR and VR technology for tasks including customer service, meetings, and training. Furthermore, according to Capgemini, 43% of manufacturing companies believe VR will become widely used in their businesses during the next three years, and 38% believe it will do so within the next three to five years (Akash, 2021).

The primary driver of Augmented Reality (AR) in Industry 4.0 is the quest for enhanced efficiency, productivity, and agility in manufacturing processes. Industry 4.0 represents the fourth industrial revolution characterized by the integration of advanced digital technologies into manufacturing, including the Internet of Things (IoT), Artificial Intelligence (AI), Big Data Analytics, and Robotics. Augmented Reality serves as a pivotal component in this paradigm shift by bridging the physical and digital realms, offering real-time contextual information overlaying physical objects or environments. AR enables manufacturers to achieve a number of important goals in the Industry 4.0 environment (Esengün and İnce, 2018; Machała et al., 2022; Masood and Egger, 2019; Santi et al., 2021; Sharma et al., 2022). These goals are described as follows.

1. **Higher Productivity:** AR provides employees with immediate access to pertinent data, directions, and advice that is displayed right in their field of vision. As a result, less manual reference materials are needed, and jobs take less time to finish, increasing productivity all around.
2. **Analytical Decision Making:** Within manufacturing operations, data-driven decision-making processes are aided by insights and analytics produced by AR. Through the utilization of contextual data overlays, production measurements, and KPIs, stakeholders can make well-informed decisions on the allocation of resources and process optimization.
3. **Better Quality Control:** AR enables real-time quality evaluations by superimposing digital data over physical objects, enabling personnel to quickly spot flaws, irregularities, or deviations. This makes it possible for proactive intervention to uphold requirements for product quality.
4. **Agile Manufacturing:** AR enables quick reconfiguration of assembly lines and manufacturing processes to adapt to shifting product configurations or production needs. With dynamic AR-guided instructions, employees may swiftly adjust to new jobs or workflows.
5. **Simplified Onboarding and Training:** AR-based training programs give staff members engaging, interactive learning experiences, which speeds up the onboarding process and lowers training expenses. In a risk-free virtual environment, workers can envision complex procedures and mimic operations.

6. **Remote guidance and cooperation:** Augmented Reality (AR) allows professionals to assist field technicians or workers who are located in different locations remotely. Real-time resolutions and teamwork are facilitated by augmented reality (AR), which shares live video feeds and adds annotations or instructions to enhance them.

Overall, the integration of augmented reality in Industry 4.0 is driven by the imperative to leverage advanced technologies to achieve operational excellence, enhance flexibility, and stay competitive in an increasingly digitalized and interconnected manufacturing landscape.

3.3 Types of Augmented Reality

In this section, the various types of Augmented Reality are discussed (Dhanalakshmi et al., 2021). Refer to Figure 3. Broadly, AR can be categorized as Marker based and Markerless.

Marker based AR: It employs a visual cue to initiate a new, interactive experience. The most frequent markers are two-dimensional QR codes. A device's camera scans a machine-readable barcode and displays visual effects. Marker-based augmented reality was considered groundbreaking, but it has serious limitations. Typically, the technology is only compatible with mobile devices like smartphones and tablets. Furthermore, users must download the app or software in order to access the information, so it lacks the instant gratification that is expected in today's digital world.

Markerless Augmented Reality: Markerless augmented reality creates visual effects without the necessity for photo identification. Instead, the technology uses a device's camera, location software, and accelerometer to identify positioning information, such as the orientation of different objects and their distance apart. Following that, it employs simultaneous localization and mapping (SLAM) to grasp the surrounding three-dimensional environment. The Augmented Reality content is then shown on top of the actual worldview, making it viewable from all angles and locations. Developing markerless AR is surely difficult. Nonetheless, the technology is increasingly overtaking its marker-based equivalent as the preferred option among forward-thinking manufacturers and aggregators. Markerless AR is classified into four types: projection-based, location-based, overlay, and contour-based.

Projection-based augmented reality (AR) creates 3D images by projecting immersive light onto a flat surface. It then applies SLAM to detect human involvement with the augmentation. For example, projection-based AR can be utilized to produce holograms for commercial or recreational purposes.

Location-Based Augmented Reality involves displaying content in a specific physical environment. It maps the actual environment and determines visual locations in the vicinity. When the smartphone recognizes a match with the mapped location, it overlays the digital image accordingly. Pokémon Go is the most well-known example of the same.

Overlay markerless AR replaces a non-augmented view of an object with an improved virtual representation of the same item, which includes many visual perspectives. The Nextech Configurator is a prime example of the same. The color and layout of a 3D product model can be changed by using the product configurator tool.

Then there's contour-based augmented reality (or outlining AR), which uses SLAM to outline the silhouettes of objects and simulate realistic human contact. It may, for example, be used to create a safe-driving app for nighttime driving, poorly lighted roads, severe precipitation, and other low-visibility conditions.

Figure 3. Types of augmented reality

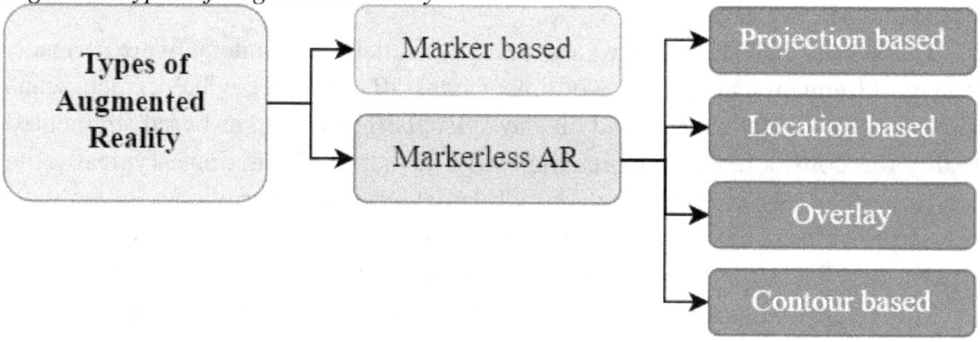

3.4 Working of Augmented Reality

The working of Augmented Reality involves the seamless integration of hardware sensors, software algorithms, and user interfaces to enhance the perception of reality by overlaying digital information onto the physical world, creating immersive and interactive experiences for users across various applications and domains. The process of augmented reality involves the following steps (Porter and Heppelmann, 2017), as shown in Figure 4.

Step I. An AR-capable device equipped with a camera, such as smart glasses, a tablet, or a smartphone, analyzes live video to recognize physical objects or the surrounding environment, like machinery or warehouse layouts.

Step II. A digital twin, essentially a cloud-hosted 3D replica of the object, bridges the gap between the real and virtual realms. It gathers data from the physical object and the digital environment.

Step III. The AR device retrieves relevant information about the object from the cloud and overlays digital content onto the physical object using markers or tracking mechanisms like GPS, accelerometers, orientation sensors, and barometric sensors. This results in a hybrid 3D interface blending reality and virtual elements.

Step IV. Utilizing real-time data from the object, users can interact with the environment by moving around and issuing commands to the cloud via touchscreen, voice commands, or gestures.

Figure 4. Working process of augmented reality

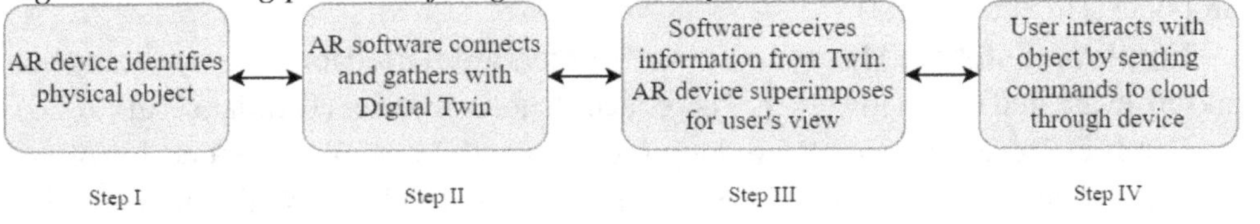

4. AUGMENTED REALITY IN MANUFACTURING

In this section, we discuss Augmented reality in terms of manufacturing providing applications, method of implementation, case examples from industries, advantages and implementation challenges of AR in manufacturing

4.1 Applications of Augmented Reality in Manufacturing

Augmented Reality (AR) has significantly influenced the manufacturing industry, revolutionizing various operational aspects. AR has streamlined tasks and transformed manufacturing processes in the following key areas (Fite-Georgel, 2011; Röltgen and Dumitrescu, 2020).

Spatial Computing: Digital representations of real-world data can be interactively experienced through spatial computing, which is made possible by augmented reality. AR makes it easier for manufacturers to comprehend certain materials and check completed goods for defects throughout production (Uva et al., 2018).

Training and Onboarding in Manufacturing: AR simplifies the learning curve for new joiners in manufacturing units, creating a safer environment for training in hazardous conditions. For instance, Ather utilizes an AR-based application to provide systematically curated guides, manuals, and documents for training purposes. AR simulates real-world scenarios, offering performance feedback to trainees for continuous improvement (Zigart et al., 2023).

Client Support: Augmented reality improves customer service by allowing manufacturers to efficiently lead customers through issue resolution processes.

Shop Floor Issues & Resolution: With AR, manufacturing units can promptly address shop floor issues by notifying the maintenance team in real time, ensuring minimal impact on production processes (Dimitris Mourtzis et al., 2017).

Quality Inspection and Control: By facilitating collaborative inspections, displaying defects, and providing assistance on resolution techniques, augmented reality (AR) improves quality control procedures and results in higher-quality products (Ho et al., 2022).

Employee Safety in Manufacturing: Workers are alerted for safety requirements and are evaluated which lowers the risk of accidents or injuries (Damiani et al., 2020).

Product Development: By reducing the time and resources needed for prototyping, AR speeds up the process of developing new products. While guaranteeing adherence to safety regulations, it expedites the ideation, development, and introduction of goods to the assembly line (Ong et al., 2011).

Maintenance: AR mitigates machine breakdowns and downtime by swiftly identifying potential flaws, significantly reducing the time spent on manual inspection. Further, when AR is used frequent breakdown of the machines is prevented, which eventually lowers maintenance costs and boosts productivity (D. Mourtzis et al., 2017).

Assembly Line Operations: In an assembly line, augmented reality (AR) ensures the timely completion of tasks while eliminating potential faults, provides visual aid using 3D images or animations, and guides operators step-by-step through complex assembly operations (Wang et al., 2016).

4.2 Deploying AR in a Manufacturing Setting

When augmented reality (AR) is integrated into workplaces a well-thought-out plan is essential that covers system integration, hardware selection, software development, and intensive training programs. Businesses can improve productivity, strengthen quality control, speed training procedures, and foster creativity in a field that is evolving quickly. An established manufacturing company can implement augmented reality by following these steps (Brunner et al., 2020; Butt, 2020; Osborne and Mavers, 2019; Schein and Rauschnabel, 2023).

Step I. **Develop use cases and describe them.** The first step in integrating augmented reality (AR) into a productive manufacturing organization is to design and specify the technology's use cases. To achieve this task, corporate identify the operations that can profit from or be improved by augmented reality. An evaluation of the company's current technology and procedures is required to achieve this task.

Step II. **Construct a Model.** The next step in AR deployment is to develop a prototype for the AR system and test it in a controlled environment. The team may find and fix any problems or defects with the prototype testing before the system is put into use.

Step III. **Create and implement the AR system.** The AR system needs to be designed, developed, and installed after the prototype has been tested and approved. It comprises identifying the hardware needed to run the system, developing and constructing original software solutions to offer the needed functionality, and incorporating them into the business.

Step IV. **Train the employees.** After establishing the AR, staff must learn how to use it efficiently. It may include formalized training or hands-on practice with simulated system components.

Step V. **Monitor and maintain.** The final stage of AR implementation involves monitoring and maintaining the system as needed. It entails regularly monitoring user feedback and modifying settings to enhance performance. In addition, regular maintenance is required to guarantee that the system operates optimally.

4.3 Current Application of AR in Manufacturing Industries

In this section, we discuss five real-life case examples of how augmented reality has been applied and transformed the processes in manufacturing industries (Fedko, 2020). These cases are discussed as follows.

Case Example 1

GE Aviation manufactures engines and avionics for worldwide airlines, including those that power Boeing and Airbus aircraft. GE Aviation experimented with using wearable augmented reality (AR) to assist its engine assembly technicians in reducing errors and working quickly. The smart glasses deliver assembly information hands-free, physically putting instructions in front of technicians' eyes - without halting or glancing away from their activities to consult printed reference materials or a computer screen. The GE Aviation pilot was designed to help technicians use a Wi-Fi based torque wrench to secure the bolts. However, the company discovered that technicians accomplished the tasks quicker and completed jobs more rapidly. This improvement in efficiency was due to the fact that they no longer had to leave the workstation to review hard copies of manuals or pick up equipment.

Case Example 2

Cisco Systems is a technology leader that employs manufacturing augmented reality to improve worker safety. They created AR-enabled safety glasses that give workers real-time information and safety alerts, allowing them to operate more effectively.

Case Example 3

Siemens AG is a major provider of industrial Augmented Reality products and services. Siemens has created a number of products that use augmented reality to improve building processes and worker safety. For example, they have used AR-enabled eyewear to provide workers with access to monitoring and performance information, as well as help with component assembly and maintenance.

Case Example 4

Dassault Systèmes is a market leader in 3D design and product engineering, and they make substantial use of Augmented Reality industrial applications in their manufacturing processes. Their 3D experience technology enables employees to access critical information via AR headsets or smartphones. It contains engineering drawings, part inventories, technical documentation, ergonomics information, and interactive assembly instructions.

Case Example 5

Lockheed Martin has implemented augmented reality (AR) into its manufacturing processes to improve speed and accuracy. Their revolutionary augmented reality technology allows employees to discover possible difficulties before they become serious problems, leading to enhanced productivity, greater quality control, and shorter production times.

4.4 Advantages of AR in Manufacturing

In the previous section, numerous applications for AR in manufacturing were suggested. AR implementation in industrial and manufacturing settings offers the following benefits overall (Hastings, 2018).

Enhanced Product Development: Utilizing AR during the design phase enables designers to meet evolving consumer demands for shorter product lifecycles and lower prototyping costs.

Simplified Procedures: Visualized workflows with step-by-step instructions aid in predictive troubleshooting, minimizing errors leading to rework, and expediting intricate assembly tasks.

Optimized Warehouse Management: AR facilitates inventory management, guides product selection, reduces downtime by aiding technicians in diagnosing and resolving issues, and improves employee training efficiency.

Increased Workforce Engagement: The immersive 3D experience and hands-on learning through AR are still novel for workers, enhancing their involvement in tasks.

Risk Reduction: AR not only replicates real-world equipment and processes but also provides a safe learning environment for workers to practice tasks virtually, improving safety, inspection, training, and workflow.

4.5 Implementation challenges of AR in Manufacturing

The advantages of the implementation of AR in manufacturing are that it is robust and can greatly improve the manufacturing organizational system. Implementing Augmented Reality (AR) in manufacturing within the context of Industry 4.0 presents several challenges, as listed below (Brunner et al., 2020; Eswaran and Bahubalendruni, 2022; Masood and Egger, 2019)

Regulations for Conformance: Implementing AR can be challenging when it comes to following rules and regulations, especially in highly regulated sectors like aircraft, pharmaceuticals, and healthcare. Complying with safety, quality, and data protection standards increases the intricacy of deploying augmented reality.

Technological Compatibility: Ensuring that AR systems are compatible with existing manufacturing equipment and software can be challenging, requiring integration with diverse hardware and software platforms.

Upkeep and Assistance: Devoted resources and specialized knowledge are needed to provide continuous maintenance, upgrades, and technical support for AR systems in manufacturing settings. Manufacturers have constant problems in ensuring the functionality, dependability, and uptime of AR systems to minimize disruptions to production activities.

Inclusion with Current Processes: It takes careful planning and coordination to integrate AR seamlessly into current manufacturing workflows and processes without interfering with daily operations. Maximizing the advantages of augmented reality while maintaining compatibility with current workflows, systems, and production schedules can be difficult to do.

Network Connectivity: AR applications often rely on stable network connections for real-time data streaming and remote support. In manufacturing environments with limited connectivity or sensitive data, ensuring consistent network access can be a challenge.

Data Security and Privacy: AR systems collect and transmit sensitive data, including proprietary manufacturing processes and equipment information. In networked production environments, safeguarding this data against hacks, illegal access, and interception is crucial but difficult.

Environmental Factors: The performance of augmented reality systems, including sensors and displays, can be impacted by manufacturing environments that are dusty, noisy, or have inconsistent illumination. It might be technically difficult to overcome environmental obstacles and guarantee dependable functioning in a variety of manufacturing settings.

Coaching to Fill Workforce Skill Gaps: Training personnel to use AR systems successfully and interpret augmented information can take time and necessitates workforce upskilling. It can be difficult to close the talent gap and ensure worker competency, particularly in high turnover industries.

Deployment costs: Deploying AR systems in production need large upfront investments in hardware, software development, and employee training. Balancing these costs against possible benefits and ROI can be difficult for firms, particularly small manufacturers with limited resources.

Interface Design and User Experience: Creating intuitive user interfaces and immersive experiences in AR applications is critical to user acceptance and effectiveness. Ensuring that AR systems are user-friendly and provide value-added functionality without overwhelming users can be difficult, necessitating iterative design and testing procedures.

5. CONCLUSION

The study concludes by highlighting the revolutionary impact of Augmented Reality (AR) technology on manufacturing within the framework of Industry 4.0. Examining AR in detail—including its introduction, types, implementation strategies, benefits, drawbacks, applications, and methodology—the study offers important insights for industry stakeholders. AR technology's overview highlights how it may combine the real and virtual worlds, changing conventional industrial processes and boosting productivity. By offering a thorough analysis of the many AR technologies and their underlying concepts, the chapter clarifies the many applications and benefits of AR in production. For companies looking to integrate AR into their operations, the logical approach provided for doing so in manufacturing contexts is helpful. Furthermore, real-world examples of AR applications in manufacturing demonstrate its usefulness in enhancing productivity, safety, quality, and training procedures. Despite the various benefits, the study recognizes the frequent problems and constraints to AR application in manufacturing, such as financial concerns, technology limits, and aversion to change. By addressing these obstacles and providing practical answers, the study provides a thorough path for using AR technology in manufacturing environments. In short, this study adds to the increasing body of knowledge on AR in manufacturing by providing useful insights into its potential, limitations, and practical implementation tactics. By embracing AR technology, manufacturers can unlock new opportunities for innovation, efficiency, and competitiveness in the Industry 4.0 era.

AR technology is still not sophisticated enough to meet industrial requirements for robustness and dependability. There is high fragmentation across hardware, software, and AR solutions, which leads to high complexity in selecting and developing AR systems (Palmarini et al., 2018). However, the use of AR technology in manufacturing has advanced to the point where hardware and software implementation has progressively attained the necessary maturity for long-term industrial AR solutions. User acceptance, on the other hand, is now impeding shop-floor adoption, which will ultimately affect how well AR technologies are integrated into industrial processes. Therefore, all pertinent components of AR systems must be considered when developing a long-term AR-based solution for the manufacturing sector. Industry predictions and research both concur that it will take a number of years for augmented reality to become a widely used technology, even if businesses have begun implementing it as an enterprise tool. Companies anticipate increased levels of efficiency from their employees by putting virtual work-related information into their field of view. While the idea of pervasive AR is rapidly taking shape, the widespread deployment of AR applications in the manufacturing industry is still a ways off.

REFERENCES

Akash, S. (2021). AR and VR in Manufacturing is Predicted to Touch US$2.8 Billion by 2023 [WWW Document]. Analytics Insight. URL https://www.analyticsinsight.net/ar-and-vr-in-manufacturing-is-predicted-to-touch-us2-8-billion-by-2023/

Alcácer, V., & Cruz-Machado, V. (2019). Scanning the Industry 4.0: A Literature Review on Technologies for Manufacturing Systems. *Engineering Science and Technology, an International Journal*, *22*(3), 899–919. DOI: 10.1016/j.jestch.2019.01.006

Azuma, R., Baillot, Y., Behringer, R., Feiner, S., Julier, S., & MacIntyre, B. (2001). Recent advances in augmented reality. *IEEE Computer Graphics and Applications*, *21*(6), 34–47. DOI: 10.1109/38.963459

Bajic, B., Rikalovic, A., Suzic, N., & Piuri, V. (2021). Industry 4.0 Implementation Challenges and Opportunities: A Managerial Perspective. *IEEE Systems Journal*, *15*(1), 546–559. DOI: 10.1109/JSYST.2020.3023041

Bellalouna, F. (2021). The Augmented Reality Technology as Enabler for the Digitization of Industrial Business Processes: Case Studies. *Procedia CIRP*, *98*, 400–405. DOI: 10.1016/j.procir.2021.01.124

Blanco-Novoa, O., Fernandez-Carames, T. M., Fraga-Lamas, P., & Vilar-Montesinos, M. A. (2018). A Practical Evaluation of Commercial Industrial Augmented Reality Systems in an Industry 4.0 Shipyard. *IEEE Access : Practical Innovations, Open Solutions*, *6*, 8201–8218. DOI: 10.1109/ACCESS.2018.2802699

Botto, C., Cannavò, A., Cappuccio, D., Morat, G., Sarvestani, A. N., Ricci, P., . . . Saturnino, A. (2020, March). Augmented reality for the manufacturing industry: the case of an assembly assistant. In 2020 IEEE Conference on Virtual Reality and 3D User Interfaces Abstracts and Workshops (VRW) (pp. 299-304). IEEE.

Brunner, K. E., Perey, C., Rauschnabel, P. A., & Sage, M. (2020). Implementation of Augmented Reality in Manufacturing: A Case Study Exercise. In Jung, T., Tom Dieck, M. C., & Rauschnabel, P. A. (Eds.), *Augmented Reality and Virtual Reality, Progress in IS* (pp. 85–97). Springer International Publishing., DOI: 10.1007/978-3-030-37869-1_8

Butt, J. (2020). A Strategic Roadmap for the Manufacturing Industry to Implement Industry 4.0. *Designs*, *4*(2), 11. DOI: 10.3390/designs4020011

Chatzopoulos, D., Bermejo, C., Huang, Z., & Hui, P. (2017). Mobile Augmented Reality Survey: From Where We Are to Where We Go. *IEEE Access : Practical Innovations, Open Solutions*, *5*, 6917–6950. DOI: 10.1109/ACCESS.2017.2698164

Damiani, L., Revetria, R., & Morra, E. (2020). Safety in Industry 4.0: The Multi-Purpose Applications of Augmented Reality in Digital Factories. *Advances in Science, Technology and Engineering Systems Journal*, *5*(2), 248–253. DOI: 10.25046/aj050232

Dargan, S., Bansal, S., Kumar, M., Mittal, A., & Kumar, K. (2023). Augmented Reality: A Comprehensive Review. *Archives of Computational Methods in Engineering*, *30*(2), 1057–1080. DOI: 10.1007/s11831-022-09831-7

Deac, C. N., Deac, C. N., Popa, C. L., Ghinea, M., & Cotet, C. E. (2017). Using Augmented Reality in Smart Manufacturing, in: Katalinic, B. (Ed.), DAAAM Proceedings. DAAAM International Vienna, pp. 0727–0732. DOI: 10.2507/28th.daaam.proceedings.102

Dhanalakshmi, R., Mai, C. D., Latha, B., & Vijayaraghavan, N. (2021). AR and VR in Manufacturing. In Palanikumar, K., Natarajan, E., Sengottuvelu, R., & Davim, J. P. (Eds.), *Futuristic Trends in Intelligent Manufacturing, Materials Forming, Machining and Tribology* (pp. 171–183). Springer International Publishing., DOI: 10.1007/978-3-030-70009-6_11

Egger, J., & Masood, T. (2020). Augmented reality in support of intelligent manufacturing – A systematic literature review. *Computers & Industrial Engineering*, *140*, 106195. DOI: 10.1016/j.cie.2019.106195

Esengün, M., & İnce, G. (2018). The Role of Augmented Reality in the Age of Industry 4.0. In *Industry 4.0: Managing The Digital Transformation, Springer Series in Advanced Manufacturing* (pp. 201–215). Springer International Publishing., DOI: 10.1007/978-3-319-57870-5_12

Eswaran, M., & Bahubalendruni, M. V. A. R. (2022). Challenges and opportunities on AR/VR technologies for manufacturing systems in the context of industry 4.0: A state of the art review. *Journal of Manufacturing Systems*, *65*, 260–278. DOI: 10.1016/j.jmsy.2022.09.016

Fedko, D. (2020). The Ultimate Guide to Augmented Reality (AR) in Manufacturing Industry [WWW Document]. Wear Studio Blog. URL https://wear-studio.com/ar-in-manufacturing/

Fite-Georgel, P. (2011). Is there a reality in Industrial Augmented Reality? in: 2011 10th IEEE International Symposium on Mixed and Augmented Reality. Presented at the 2011 IEEE International Symposium on Mixed and Augmented Reality, IEEE, Basel, pp. 201–210. DOI: 10.1109/ISMAR.2011.6092387

Hastings, W. (2018). *Augmented reality for manufacturing: bringing digital transformation to skilled workers*. ARC White Pap.

Ho, P. T., Albajez, J. A., Santolaria, J., & Yagüe-Fabra, J. A. (2022). Study of Augmented Reality Based Manufacturing for Further Integration of Quality Control 4.0: A Systematic Literature Review. *Applied Sciences (Basel, Switzerland)*, *12*(4), 1961. DOI: 10.3390/app12041961

Kipper, G., & Rampolla, J. 2013. Augmented reality: an emerging technologies guide to AR, First Edition. ed. Syngress, Waltham, Mass.

Levialdi, S., & AVI (Conference). (2000). AVI 2000: Proceedings of the Working Conference on Advanced Visual Interfaces, Palermo, Italy, May 23-26, 2000. Association for Computing Machinery.

Machała, S., Chamier-Gliszczyński, N., & Królikowski, T. (2022). Application of AR/VR Technology in Industry 4.0. *Procedia Computer Science*, *207*, 2990–2998. DOI: 10.1016/j.procs.2022.09.357

Masood, T., & Egger, J. (2019). Augmented reality in support of Industry 4.0—Implementation challenges and success factors. *Robotics and Computer-integrated Manufacturing*, *58*, 181–195. DOI: 10.1016/j.rcim.2019.02.003

Mendoza-Ramírez, C. E., Tudon-Martinez, J. C., Félix-Herrán, L. C., Lozoya-Santos, J. D. J., & Vargas-Martínez, A. (2023). Augmented Reality: Survey. *Applied Sciences (Basel, Switzerland)*, *13*(18), 10491. DOI: 10.3390/app131810491

Milgram, P., Takemura, H., Utsumi, A., & Kishino, F. 1995. Augmented reality: a class of displays on the reality-virtuality continuum, in: Das, H. (Ed.). Presented at the Photonics for Industrial Applications, Boston, MA, pp. 282–292. DOI: 10.1117/12.197321

Mourtzis, D., Vlachou, A., & Zogopoulos, V. (2017). Cloud-Based Augmented Reality Remote Maintenance Through Shop-Floor Monitoring: A Product-Service System Approach. *Journal of Manufacturing Science and Engineering*, *139*(6), 061011. DOI: 10.1115/1.4035721

Mourtzis, D., Zogopoulos, V., & Vlachou, E. (2017). Augmented Reality Application to Support Remote Maintenance as a Service in the Robotics Industry. *Procedia CIRP*, *63*, 46–51. DOI: 10.1016/j.procir.2017.03.154

Nee, A. Y. C., Ong, S. K., Chryssolouris, G., & Mourtzis, D. (2012). Augmented reality applications in design and manufacturing. *CIRP Annals*, *61*(2), 657–679. DOI: 10.1016/j.cirp.2012.05.010

Ong, S. K., Zhang, J., Shen, Y., & Nee, A. Y. C. (2011). Augmented Reality in Product Development and Manufacturing. In Furht, B. (Ed.), *Handbook of Augmented Reality. Springer New York, New York* (pp. 651–669)., DOI: 10.1007/978-1-4614-0064-6_30

Osborne, M., & Mavers, S. 2019. Integrating Augmented Reality in Training and Industrial Applications, in: 2019 Eighth International Conference on Educational Innovation through Technology (EITT). Presented at the 2019 Eighth International Conference of Educational Innovation through Technology (EITT), IEEE, Biloxi, MS, USA, pp. 142–146. DOI: 10.1109/EITT.2019.00035

Palmarini, R., Erkoyuncu, J. A., Roy, R., & Torabmostaedi, H. (2018). A systematic review of augmented reality applications in maintenance. *Robotics and Computer-integrated Manufacturing*, *49*, 215–228. DOI: 10.1016/j.rcim.2017.06.002

Plakas, G., Ponis, S. T., Agalianos, K., Aretoulaki, E., & Gayialis, S. P. (2020). Augmented Reality in Manufacturing and Logistics: Lessons Learnt from a Real-Life Industrial Application. *Procedia Manufacturing*, *51*, 1629–1635. DOI: 10.1016/j.promfg.2020.10.227

Porter, M., & Heppelmann, J. (2017). How Does Augmented Reality Work? *Harvard Business Review*, ●●●, 58.

Reinhart, G., & Patron, C. (2003). Integrating Augmented Reality in the Assembly Domain - Fundamentals, Benefits and Applications. *CIRP Annals*, *52*(1), 5–8. DOI: 10.1016/S0007-8506(07)60517-4

Röltgen, D., & Dumitrescu, R. (2020). Classification of industrial Augmented Reality use cases. *Procedia CIRP*, *91*, 93–100. DOI: 10.1016/j.procir.2020.01.137

Sääski, J., Salonen, T., Hakkarainen, M., Siltanen, S., Woodward, C., & Lempiäinen, J. (2008). Integration of Design and Assembly Using Augmented Reality. In Ratchev, S., & Koelemeijer, S. (Eds.), *Micro-Assembly Technologies and Applications, IFIP — International Federation for Information Processing* (pp. 395–404). Springer US., DOI: 10.1007/978-0-387-77405-3_39

Santi, G. M., Ceruti, A., Liverani, A., & Osti, F. (2021). Augmented Reality in Industry 4.0 and Future Innovation Programs. *Technologies*, *9*(2), 33. DOI: 10.3390/technologies9020033

Schein, K. E., & Rauschnabel, P. A. (2023). Augmented Reality in Manufacturing: Exploring Workers'. *IEEE Transactions on Engineering Management*, *70*(10), 3344–3357. DOI: 10.1109/TEM.2021.3093833

Sharma, A., Mehtab, R., Mohan, S., Mohd Shah, M.K., 2022. Augmented reality – an important aspect of Industry 4.0. IR 49, 428–441. DOI: 10.1108/IR-09-2021-0204

Uva, A. E., Gattullo, M., Manghisi, V. M., Spagnulo, D., Cascella, G. L., & Fiorentino, M. (2018). Evaluating the effectiveness of spatial augmented reality in smart manufacturing: A solution for manual working stations. *International Journal of Advanced Manufacturing Technology*, *94*(1-4), 509–521. DOI: 10.1007/s00170-017-0846-4

Valentini, P. P. (2009). Interactive virtual assembling in augmented reality. *International Journal on Interactive Design and Manufacturing*, *3*(2), 109–119. DOI: 10.1007/s12008-009-0064-x

Wang, X., Ong, S. K., & Nee, A. Y. C. (2016). A comprehensive survey of augmented reality assembly research. *Advances in Manufacturing*, *4*(1), 1–22. DOI: 10.1007/s40436-015-0131-4

Wiedenmaier, S., Oehme, O., Schmidt, L., & Luczak, H. (2003). Augmented Reality (AR) for Assembly Processes Design and Experimental Evaluation. *International Journal of Human-Computer Interaction*, *16*(3), 497–514. DOI: 10.1207/S15327590IJHC1603_7

Xu, L. D., Xu, E. L., & Li, L. (2018). Industry 4.0: State of the art and future trends. *International Journal of Production Research*, *56*(8), 2941–2962. DOI: 10.1080/00207543.2018.1444806

Yuan, M. L., Ong, S. K., & Nee, A. Y. C. (2008). Augmented reality for assembly guidance using a virtual interactive tool. *International Journal of Production Research*, *46*(7), 1745–1767. DOI: 10.1080/00207540600972935

Zheng, P., Wang, H., Sang, Z., Zhong, R. Y., Liu, Y., Liu, C., Mubarok, K., Yu, S., & Xu, X. (2018). Smart manufacturing systems for Industry 4.0: Conceptual framework, scenarios, and future perspectives. *Frontiers of Mechanical Engineering*, *13*(2), 137–150. DOI: 10.1007/s11465-018-0499-5

Zigart, T., Stürzl, F., Niedermayr, D., Wolfartsberger, J., & Sorko, S. R. 2023. Comparative Evaluation of Virtual Reality and In-Person Onboarding for Assembly Trainings in Manufacturing, in: 2023 IEEE International Symposium on Mixed and Augmented Reality Adjunct (ISMAR-Adjunct). Presented at the 2023 IEEE International Symposium on Mixed and Augmented Reality Adjunct (ISMAR-Adjunct), IEEE, Sydney, Australia, pp. 167–174. DOI: 10.1109/ISMAR-Adjunct60411.2023.00041

Chapter 12
AR-VR Enabled Personalized Vehicle Exploration for Urban Mobility and Smart City Development

N. Venkateswaran
Department of Master of Business Administration, Panimalar Engineering College, Chennai, India

A. Srinivasa Reddy
https://orcid.org/0000-0001-7600-0371
Department of Computer Science and Engineering (Data Science), CVR College of Engineering, Hyderabad, India

B. Uma Maheswari
https://orcid.org/0000-0001-9707-285X
Department of Computer Science and Engineering, St. Joseph's College of Engineering, Chennai, India

P. Suthanthira Devi
https://orcid.org/0000-0002-7386-821X
Department of Data Science and Business Systems, School of Computing, SRM Institute of Science and Technology, Kattankulathur, India

M. Aishwarya
Department of Science and Humanities, Prince Shri Venkateshwara Padmavathy Engineering College, Chennai, India

J. K. Periasamy
https://orcid.org/0009-0008-5822-1056
Department of Computer Science and Engineering, Sri Sai Ram Engineering College, Chennai, India

ABSTRACT

This chapter explores the use of Augmented Reality (AR) and Virtual Reality (VR) for the development of immersive showrooms, where vehicles, city-wide progressive mobility, smart city development, etc. can be personalized and experienced. AR-VR

DOI: 10.4018/979-8-3693-7287-6.ch012

tech can generate immersed experiences such as real-time customizing, interactive test drives, and even position the client's own vehicle into his area to study how it will function on his street, for example. This all reduces the need for physical showrooms, speeds up decision-making processes, and promotes sustainable practices. Additionally, the chapter scopes the AR-VR potential in enabling urban mobility solutions by delivering real-time scenarios simulation to optimize vehicle designs and enhance the sustainability targets. By bringing together cutting-edge technology with responsive user experiences, it sets a new bar for smart city infrastructure and next-gen mobility systems.

INTRODUCTION

A new era in urban mobility development has been enabled by the convergence of technology and urbanization. Among this revolutionary way of going through the Metaverse, Augmented Reality (AR) and Virtual Reality (VR) are some of the most defensive high-tech that cannot wait for enhancing our way of living in our personal home. New use cases and contexts for urban mobility and smart cities: augmented and virtual reality showrooms (public access) to browse customizable cars. Though the automobile sector is substantially contributing to climate change and working towards developing a new sustainable urban ecosystem, with the incorporation of AR-VR technology, the auto manufacturers are redefining customer experience and providing an enhanced mobility solution(Gabbard et al., 2019).

They're moving from gaming and entertainment into fields like education, healthcare, real estate, and retail. The automotive industry has already made a leap towards these technologies as businesses look to innovate and revolutionize business models and the customer experience. Among the biggest innovations, the immersive showrooms enable users to view cars in their tailored, interlinked configurations—and all of this without the need for a showroom. That also allows your customers to finalize their configuration and virtually experience their future cars and test in simulated prombs at the touch of a button, all without needing to travel into dealerships or other physical touchpoints(Aslan & Yıldız, 2020).

The biggest advantage of AR-VR-enabled showrooms is that they provide a personalized end-to-end experience. Instead of being limited in the same way as regular showrooms that require costly physical space, immersive showrooms push all these restrictions and matters to the side and take customer satisfaction to the next level. With AR, users are able to overlay digital images of vehicles in their environment, allowing them to visualize what a car would look like in their driveway or what a vehicle would look like rolling down their streets. The next level up from this is VR—where users not only experience a full virtual environment but interact with vehicles in the number of ways they'd do in real life: not only exploring the features of the vehicle on offer but taking it for a simulated test drive, seeing how it would look and feel under a variety of simulated conditions(Boboc et al., 2020).

When considering the rise of eventful transport in smart cities, this is a very on-point way to experience an engaging and context-specific journey. However, with the sharp rise of urban density and pollution, there is a demand for a mobility solution that must be answered. AR-VR technologies help break these barriers by enabling city planners and vehicle manufacturers to create solutions that are data-driven and customer-centric. Immersive technologies can deliver sequential learning that includes evaluating electric or autonomous vehicle performance in real-world traffic conditions as well as working through various traffic scenarios. Not only does that give consumers a little more power, but it also provides manufacturers and city planners with useful data that can improve vehicle design and urban infrastructure(Yıldız, 2020).

AR-VR integrated showrooms also provide a significant advantage in sustainability. In addition, by decreasing the necessity for physical infrastructure, these platforms decrease resource consumption and carbon footprints, Wimmer believes. A conventional showroom requires land, energy, and materials to build and operate, all of which will ruin the planet. Furthermore, immersive showrooms leverage digital spaces, which is an eco-friendlier option and adheres to the principles of smart city development. Virtual test drives not only reduce the fuel and emissions generated through individual test drives but also work to reduce physical test drives entirely, further impacting sustainability metrics(Dalle Mura & Dini, 2021).

All of these are classified into two main categories. The first is the mass production facilities with very low manufacturing costs. Operating physical showrooms is extremely expensive due to the real estate, staffing and upkeep involved. Immersive showrooms alleviate some of these costs by creating a virtual environment where consumers can shop and purchase vehicles without the need for on-site resources. That change not only creates operational efficiencies but also opens up new revenue opportunities through worldwide coverage. Without the layers of geography getting in the way of interaction, customers of the world can observe what a brand has to offer, extending the reach of a brand and resulting in brand loyalty(Dalle Mura & Dini, 2021).

While the applications are myriad, AR-VR adoption in captive car showrooms and urban mobility comes with challenges. The primary barrier to the widespread adoption of these technologies is the high capital cost associated with development and deployment. Creating quality virtual environments and ways to interact are sophisticated and often require investment in software and hardware, not to mention expertise that smaller companies may not have the budget for. Moreover, to gain widespread adoption, it needs a frictionless and user-friendly experience. However, if there are technical problems or poorly designed interfaces that repel users, the potential of these breakthrough solutions can be lost(Manimuthu et al., 2022).

Leveraging AR-VR solutions to existing urban mobility systems and smart city architecture. Complex ecosystems of stakeholders, beneficiaries, governing frameworks, and infrastructural constraints inhabit urban settlements. For immersive technologies to realize their potential, strong partnerships between technology suppliers, automotive companies, urban planners, and government entities will need to be built. Standardized platforms, cybersecurity, and privacy issues need to be addressed to realize interoperability within or between different systems(Tripathi & Talukder, 2023).

User adoption and behavioral challenges also significantly drive the effectiveness of AR-VR-enabled showrooms. While younger generations who are relatively well educated about technology might adopt easily, older or less tech-savvy generations may find these technologies intimidating—or futile. Bridging this gap will need focused educational and awareness initiatives that highlight the benefits and simplicity of utilizing immersive platforms. With the provision of interfaces that are easy to use and guided journeys, the adoption rates should continue to scale across different audience slopes(Bale et al., 2024).

Examples of AR-VR in Urban Mobility & Smart Cities Beyond car dealerships, these technologies may also be adopted in public transport systems, urban planning, and infrastructure design. For instance, this technology offers AR-VR realistic simulation of pedestrian and traffic flows and optimization of routes for public transport. Of course, at the planning of new intersections and districts of the cities It allows sustainable urban design. Smart cities, fueled by big data, can shape a more sustainable future through data-driven decisions and community engagement(Gouda & Tiwari, 2024).

AR-VR-enabled immersive showrooms are thus a game-changing innovation at the juncture of technology, mobility, and urban development. Such platforms are ideally positioned to not only change how customers relate to vehicles but also fulfill broader themes of sustainability and urbanization with tailored, interactive, and environmentally friendly engagement. There are some obstacles to implementing and adopting these technologies, but it brings long-term benefits that far outweigh their initial costs and places them at the core of future smart city ecosystems. As these AR-VR technologies continue to develop, so do their potential applications in urban mobility solutions, providing a credible pathway to a hyper-connected, optimized, and sustainable future.

Objectives and Scope

This chapter attempts to investigate the future-oriented potential of augmented reality (AR) and virtual reality (VR) technology to change both vehicle showrooms and urban mobility towards smart city development. It inspects AR-VR technologies to deliver personalized, immersive customer experiences with vehicles that help customers virtually explore and experience vehicles, customizing them and providing interactive simulation. Additionally, it explains how being physically dependent on showrooms affects the environment, economy, and urban mobility and provides a walkthrough into the radical improvements the AR-VR industry can bring to such surrounding problems through simulated world experiences. The chapter defines the scope of how AR-VR could be utilized to enable data-driven sustainable mobility solutions by specifying the technical aspects, challenges, and opportunities.

CONCEPTUAL FRAMEWORK

AR-VR in Automotive Showrooms

Car dealerships are using augmented reality (AR) and virtual reality (VR) to provide customers with an interactive retail experience. She says AR provides customers the ability to visualize vehicles in their own physical environments, with digital content layered atop the physical world in real time; VR focuses more on creating 100 percent immersive experiences, enabling users to "walk around" and explore a full vehicle in a 3D environment. Together these technologies eliminate the necessity for extensive stables, which makes it as cost-effective as a sustainable solution for manufacturers and dealers alike. Users can experience virtual test drives, real-time alterations of the color and features, and the 3D visualization of the vehicle with AR-VR platforms—a dreamy combination of ease and craftsmanship. Customers can 'visit' vehicle interiors, even virtually 'sit' behind the wheel and make informed decisions—all from the comfort of their own home. The analytics is used to tailor the customer experience, so the showroom suggests things according to likes and habits. Additionally, AR-VR technology is employed in sales team's product training and in marketing as well to engage the audience with interactive ads. While urban wear and tear takes the front seat, AR-VR-enabled showrooms tap into the rising levels of appetite for smart and sustainable solutions, in step with the blueprint of the future city and paving the way for new champions in the automotive retail space(Hadas et al., 2018; Heydaryan et al., 2018).

Personalized Vehicle Exploration: Definition and Scope

Personalized vehicle exploration also utilizes advanced digital technologies to tailor the vehicle's selection and personalization process to all of the customers preferences. Both AR and VR converge around these abilities by providing customers virtual access to cars to customize everything from color and trim to engine specs to add-ons in-house. Its reach goes well beyond what we think of as customization by surrounding customers with virtual environments that mimic the look, feel, and handling of the setups they choose. Our algorithms of personalized vehicle/vehicle exploration can use data to inform consumer preferences, allowing manufacturers to shape their design and manufacturing process. Finally, it opens the crossover between real and digital experiences by allowing users to see how a vehicle occupies a handful of predetermined places throughout the city, such as parking lots, open surfaces, or smart city infrastructure. This leads to improved user satisfaction since it allows a more personalized and intuitive exploration experience and a closer customer-product relationship. Interested in the possibilities as an architecture as well—Internet-connected devices updating in real time, records being carried on the fly to existing databases, or inclusion in the future city infrastructure. Customized Vehicle Exploration—As a major trend in urban mobility innovations, vehicle personalization through AR-VR develops the way customers experience vehicles in the kinds of convenience, effectiveness, and sustainability.

AR-VR IN SHOWROOM EXPERIENCES

Augmented reality (AR) and virtual reality (VR) showroom experience: From dealer showroom to an immersive experience via AR and VR: A personalized, augmented (vehicle) and virtual (virtual showroom) with (vehicle) experience, rightsized to our engagement experience. With this technology, customers have access to a visualization of vehicles with customized configurations, as well as driving experience simulations in a virtual environment. AR-VR is changing the staple dealership procedure by combining AR's power to visualize digital features in real-world settings and VR's power to create wholly engaging environments with futuristic choices for technology-smart clients(Lee et al., 2023; Piromalis & Kantaros, 2022). Figure 1 depicts the use of AR-VR technologies in showroom interactions, enhancing the overall experience in the showroom setting.

Figure 1. AR-VR experience in showroom interactions

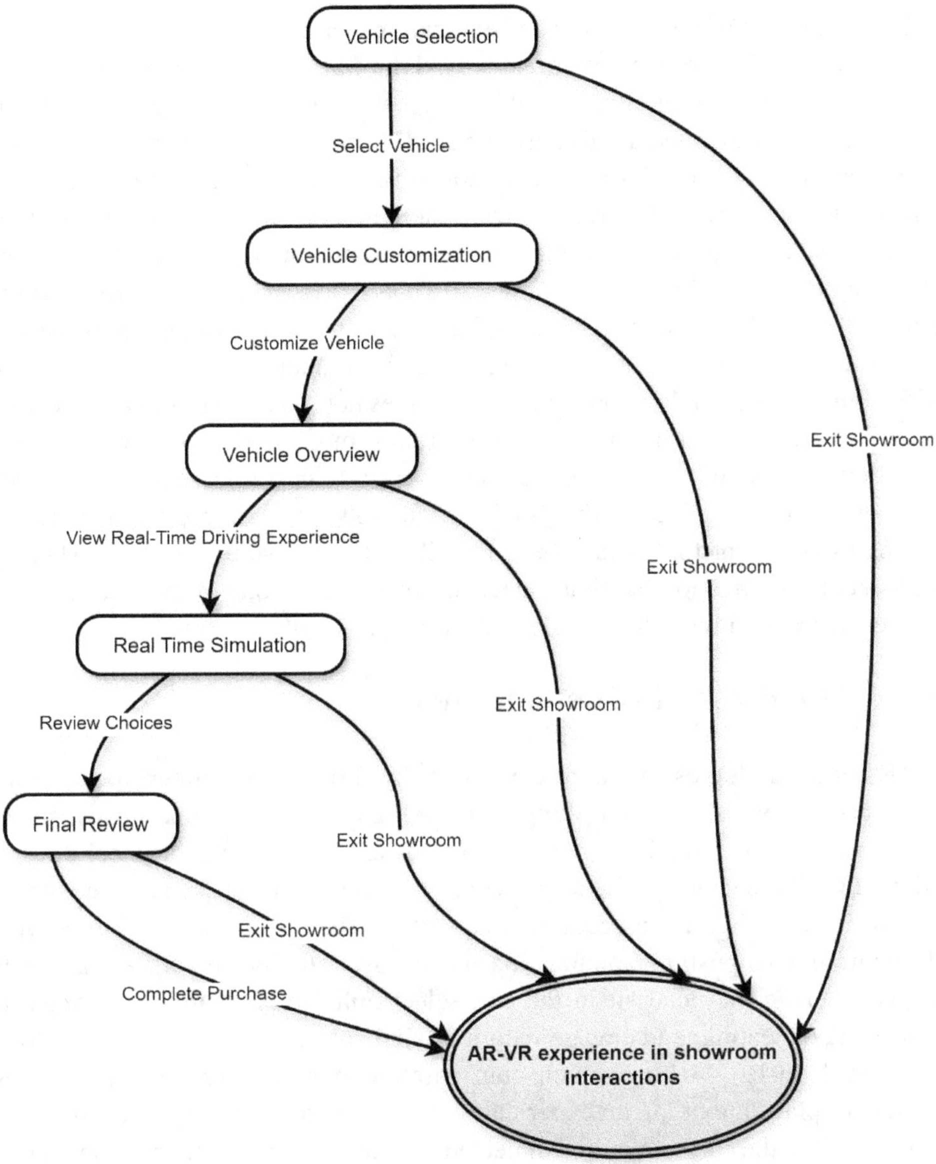

Immersive Visualization Techniques

By making use of the immersive visualization techniques available through AR and VR, automotive companies have changed the way customers view and interact with vehicles. These technologies enable prospective buyers to engage with vehicles in high-resolution 3D models, giving them an all-encompassing perspective of every angle, part, and function. Using AR, customers can project a life-size avatar of the vehicle into their physical space, allowing them to visualize how the car would fit into their driveway, parking space, or urban environment. Virtual reality takes this a step higher by creating virtual showrooms that allow customers to look around vehicles that can be placed in a variety of areas, like paved highways, city streets, or rugged off-road environments that are realistically simulated(Lee et al., 2023; Stark, 2022). Immersive visualization development does not require a physical showroom space, providing the ability for dealerships to show more models, variants, and configurations without needing more showroom space. At customer request, customers can zoom in, for instance, on the dashboard, upholstery, and infotainment system designs to an extent that would be impossible through static displays. Thereby, this level of visualization facilitates decision-making, as none of the visualizations requires interaction with the vehicle to become physically accurate.

Personalized Vehicle Customization

AR-VR technologies enrich the experience by allowing consumers to customize each aspect of a vehicle to their liking. Using AR interfaces, customers can instantly change vehicle specifications, including the exterior paint colors, wheel designs, and lighting features on mobile devices or AR glasses. Changes can be seen in real-time, which can deliver an accurate image of the final product. In a similar vein, VR environments push personalization further by virtually immersing customers in spaces where they can sit in the car, select upholstery materials, change the positioning of seating, and engage with infotainment systems(Besnea et al., 2019; Gerber et al., 2019). Making vehicle configuration more of a figurative art and less of a record played, making an experience as opposed to capturing, creating to get the most out of the car, so a more connected appointment experience to the product and the customer. In addition, AR-VR platforms are also integrated with AI-based analytics, which recommend configurations based on customer preferences, purchase history, and market trends, with personalized recommendations that align with lifestyle and requirements. These brands simply don't have decades of input from millions of consumers to inspire their models, but the direct connection that the PC offers creates a more tailored, engaging, and fun experience that is enforceable consumer-focused.

Interactive Test Drive Simulations

However, interactive test drive simulations are one of the most innovative use cases of AR and VR in showroom experiences. Conventional test drives are constrained by logistical issues, ranging from vehicle availability to time pressures and environmental conditions. AR-VR technologies offer solutions to these challenges in customers by allowing them to experience a hyper-realistic driving environment in their synchronized virtual world(Hadas et al., 2018; Heydaryan et al., 2018). Using virtual reality headsets and motion-sensing technology, users can experience driving a vehicle through various terrains, weather conditions, and traffic scenarios. Combined with these simulations, customers get a complete idea of how the car is, its performance, handling, features, and basis for making the decision to purchase. Adding to this, AR also provides heads-up displays (HUDs) that show real-time information on the vehicle's performance, like speed, fuel efficiency, and navigation, to be projected in the customer's field of view as the simulation is running. Giving consumers the ability to experience a fully simulated test drive allows them, in their own time, to acquire those things that sit under the hood as well as are as much a part of the driving experience as the actual driving experience, giving consumers the edge that only a digital experience can offer them. For example, customers can live-experience advanced driver-assistance systems (ADAS), electric vehicle (EV) functionalities, and connected car technologies. By replacing physical test drives with AR-VR technology, the automaker can contribute towards sustainability and reduction in emission levels due to fuel consumption.

AR-VR-based showroom technology improves the best shopping experience and also provides automotive manufacturers and dealers with a great experience in different domains. It lessens reliance on more extensive physical spaces and opens the door to smaller, economical, and eco-friendly showrooms first created online." Additionally, customers can tour a showroom from the convenience of their living rooms through these technologies, eliminating geographical limitations and tapping into a completely new demographic of prospective buyers. AR-VR integrated customer relationship management (CRM) systems also use this information to track interactions, preferences, and feedback from these users, which dealerships can use to refine their offerings and services.

Thus, AR-VR in showroom experiences makes one ever used to the touch and feel of a car pre-purchase journey. My future-ready immersive touch experiences are sustainable. Such immersive visualization enables customers to configure bespoke vehicles to their taste and carry out virtual test drives before making an informed purchase decision, all of which are hugely customer-centric. These technologies are vital for automotive manufacturers and dealers to address the changing needs of the modern shopper and the innovation agenda for retailing automobiles alongside smart city practices that can make urban mobility the focus.

Figure 2. Processes where AR and VR are integrated into urban mobility

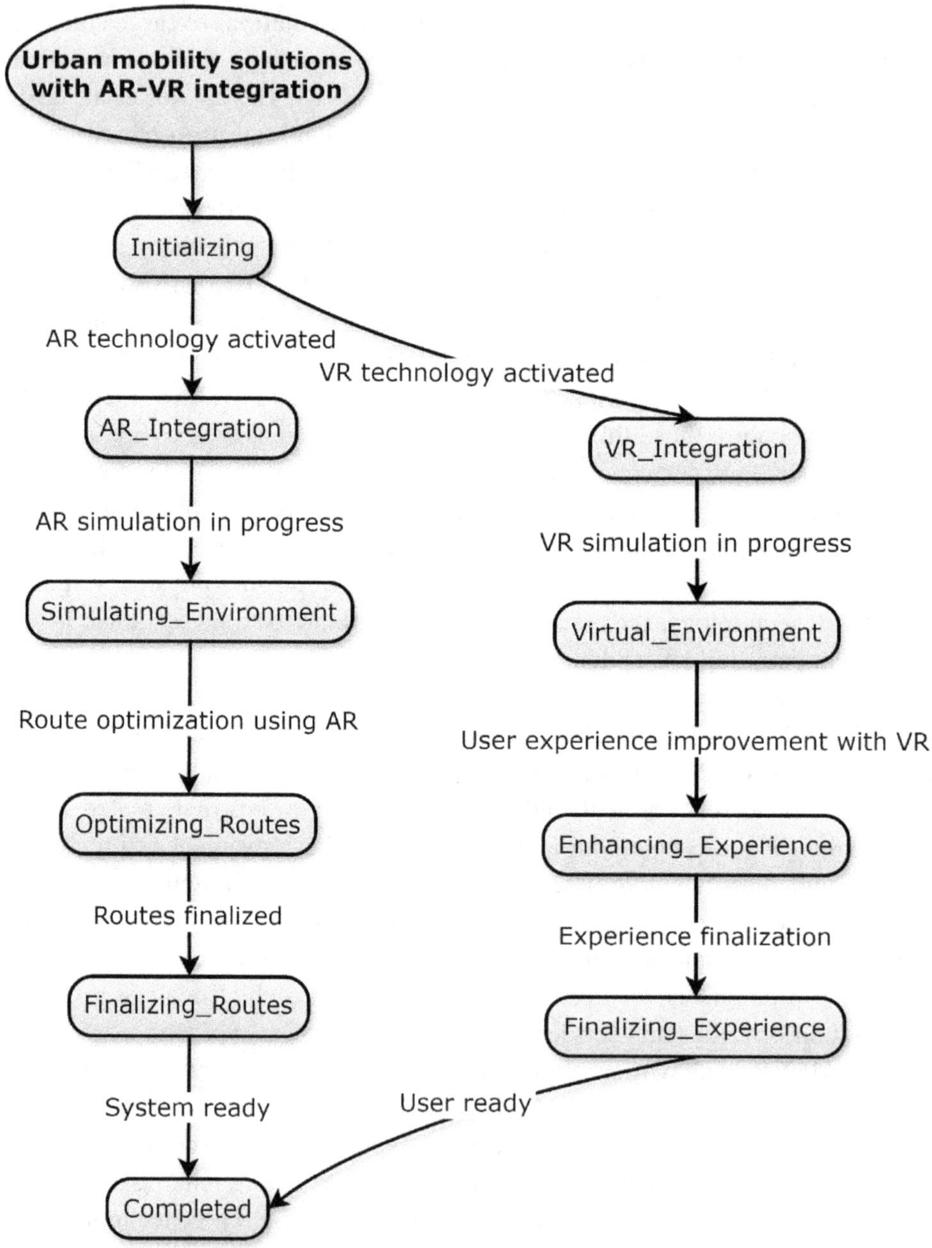

URBAN MOBILITY SOLUTIONS WITH AR-VR

Augmented Reality, or AR, and Virtual Reality, or VR, technologies are redefining urban mobility, from the way that metropolitan areas rethink their transportation networks to the way that vehicles are conceived and constructed—and everything, including the way a more sustainable future is created, in between. Adopting these technologies brings enormous gains in user experience and systems performance via real-time immersive simulations and optimizations. Augmented and Virtual Reality Impacts on Urban Mobility next to Impacts of AR and VR in Urban Mobility: As cities become smarter, the impact of AR and VR in urban mobility will be greater, which will lead to innovations in smart vehicles, sustainable transport systems, and sustainable initiatives. By enabling the most unprecedented degrees of interaction, visualization, and data analysis, AR and VR are helping cities and manufacturers develop more efficient, user-friendly, and sustainable transportation technologies(de Freitas et al., 2022; Kim & Oh, 2022). The integration of AR and VR into urban mobility processes is depicted in Figure 2.

Enhancing Mobility through Real-Time Simulations

Illustrated in the second chapter of this report, these dynamic environments, supported by AR and VR, are critical to urban mobility, as they facilitate the transfer of information and enable participants to better comprehend the complexity involved in transportation systems. Through transportation scenarios, including future developments, city planners, automotive designers, and even the public can preview how transportation networks will perform before they are built and what a transit world will do in many situations. AR and VR are enabling the user to experience real-time simulation of patterns in traffic flows, pedestrian movement, and public transportation routes that can inform urban planning and infrastructure design or transportation policies(Wang et al., 2023).

As an example, city planners can use VR simulations to see how traffic patterns might change when they add new roads, public transportation systems, or carpool lanes. Some of the techniques commonly used in this domain would be data mining, pattern recognition, etc., that can identify bottlenecks, traffic jams, and accident-prone areas, which, over time, we will be able to take data-driven decisions that can help make us more efficient and safer. On the consumer side, real-time 3D simulations let drivers see routes in virtual city environments as they adapt to current traffic, construction, and accident locations. This kind of dynamic interaction levels allow users to decide where to go and when, preventing congestion and ultimately improving mobility. AR also overlays digital information on a driver's real-world perspective: it can provide navigation directions, alerts about road hazards, and information about nearby parking spots, enhancing the user experience while also helping to address issues like traffic congestion and parking availability.

Moreover, the AR-VR technologies enable real-time virtual simulations of the vehicle performance across a variety of urban conditions. These simulations provide automakers with the ability to test vehicles in a virtual cityscape that closely resembles real-life environments, which include a wide range of road conditions, weather patterns, and urban infrastructure. Such intelligence assists manufacturers in tailoring vehicle components, such as the suspension systems, braking, fuel economy, and electric vehicle range for metropolitan areas, which makes their designs suitable for city behavior. Such simulations may also validate autonomous vehicle systems, such as lane-change assist, braking, and other collisions that may only reach drivers beyond metropolitan areas.

Vehicle Design Optimization for Urban Environments

With the evolution of cities and the increasing demand for urban mobility, vehicle design optimization has emerged as a key factor in meeting the demands of urban environments. AR and VR technologies are at the forefront of this optimization, providing solutions to the challenges of urban living, from congestion to limited parking to environmental impact to new shared mobility options. Manufacturers can use VR to test different drives and environments across all their models, providing a realistic idea of how a vehicle will perform in real-world situations. This enables designers to safely run new vehicle models through virtual city streets—narrow lanes, crowded intersections, and high-traffic areas. These simulations are essential in fine-tuning vehicle size, turning radius, and handling so that whatever's on the road can slide through constrained urban streets with no trouble. This optimization goes well beyond the vehicle itself and also has to cover features such as parking assistance systems, autonomous driving, or energy-efficient designs that become very laudable parameters for urban mobility systems(Ejichukwu et al., 2024).

AR, by contrast, lets designers and consumers place different car configurations into a real-world city context and see how the car will actually look where they plan to drive it. Through AR, for instance, consumers can request that a car be delivered as if it were parked in their driveway or in the local environment in which they typically drive an EV. This visualization aids making choices towards manufacturers and customers; one of the reasons is to understand if the vehicle is adaptable to its environment (urban or not) and if it fits the lifestyle of a modern, tech-savvy, city-dwelling consumer that looks for functionality/style/sustainability in a vehicle. Additionally, it allows for design validation of new vehicles by testing them in Phoenix, Arizona, without the need to build and develop prototypes. Manufacturers, for example, can model the behavior of battery-powered vehicles with varying battery capacities in crowded urban traffic, accounting for real-time charging stations and energy consumption. This optimizes vehicle designs to be sustainable, considering how efficient, environmentally friendly, and practical they are given the current urban landscape.

Role in Sustainable Transportation

With the aim to accomplish the target to reduce carbon emissions, ameliorate air quality, and incentivize the use of sustainable transport modes, the segment of AR and VR in sustainable transportation is being highly pronounced. Additionally, AR and VR can optimize transportation systems and vehicles to reduce waste and energy consumption, and therefore environmental impact. Information technologies based on these technologies can map sustainable transportation systems, electric vehicles (EVs), and sharing transportation systems, or smart grid types of systems such as green transportation networks. These advantages contribute to sustainable transportation, specifically with how AR and VR can optimize electric vehicle design and performance(Sadeghzadeh et al., 2024). VR simulations enable manufacturers to test how efficient an EV battery is at charging times in diverse urban situations, from heavy traffic and steep slopes to adverse weather. The insights earned along the way will serve to help improve vehicle performance, but will also ensure that EVs are fitted to life in our cities, helping to allay fears over range anxiety and charging infrastructure. VR simulations can also model how EVs will fit into our existing transportation networks to help identify where we can shore up the infrastructure (like charging stations) or where we may need to upgrade the grid to better accommodate a surge in the demand for electric cars and toners, he said.

AR also contributes to the sustainability of transport systems. Using AR navigation, drivers can be guided on the most fuel-efficient routes, thus resulting in lower fuel consumption and emissions. AR can provide real-time information about the fastest routes—enabling the detours through traffic jams—and recommend alternative paths that use less fuel, for example. Through AR, for instance, end-users can visualize, in their path, people and vehicle sharing that include stations for bike-sharing or even e-scooters that endorse alternate mobility alternatives to private cars. Thus, AR and VR technologies are changing the aspect of urban mobility by enhancing real-time simulators, optimizing the design of vehicles to suit urban conditions, and promoting sustainable travel habits. From rich, immersive experiences to data-centric insights, these technologies are leading the charge towards pedestrian-friendly, adaptable, greener, safer, and smarter solutions and services for the most legitimate future of living in urban environments. By leveraging the physical urban landscape to enhance the immersive experience, AR and VR will become key players in era II of urban mobility evolution: as enablers of shaping and building smarter, more livable cities and also as connectors of seamless travel experiences through which users explore the city around them.

IMPACT ON SMART CITY DEVELOPMENT

AR and VR have completely changed the game for smart cities, redefining how urban environments are designed, implemented, and experienced by residents. They enable more efficient use of resources, intelligent transportation systems, and greater sustainability—all of which are essential to a smart city that works. By reducing the volume of physically necessary infrastructure, supporting eco-automation, and literally linking to the Internet of Things (IoT) and smart systems, AR and VR become an integral part of the development of future urban landscapes. These solutions improve residents' quality of life while creating a more efficient, sustainable, and connected city(Boopathi, 2024; Jha et al., 2024; Maguluri et al., 2023). The use of AR and VR technologies has a significant impact on the development of smart cities, as depicted in Figure 3.

Figure 3. Impact of AR and VR technologies on smart city development

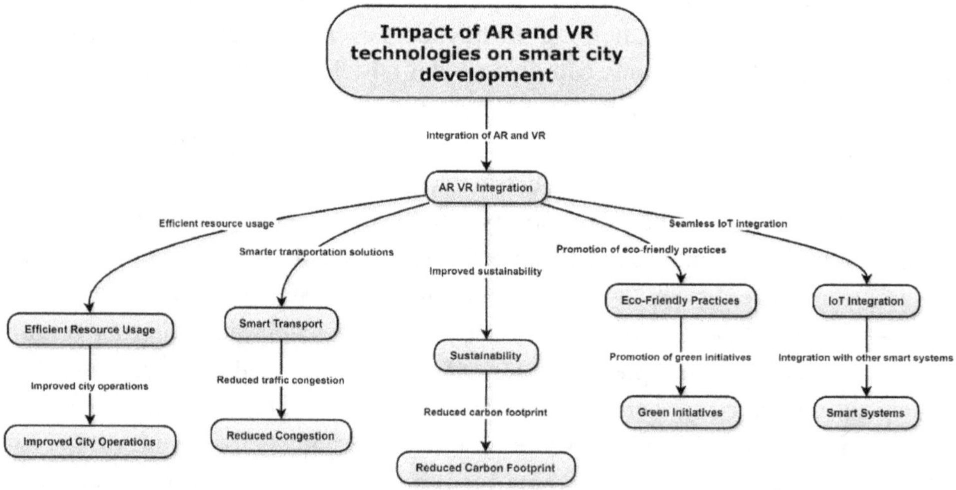

Reducing Physical Infrastructure Needs

One major benefit of AR and VR technologies in smart city development is the possible reduction of physical infrastructure. Since conventional cities occupy vast physical space for their roads, transport systems, utility networks, and commercial spaces, this often results in urban sprawl, inefficiencies, and environmental degradation. Bringing in AR and VR, urban centers can achieve virtualized many, if not all, of these elements, drastically reducing the physical space required to accommodate urban revitalization and allowing for more sustainable solutions(Boopathi, 2024; Molakatala et al., 2024).

The most conspicuous case could be in vehicle sales and transportation, where AR & VR technologies have made it possible to build virtual showrooms, greatly reducing the need for big physical dealership spaces. Viewing cars and other vehicles from home, consumers can browse and customize vehicles or even test-drive virtual vehicles in virtual reality, cutting down on the amount of physical real estate set aside for traditional showrooms as well as reducing transportation-based emissions from physical test drives. This vast reduction in physical concoction helps conserve invaluable urban spots, in addition to benefiting the objective of reducing our environmental jackpot at the footings of big facilities on the retail and transportation. Additionally, mind you that we go forward in construction and city planning, as virtual simulations allow city planners to observe urban blocs as a whole, their transport networks, as well as the public services as much as their monuments, before they even start building. This leads to better decision-making that is justified by the need for and impact of infrastructure projects. VR, for instance, can simulate how a new public transportation line will affect traffic patterns, pedestrian movement, and services distribution, Heller said. These simulations greatly reduce trial and error compared to typical large-scale construction projects, thus saving money spent on wasted space, resources, and unneeded renovation of already separated infrastructure.

Promoting Eco-Friendly and Efficient Practices

The concept of smart cities and its potential to transform all the facets of urban life, including transport, education, healthcare, and energy, is finally coming to fruition, thanks to trends such as urban sustainability being furthered by AR and VR across industries. It can enable streamlined management of resources, reduction of waste generation, and promotion of energy-efficient behaviors—activities that are complementary and aligned with the overall environmental goals of a smart city(Molakatala et al., 2024; Sharma et al., 2024; Sonia et al., 2024).

For transportation, AR and VR help to optimize the movement of people and goods and can provide real-time simulations of strategies to minimize congestion, emissions, or visual pollution. Theories and VR-based traffic simulation potential use cases Theories and VR-based traffic simulation potential use cases City planners can leverage these simulations to make data-driven decisions on where to deploy sustainable transportation infrastructure, helping ensure that new projects are aligned with sustainability objectives.

Additionally, AR navigation systems change the way human beings communicate with their environment. AR promotes eco-friendly alternative modes of transportation because it provides real-time information about the nearest public transport options, bike-sharing stations, and electric vehicle charging points. AR interfaces can tell users the fastest way to their destination while also recommending alternative routes, including cycling paths or public transport, which aids in reducing road congestion and individual travel carbon footprints. AR and VR paired with data systems play a significant role in reducing the global ecological footprint of urban transportation by enabling transport networks to align with the principles of sustainability.

AR and VR also needed green designs for buildings and infrastructure. For instance: In energy-efficient building simulations in VR, it's a means by which to be able to test out different designs, different materials, and different energy systems, and before you even have to pay to construct it, which is pretty darn cool." That data helps architects and city planners create optimized building layouts that can help reduce energy consumption, waste, and overall carbon footprints. Virtual reality also allows for virtual inspections that replace physical site visits, reducing construction-related travel and carbon emissions."

Integration with IoT and Smart Systems

Interconnected, responsive urban development's core feature of smart cities is the integration of smart devices, sensors, and other digital technologies into an interconnected, responsive urban environment. - AR/VR: Working with Internet of Things (IoT) devices, these platforms are often integrated into the platform to provide intuitive interfaces for controlling and monitoring these devices while visualizing data in real time. This use is beneficial for the enhancement of an urban management model and the environmental quality of life(Malathi et al., 2024; Molakatala et al., 2024).

When integrated within transportation systems, IoT-enabled devices can capture data related to vehicle performance, road conditions, and traffic congestion, which can then be visualized by the AR interfaces. It gives drivers live information about the availability of parking spaces, local EV charging points, and even road traffic accidents so that they can make better choices about their journey. For example, city planners are using VR in conjunction with IoT sensor data to model traffic flows and explore correlation and impact as a means to generally reduce congestion and improve public transport routes and improve mobility as a whole.

AR and VR also supplement smart city infrastructure by offering dynamic, interactive means to manage public utilities and services. Through AR, maintenance workers are also presented with real-time, context-relevant information about the project—whether it's water lines, power grids, or sewer systems—with which they can locate problems and respond to them faster. VR can also offer training environments for utilities and emergency services, with power outages, water leaks, and other simulated scenarios preparing them for real-life events.

So overall the AR and VR technologies will play a transformational role in the total development of smart cities as they will help in bringing efficiency, sustainability, and enhanced quality of life for the urban citizens. Combined, these technologies help create smarter, more sustainable, and resilient cities by responding more intelligently to the needs of their inhabitants, minimizing the need for the extensive physical infrastructure necessary to interact with such systems, and more readily promoting environmentally conscious behavior that is naturally integrated with IoT and other smart systems. As the next generation of urban environments becomes smarter and more connected, AR and VR technologies are set to play an ever-greater role in both their design and evolution.

APPLICATIONS IN URBAN MOBILITY

Integration of AR and VR into urban mobility systems When we extensively study AR and VR applications to urban mobility systems, we tend to get into wider discussions regarding AR and VR systems that enhance user experiences, enable better decision-making, and optimize the urban mobility system as a whole. These technologies will have a profound effect on how people will engage with their surroundings and how they will experience future mobility services, providing an opportunity to tackle many of the major challenges of urban mobility, such as congestion (and its management), accessibility (in terms of vulnerable groups), sustainability, and efficiency(Jha et al., 2024; Sundaramoorthy et al., 2024). The enrichment of customer experience, in-situ scenario simulations, and the data-driven optimization of flows between mobility actors through AR and VR approaches allow mobility players to introduce tap-into technology to smarter and more fluid urban mobility deployments.

Figure 4. Interaction between various applications of AR and VR technologies in urban mobility

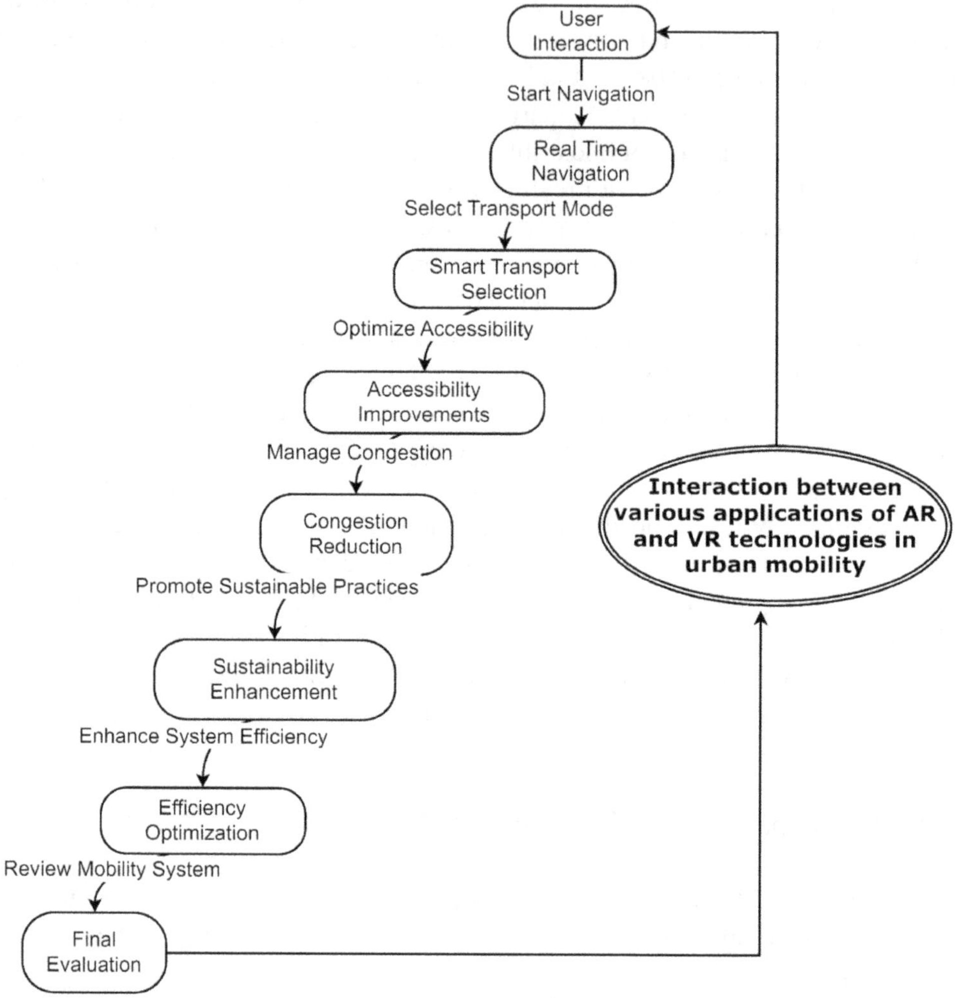

Enhanced Customer Experience

The entire customer experience inside urban mobility systems is being reinvented through AR and VR. Across more advanced cities in an increasingly connected society and a wider variety of transportation options, users are demanding not only smart, connected but personalized, effective and seamless means of threading their everyday urban existence together. AR and VR may serve as a solution to these demands as they can provide intuitive and immersive interfaces wherever their users are, enabling them to access the respective corresponding mobility services in real-time, resulting in better convenience and gratification(Naveeenkumar et al., 2024; Ravisankar et al., 2024).

AR can also enhance the experience of the commuter such as giving details about the schedule of the bus or train in real-time, the density of the crowd, or next stop. Mobile-powered augmented reality applications grant real-time information to passengers, giving them a mobile-powered guide that allows users to know the fastest route available, or where the nearest bus stop is located, or look for an available parking place. This data helps users to be well prepared for their trips and also reduces the waiting time, contributing to a smooth and delightful process. Applying for a VR employee to make-believe travel routes when you leave for another outing assists travelers in understanding their experience better and picturing transit associations.

With AR, users of car-sharing and ride-hailing services can see the cars nearby, the type of vehicle, fuel efficiency, and a cost estimate on their phone screen or AR glasses. This level of visibility allows more control over the choice of ride and helps build trust with the customers. Similarly, VR can provide an experience akin to being in various types of shared cars, enabling users to get a feel for the ride quality, comfort, etc. As augmented and virtual reality have their presence in urban mobility, taking off the glasses, at least, and putting on the glasses will elevate the level of customer satisfaction and convenience, and Urban mobility services can pack a much more attractive and consumer-friendly outfit.

Real-World Scenario Simulations

AR and VR technologies are fundamentally transforming the field of real-world scenario simulations in the sense that they enable city planners and the people who live there to understand how urban mobility systems would behave in the real world. Real-time simulations give key decision-makers the power to predict the effects on areas of new policies, transportation infrastructures, or alternative mobility solutions before they are implemented and minimize the risk of potentially costly errors. Simulating urban mobility systems can help to gain detailed insights on traffic patterns, pedestrian flows, and environmental(Sekhar et al., 2024).

VR predictions of how traffic would flow, for example, when adding new roads or bus lanes or implementing a bike-sharing program, could be used by planners at a city or transportation authority. This would allow simulating mapping interconnected worlds and testing various urban mobility options in relation to how they might affect congestion, air quality, and efficiency, for instance. Backed by data through October 2023, these simulations allow for the testing of infrastructure interventions that might exacerbate traffic bottlenecks or aid in developing better access to public transit. By simulating a variety of scenarios and testing numerous options in a virtual environment, cities are capable of making better-informed, data-based decisions on how to optimize their mobility systems and their entire urban design.

With regards to user experience, VR simulations provide users with an opportunity to "try" out various transport types in simulated urban environments before making choices in the real world. It can help to test a new metro line or bus or bike-share service in a virtual environment to see how comfortable and then how fast services are likely to be, for example. These simulations give users an idea of what they may have to choose from with corresponding modes of transportation and how to assess which modes meet their specific needs. Transport providers are also fed back from them to amend their services as per user expectations and experience.

Data-Driven Optimization of Mobility Solutions

Moreover, AR & VR can also facilitate data collection and navigation, as well as improvement of smart transportation networks. The incorporation of emerging technologies like advanced analytics, artificial intelligence, and big data science allows for real-time data collection, analysis, and visualization, augmenting potential for smarter, more efficient mobility solutions(Paul et al., 2024).

AR and VR Optimize Urban Mobility with Data Overlays, Predictive Models For instance, augmented reality (AR) technology research argues that AR can be applied for navigation systems so that drivers and commuters will gain access to real-time data traffic flow, road condition, and accidents while moving around the city. Such a degree of access to live data provides users with the ability to make the right call between the routes and avoid traffic jams as well as identify alternative routes to their destinations to make trips shorter. In contrast, AR and VR can provide that driver with additional, contextual data, suggesting the most efficient driving route given dynamically changing traffic conditions throughout the day. That cuts down on delays and makes the most of individual commutes.

AR and VR also support the organization of public transport by collecting detailed and real-time data to guide operational decisions. One way to do this is through VR simulations, which can be used to study the flow of people through stations to determine where crowding might form, thus allowing designers to anticipate features of the station, like platforms or train schedules, that should be optimized. When the day comes that cars communicate with each other and infrastructure, in combination with a swarm of IoT and sensor data, VR models will also be able to predict before it happens where congestion may occur and allow for the proactive deployment of additional services to meet demand. Likewise, data collected through ridesharing and car-sharing aggregators can be mined in virtual environments to forecast demand and optimize the fleet of cars, assuring that the amount of cars will be enough during the most congested periods while decreasing idle times compared to other periods and optimizing fuel use.

Hence, AR and VR technologies are likely to have a drastic impact on urban mobility by enabling new solutions that can improve the customer experience while simultaneously allowing scenario simulations and data-driven optimizations. By enabling cities to leverage immersive, real-time simulations and data making the case for their mobility ecosystem, these technologies allow cities to minimize inefficiencies within transportation systems and offer mobility experiences that are more tailored and convenient to citizens. As these technologies are continuously being developed, their ability to revolutionize urban mobility will grow further, providing smarter, more efficient, and sustainable cities while also enhancing the quality of life for cities' inhabitants.

CASE STUDIES

Implementing AR and VR in Urban Mobility Solutions

A growing urban center that had problems with its public transport system. Traffic jams, parking issues, and lack of real-time data proved to be a challenge for both commuters and city planners. Consequently, the city's transport authority decided to deploy augmented reality (AR) and virtual reality (VR) technologies for stimulating mobility, enhancing infrastructure, and improving user experience. It was also supposed to make for a much smoother, cleaner, and data-optimized urban transport system(Boopathi et al., 2022).

Challenge: City The doomed public transportation network was plagued with congestion, schedule inefficiencies, and crumbling infrastructure. Traffic jams occasionally bedeviled commuters along the corridor, but they had scant sightlines to potential alternate routes. Transportation providers were struggling to optimize their services—and dissatisfied urban mobility users—but urban mobility solutions were sprouting on disparate, unconnected screens. Concurrently, greater demand is emerging for sustainable, liftable, and user-friendlier modes of transportation.

Solution To address these issues, City deployed AR and VR technologies in several ways:

a) **AR for Navigation:** The city integrated AR navigation solutions on each commuter's device. It offered users live traffic threats, public transport info, and parking availability. An intelligent augmented-reality interface provided basic direction—including back routes, warnings of pending road closures, and suggestions for utilizing shared transport, such as bikes and electric scooters.

b) **VR Simulations for Urban Planning:** The city transport authority used VR simulations to visualize how the new transport policies, infrastructure changes, and mobility solutions will affect low-income communities. City planners could experiment with different setups, such as what if they added more bus routes to the system or used smart parking solutions, in a virtual model of the city so they make the best decisions without the associated risks of making physical changes to the infrastructure.
c) **Data-Driven Fleet Optimization:** AR collected real-time data through the IoT devices installed into the public transport vehicles and used the information to monitor the vehicles regularly, their fuel consumption, the number of passengers, etc. Drivers had a ton of information available, with real-time updates fed through what amounted to an augmented reality display in their car, providing data about vehicle health and the best courses through the unpredictable external environment, taking into account not just traffic but environmental conditions. This increased uptime and efficiency and reduced carbon emissions.

This made the city better able to leverage its urban mobility system with its AR and VR features. Traffic congestion decreased by 25%, and commuters had an average travel time 30% reduced. Times when you boarded as a commuter on public transport service were catered to better as it got more efficient, roads got optimized, and routes cut in distance but fewer kilometers. Additionally, our findings indicated that AR navigation can also promote non-motor modes of transportation such as bike-sharing (20%), thus contributing to more sustainable urban mobility. "This enabled us to get to our VR simulations" that were not only a basis for more informed city building but also quicker decisions, he added, so "investments about infrastructure" were affordable as well as aligned with the decades-long evolution of the city.

Some of the most interesting applications of AR and VR technologies to urban mobility are in the city strategy report on the role that AR and VR play in enhancing transportation systems. They are, in fact, essential for the development of optimal, efficient, and sustainable cities: AR and VR answer customer experience needs, provide real-time data, and enable superior mobility solutions.

CHALLENGES

The coming boom: The use of Augmented Reality (AR) and Virtual Reality (VR) in urban mobility systems brings tremendous benefits to the mobility experience and efficiency of successful mobility strategies in urban environments. There are technical and implementational barriers and cost and accessibility issues. Answers to these questions are important not only for successful transmission of significant data but also for ahead investments of time, energy, and resources in AR & VR, which will equally help develop intelligent & most effective modes of urban mobility(Boboc et al., 2020; Gabbard et al., 2019).

Technical and Implementation Barriers

Many cities have legacy systems in place, older public transportation networks and road infrastructure, as well as older fleet management solutions that don't easily integrate with new, advanced technologies. Utilizing AR and VR, the systems need to be enhanced and integrated with new data-sharing platforms, IoT devices, and real-time monitoring systems. This inevitably requires extensive time and technical expertise and coordination of multiple departments and stakeholders, from government authorities to technology vendors to transport providers. And, like any next-generation consumer experience that relies on a special app that connects to the internet, the performance of AR and VR applications is very much tied to a well-performing and, in the case of densely populated buildings, also stable and reliable network connection. In the event where connectivity is initial or potentially weak, these technologies are compromised, leading to system lag or partial data presentation, navigation errors, and in real-time scenarios where data is significantly influenced, transitional data will be absent. For AR and VR tools to work properly, there needs to be a clear defensive investment in the backbone of cities—particularly a 5G or fiber-optic infrastructure backbone—which can be costly and lengthy to build out.

The development and implementation of VR simulations on urban mobility come with their own challenges too. Simulating real-world conditions, such as traffic patterns, environmental factors, and human behavior, involves large amounts of data collection and advanced algorithms. Must make sure these simulations are accurate and intuitive to use, which can be technically challenging (and needs to be continuously refined to adapt to dynamic aspects of urban behavior).

Cost and Accessibility Concerns

The cost of developing, deploying, and maintaining AR and VR in urban mobility is another big hurdle to widespread adoption. To have these technologies in position, considerable investments are required on the parts of hardware, software, capital expenditure, infrastructure, and experienced persons. These costs to be paid up front may be even too high for an average city—especially in developing countries or in ones that are in fact poor. And along with the costs of long-term installation, ongoing updates and maintenance are needed to run these systems. However, access is the key issue. For the most part, AR and VR applications such as this can provide a much more immersive and efficient experience for the user; however, not all residents have access to the required hardware or technology to experience these solutions. Consider the high-quality VR meditation simulations, which will call for costly headsets that not everyone can afford. Similarly, AR-empowered versatile applications (applications) require cell phones with significant capacities, for example, GPS, camera highlights, and high-level handling force, which may not generally be promptly available to all, particularly to low-salary or underserved people.

This creates a significant gap in access, leaving a portion of the overall population out of the equation when it comes to leveraging the advanced capabilities of AR and VR technologies. Cities will also need to figure out how to make each of these technologies cheaper and have a low option in each category so that those without access to devices that can handle these new technologies will still have the ability to engage. AR- and VR-based services could be offered free of charge through public libraries, transportation hubs, community centers, etc. This would bridge some of the equity gap and ensure equal access to these new mobility solutions.

In a more complex way, AR and VR technologies can address the role for both in the conceptualization and planning for the mobility option; however, it depends on these technical, financial, and accessibility drivers to be introduced. This is a concern that cities must leave challenging integration that or risk connectivity, cost, and accessibility of devices that makes the penchant deployment in need of help for all residents.

FUTURE TRENDS IN AR-VR FOR URBAN MOBILITY

In the summation of AR and VR technologies in urban mobility, there are many emerging prospects, summarizing ways in which cities can be revamped for improved transport means and enhanced life quality of residents. A few trends are emerging that could significantly alter urban mobility solutions in the coming years, as these technologies continue to improve(Bale et al., 2024; Gouda & Tiwari, 2024; Tripathi & Talukder, 2023). Future trends of urban mobility using AR-VR technology are shown in Figure 5.

Figure 5. Future trends in AR-VR for urban mobility

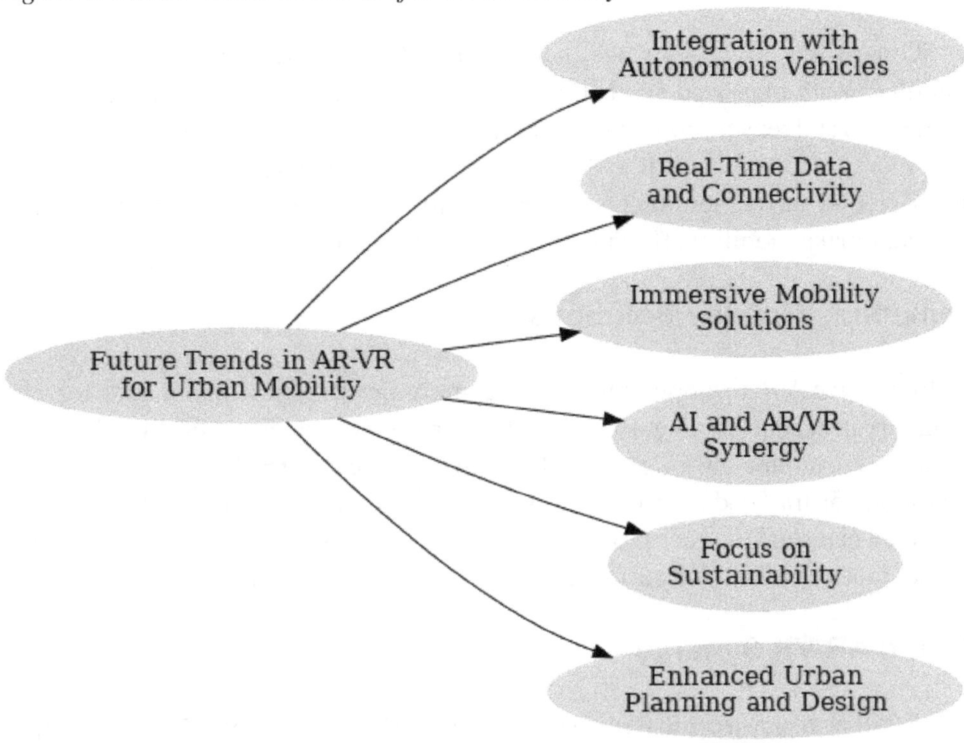

Widespread Integration with Autonomous Vehicles

Augmented Reality and the Autonomous Revolution: As we approach the dawn of autonomous vehicles (AVs), there will undoubtedly be a large part of our lives integrated with them. Using AR for AVs allows for its use to detect real-time navigation and road condition mapping, giving the vehicle a sense of certain situational awareness. VR simulations will also facilitate realistic representations of the way in which AVs might interact with pedestrians, which would enable urban planners and vehicle manufacturers to program for a desired AV behavior in the dense environment of a city: developing and governing the mutually acceptable AV behaviors that strive to create the best conditions for integration into transport infrastructure and safe interactions with other road users (pedestrians, bicycles, and other vehicles).

Real-Time Data Sharing and Enhanced Connectivity

5G networks would enhance the working of. What are AR and VR in urban mobility? With improved and extremely reliable connectivity, real-time data from IoT devices and sensors embedded in vehicles and infrastructure will likely integrate even more smoothly into AR and VR applications. Such information will allow for hyper-precise navigation, real-time updates for commuters, new ways to manage flow and make mobility efficient, and help reduce congestion.

Immersive Mobility Solutions

Immersive VR technologies will become more than just simulators for city planning, instead providing users with fully immersive environments to plan daily commutes, examine transportation alternatives, and even simulate adverse weather conditions or traffic disruptions. This change will allow commuters to better see how they can shift to other modes of transportation (like e-bikes or shared SUVs) to take more efficient and sustainable ways to get to their destination.

AI and AR/VR Synergy

AI will also play a critical role in accelerating the potential of AR and VR applications across industries by offering predictive analytics and providing personalized solutions on how to navigate in mobility. For instance, AI-powered AR interfaces could present real-time explanations in line with user preferences, ensuring optimal navigation and transportation options while simultaneously minimizing environmental consequences and length of transit.

Thus, with the ongoing technological developments, AR and VR will play transformational roles in urban mobility and create smarter, safer, and more sustainable transportation systems.

CONCLUSION

The Augmented Reality (AR) and Virtual Reality (VR) into urban mobility systems will revolutionize the future of smart cities. Augmented reality helps with customer experience by offering immersive experiences, providing real-time navigation, and tailoring transportation solutions. These technologies are also used by city planners, who can visually represent urban infrastructure and optimize it through various scenarios. AR and VR facilitate more informed decision-making and sustainable development by simulating dynamic data-driven scenarios that can help anticipate the effects of different transportation policies and designs. Nevertheless, these technologies' successful implementation has to juggle several hurdles, including technical integration with existing systems, high deployment costs, and guaranteed access for all urban residents. Despite noticeable barriers to company-road and company-consumer polarization with AR and VR technology—such as ethical usage issues, regulations, and high development costs—their transformative capabilities, from improving traffic management to fleet optimization and sustainable transport solutions, make them core contributors to the future of urban mobility.

As technology continues to progress, emerging trends such as the rise of autonomous vehicles, advanced connectivity via 5G networks, and AI-driven personalization will only augment the role of AR and VR in the evolution of urban mobility. Thus, AR and VR are helping us not only navigate the cities we live in today more effectively but also paving the way for smarter, more efficient, and more computable cities in the future. So it can be said that AR and urban transportation and VR and driving simulation all contribute a lot to the global sector of smart cities directly by providing effective, efficient, and sustainable tech to effectively illegal residing people as living condition-improving technology.

REFERENCES

Aslan, B., & Yıldız, A. R. (2020). Optimum design of automobile components using lattice structures for additive manufacturing. *Materials Testing*, *62*(6), 633–639. DOI: 10.3139/120.111527

Bale, A. S., Patil, M., Prathima, M., Telgane, S. V., Ghorpade, N., & Anish, B. (2024). Augmenting Knowledge Management with Immersive Technologies: Exploring Transformative Use Cases for AR and VR. *2024 International Conference on Advances in Computing, Communication and Applied Informatics (ACCAI)*, 1–6. DOI: 10.1109/ACCAI61061.2024.10602348

Besnea, F.-L., Resceanu, I.-C., Cismaru, S.-I., Ganea, I.-E., Pistritu, I.-L., & Bizdoaca, N.-G. (2019). Experiments regarding implementation of a virtual training environment for automotive industry. *2019 20th International Carpathian Control Conference (ICCC)*, 1–6.

Boboc, R. G., Gîrbacia, F., & Butilă, E. V. (2020). The application of augmented reality in the automotive industry: A systematic literature review. *Applied Sciences (Basel, Switzerland)*, *10*(12), 4259. DOI: 10.3390/app10124259

Boopathi, S. (2024). Advancements in Machine Learning and AI for Intelligent Systems in Drone Applications for Smart City Developments. In *Futuristic e-Governance Security With Deep Learning Applications* (pp. 15–45). IGI Global. DOI: 10.4018/978-1-6684-9596-4.ch002

Boopathi, S., Thillaivanan, A., Mohammed, A. A., Shanmugam, P., & VR, P. (2022). Experimental investigation on Abrasive Water Jet Machining of Neem Wood Plastic Composite. *IOP: Functional Composites and Structures, 4*, 025001.

Dalle Mura, M., & Dini, G. (2021). An augmented reality approach for supporting panel alignment in car body assembly. *Journal of Manufacturing Systems, 59*, 251–260. DOI: 10.1016/j.jmsy.2021.03.004

de Freitas, F. V., Gomes, M. V. M., & Winkler, I. (2022). Benefits and challenges of virtual-reality-based industrial usability testing and design reviews: A patents landscape and literature review. *Applied Sciences (Basel, Switzerland)*, *12*(3), 1755. DOI: 10.3390/app12031755

Ejichukwu, E., Tong, L., Hazime, G., & Jia, B. (2024). Enhancing Autonomous Vehicle Design and Testing: A Comprehensive Review of AR and VR Integration. *arXiv Preprint arXiv:2404.19021*.

Gabbard, J. L., Smith, M., Tanous, K., Kim, H., & Jonas, B. (2019). AR drivesim: An immersive driving simulator for augmented reality head-up display research. *Frontiers in Robotics and AI*, *6*, 98. DOI: 10.3389/frobt.2019.00098 PMID: 33501113

Gerber, M. A., Schroeter, R., & Vehns, J. (2019). A video-based automated driving simulator for automotive UI prototyping, UX and behaviour research. *Proceedings of the 11th International Conference on Automotive User Interfaces and Interactive Vehicular Applications*, 14–23. DOI: 10.1145/3342197.3344533

Gouda, G. K., & Tiwari, B. (2024). Unravelling the enablers of Industry 4.0 in Indian automobile industry amid COVID-19: An integrated TISM and fuzzy MICMAC approach. *Journal of Business and Industrial Marketing*, *39*(2), 227–243. DOI: 10.1108/JBIM-02-2022-0067

Hadas, Z., Janak, L., & Smilek, J. (2018). Virtual prototypes of energy harvesting systems for industrial applications. *Mechanical Systems and Signal Processing*, *110*, 152–164. DOI: 10.1016/j.ymssp.2018.03.036

Heydaryan, S., Suaza Bedolla, J., & Belingardi, G. (2018). Safety design and development of a human-robot collaboration assembly process in the automotive industry. *Applied Sciences (Basel, Switzerland)*, *8*(3), 344. DOI: 10.3390/app8030344

Jha, S. K., & Beevi, S. J. P., H., Babitha, M. N., Chinnusamy, S., & Boopathi, S. (2024). Artificial Intelligence-Infused Urban Connectivity for Smart Cities and the Evolution of IoT Communication Networks. In *Blockchain-Based Solutions for Accessibility in Smart Cities* (pp. 113–146). IGI Global. DOI: 10.4018/979-8-3693-3402-7.ch005

Kim, J. Y., & Oh, J. M. (2022). Opportunities and challenges of metaverse for automotive and mobility industries. *2022 13th International Conference on Information and Communication Technology Convergence (ICTC)*, 113–117.

Lee, J., Chua, P. C., Chen, L., Ng, P. H. N., Kim, Y., Wu, Q., Jeon, S., Jung, J., Chang, S., & Moon, S. K. (2023). Key enabling technologies for smart factory in automotive industry: Status and applications. *International Journal of Precision Engineering and Manufacturing-Smart Technology*, *1*(1), 93–105. DOI: 10.57062/ijpem-st.2022.0017

Maguluri, L. P., Arularasan, A., & Boopathi, S. (2023). Assessing Security Concerns for AI-Based Drones in Smart Cities. In *Effective AI, Blockchain, and E-Governance Applications for Knowledge Discovery and Management* (pp. 27–47). IGI Global. DOI: 10.4018/978-1-6684-9151-5.ch002

Malathi, J., Kusha, K., Isaac, S., Ramesh, A., Rajendiran, M., & Boopathi, S. (2024). IoT-Enabled Remote Patient Monitoring for Chronic Disease Management and Cost Savings: Transforming Healthcare. In *Advances in Explainable AI Applications for Smart Cities* (pp. 371–388). IGI Global.

Manimuthu, A., Venkatesh, V., Shi, Y., Sreedharan, V. R., & Koh, S. L. (2022). Design and development of automobile assembly model using federated artificial intelligence with smart contract. *International Journal of Production Research, 60*(1), 111–135. DOI: 10.1080/00207543.2021.1988750

Molakatala, N., Kumar, D. A., Patil, U., Mhatre, P. J., Sambathkumar, M., & Boopathi, S. (2024). Integrating 5G and IoT Technologies in Developing Smart City Communication Networks. In *Blockchain-Based Solutions for Accessibility in Smart Cities* (pp. 147–170). IGI Global. DOI: 10.4018/979-8-3693-3402-7.ch006

Naveeenkumar, N., Rallapalli, S., Sasikala, K., Priya, P. V., Husain, J., & Boopathi, S. (2024). Enhancing Consumer Behavior and Experience Through AI-Driven Insights Optimization. In *AI Impacts in Digital Consumer Behavior* (pp. 1–35). IGI Global. DOI: 10.4018/979-8-3693-1918-5.ch001

Paul, A., & Thilagham, K. T., KG, J. C., Reddy, P. R., Ravishankar, S., & Boopathi, S. (2024). Multi-criteria optimization on friction stir welding of aluminum composite (AA5052-H32/B4C) using titanium nitride coated tool. *Engineering Research Express, 6*(1), 015402.

Piromalis, D., & Kantaros, A. (2022). Digital twins in the automotive industry: The road toward physical-digital convergence. *Applied System Innovation, 5*(4), 65. DOI: 10.3390/asi5040065

Ravisankar, A., Shanthi, A., Lavanya, S., Ramaratnam, M., Krishnamoorthy, V., & Boopathi, S. (2024). Harnessing 6G for Consumer-Centric Business Strategies Across Electronic Industries. In *AI Impacts in Digital Consumer Behavior* (pp. 241–270). IGI Global.

Sadeghzadeh, A., Islam, M. B., Uddin, M. N., & Aydin, T. (2024). ARVA: An Augmented Reality-based Visual Aid for Mobility Enhancement Through Real-time Video Stream Transformation. *IEEE Access : Practical Innovations, Open Solutions, 12*, 137268–137283. DOI: 10.1109/ACCESS.2024.3462628

Sekhar, K. Ch., Ingle, R. B., Banu, E. A., Rinawa, M. L., Prasad, M. M., & Boopathi, S. (2024). Integrating VR and AR for Enhanced Production Systems: Immersive Technologies in Smart Manufacturing. In *Advances in Computational Intelligence and Robotics* (pp. 90–112). IGI Global. DOI: 10.4018/979-8-3693-6806-0.ch005

Sharma, M., Sharma, M., Sharma, N., & Boopathi, S. (2024). Building Sustainable Smart Cities Through Cloud and Intelligent Parking System. In *Handbook of Research on AI and ML for Intelligent Machines and Systems* (pp. 195–222). IGI Global.

Sonia, R., Gupta, N., Manikandan, K., Hemalatha, R., Kumar, M. J., & Boopathi, S. (2024). Strengthening Security, Privacy, and Trust in Artificial Intelligence Drones for Smart Cities. In *Analyzing and Mitigating Security Risks in Cloud Computing* (pp. 214–242). IGI Global. DOI: 10.4018/979-8-3693-3249-8.ch011

Stark, R. (2022). *Virtual product creation in industry*. Springer. DOI: 10.1007/978-3-662-64301-3

Sundaramoorthy, K., Singh, A., Sumathy, G., Maheshwari, A., Arunarani, A., & Boopathi, S. (2024). A Study on AI and Blockchain-Powered Smart Parking Models for Urban Mobility. In *Handbook of Research on AI and ML for Intelligent Machines and Systems* (pp. 223–250). IGI Global.

Tripathi, S., & Talukder, B. (2023). Supply chain performance and profitability in Indian automobile industry: Evidence of segmental difference. *Global Business Review*, *24*(2), 371–392. DOI: 10.1177/0972150919898302

Wang, H., Wang, Z., Chen, D., Liu, Q., Ke, H., & Han, K. K. (2023). Metamobility: Connecting future mobility with the metaverse. *IEEE Vehicular Technology Magazine*, *18*(3), 69–79. DOI: 10.1109/MVT.2023.3263330

Yıldız, B. S. (2020). Optimal design of automobile structures using moth-flame optimization algorithm and response surface methodology. *Materials Testing*, *62*(4), 371–377. DOI: 10.3139/120.111494

Chapter 13

Innovative Technologies and Implementation of Autonomous Vehicles in Developments of Urban City

Dharam Prakash Paswan
Department of Economics, L.N. College, B.R.A. Bihar University, Bhagwanpur, India

Brij Kishor Ram
Department of Economics, S.N.S. College, B.R.A. Bihar University, Motihari, India

Kundan Kumar
Department of Economics, B.R.A. Bihar University, Muzaffarpur, India

Anil Kumar
Department of Economics, B.R.A. Bihar University, Muzaffarpur, India

Sheobarat Ravidas
https://orcid.org/0009-0000-1194-2907
PG Department of Economics, A.N.S. College Barh, Patliputra University, Patna, India

ABSTRACT

The Autonomous Vehicle is an urban mobility sector user-generative disruptor. The chapter describes the core technology that allows AVs to function—sensory systems,

DOI: 10.4018/979-8-3693-7287-6.ch013

machine learning, and the artificial intelligence that orchestrates the autonomous process itself. Finally, how can ACT align itself with urban systems, not just at pragmatic levels, such as what the unavoidable safety issues end up being and in terms of legislation, but with ACT itself considered an infrastructure? It will also consider how A.V.s may affect traffic patterns, as well as environmental sustainability and equity. This chapter elaborates on the myriad systems that a city employs to ensure that a fleet of vehicles does what it is designed to do, arguing that many details of this framework are what innovators will need to build if they hope to succeed. And it also lays out the immediate benefits the innovations provide to the city itself.

INTRODUCTION

The autonomous vehicles (AVs) era is a turning point in the transportation industry that will transform urban mobility in the future. As technology continues to evolve, self-driving cars have transitioned from thought experiments to real-world innovations—and now seem to offer a potential solution to some of the challenges faced by modern urban hubs. Urban areas have become the focal point for AV implementation, with complexities, high density, and increasing demand for sustainable solutions. With their potential to solve a slew of challenges that confront transport systems in urban environments—traffic, safety, environmental impacts, and equitable demand-led mobility, to name but a few—autonomous vehicles arguably have more potential than any previous technology in human history(Li et al., 2015).

The revolution will be driven by a convergence of cutting-edge technologies. Sensor systems—LiDAR (Light Detection and Ranging), radar, and cameras—allow AVs to perceive their environment in real-time, and together they provide the vehicle with an unparalleled situational awareness. These sensors continuously collect immense data, which are uniquely processed with very advanced algorithms to take decisions within milliseconds. These data give a form to machine learning and artificial intelligence (AI), essential parts of self-controlling straight associates that enable the vehicle's programming to learn from past experiences and respond diversely in future circumstances, making the venture toward independent driving expectably exact and intuitive. These advancements have also enhanced the functionalities of AVs, so today self-driving vehicles can navigate complex terrains with little to no driver input(Du, 2016).

However, technological sophistication needed to design technologies and roll out autonomous vehicles in cities is just a slice of the equation. " Perhaps the biggest challenge is how AVs can coexist with existing infrastructure in any given city. Urban areas are complicated by crowded roads, a range of participants from cars and trucks to pedestrians and cyclists, and, in many cases, outdated infrastructure.

Infrastructure will need to evolve to accommodate the unique challenges of self-driving cars, meaning urban planners, policymakers, and engineers will need to work together to lay the necessary groundwork for a safe transition to autonomous mobility. Such adjustments may involve redesigning roadways, modifying traffic signals, and deploying new communication systems that allow autonomous vehicles to communicate with one another and surrounding infrastructure. Determining how to ensure AVs can operate safely within city limits presents another challenge that policymakers face, as all new laws and safety standards would need to be established(Gružauskas et al., 2018).

There are many other issues related to the integration of AVs as well but will start with safety, which is a key issue. Autonomous automobiles can handle a lot of weight in terms of protecting against human error, wherein there are lots of human-caused mishaps, like running a vehicle, being unconscious with the authority to do so, an attempt to divert during driving, etc. Remaining are critical areas of human slip-ups close to human slip-up primary mishap road. But things like pedestrian safety, interactions between AVs and vehicles driven by people, and how effectively AVs can handle unanticipated road situations are all still subjects of active scientific inquiry. These issues are compounded by the emergence of robust cybersecurity measures designed to protect the vehicle's systems from malicious attacks(Kovačić et al., 2022a). Alongside that, public acceptance of autonomous technology is a major challenge. People are hesitant about relying on machines to make decisions that affect their lives. Public education, transparency regarding testing, and evidence of the safety benefits that AVs can provide will be key to increasing public acceptance.

There are important questions to consider on top of the technical and regulatory challenges associated with the deployment of AVs in cities about their impact on society and the environment. Perhaps the most exciting benefit of AVs is their ability to reduce traffic congestion, a problem faced by cities across the globe. They can communicate with each other, sending and receiving real-time traffic data to adjust speed and position for a smoother traffic flow and optimal use of road space. Theoretically, shared vehicles could translate to fewer vehicles on the road, especially if shared mobility solutions, such as ride-hailing systems (e.g., Uber, Lyft, etc.), are integrated into the AV ecosystem. Commercial AVs could even help reduce the overall number of vehicles on the road, further decreasing emissions and making AVs a true ally in our move toward sustainable cities. Autonomous vehicles (AVs) can greatly reduce the need for energy that is spent on wasteful extra driving as well as idling on the road(Wang et al., 2022).

Environmental advantages of AVs extend beyond emissions reductions. Cities will be able to embrace renewable energy through increased electrification of autonomous vehicles, which can reduce greenhouse gases significantly. Urban transportation should find it easier to decarbonize with electrified AVs—the latter can run on power produced by renewables. Moreover, increased adoption of AVs could also affect urban design, as there may be less need for expansive parking structures: autonomous vehicles can drop people off at their destination and move on to the next job. This change may free up precious urban land for greening, social housing, or public functions and create a more livable, sustainable city(Faisal et al., 2019).

Social equity is another consideration to pay attention to when we talk about urban AV integration. The technology that goes into self-driving cars could also have substantial benefits, but it is crucial that these benefits are extended and available to all citizens, and not only those who are wealthy or have access to areas that are more innovative and technologically advanced. For AVs to be successfully deployed, they need to be accessible to underserved communities, such as low-income neighborhoods and those with disabilities(Rahman & Thill, 2023). Public transport systems utilizing autonomous vehicles could result in a cheaper, more inclusive form of transport, enhancing mobility and independence for some individuals. Furthermore, AVs could mitigate road safety issues, especially targeting vulnerable populations, like the elderly, who may struggle to traverse in legacy transport systems.

Urban areas can be even more disruptive; autonomous vehicle adoption may catalyze radical and beneficial changes to transportation systems. If technologists, urban planners, policymakers, and the public collaborate, AVs can transform our cities for the better, enhancing safety, sustainability, and inclusion. But a host of issues ranging from technological hurdles to regulatory hearings will hold back progress, and the interplay between AVs will necessitate sophistication in planning that accounts for these complexities. These advancements lay the groundwork for new, better, more sustainable, and equitable transport solutions for everyone.

Objectives

In this chapter, it examines the pillars of technology supporting AVs: sensor systems, machine learning, and artificial intelligence, and how these technologies converge to make driving vehicles without human input a reality. The purpose of the chapter is to scan the bandwidth of regulatory and infrastructure challenges related to the integration of AVs into the existing urban infrastructure. It will also examine the impact of AVs on urban mobility, particularly in areas like traffic flow, sustainability, and social equity. In addition, it seeks to illustrate the world case studies to give a concrete overview of futuristic aspirations on self-driving cars in an urban fabric, providing a wholesome insight on the future transformation of cities.

CORE TECHNOLOGIES OF AUTONOMOUS VEHICLES

AVs are considered a milestone in transportation technology and utilize several core technologies to function properly and effectively. Sensor systems, machine learning algorithms, artificial intelligence (AI), and complex data processing and communication systems are some of the critical components that drive autonomous capabilities. From real-time situational awareness and making decisions to communicating with other vehicles and infrastructure, these technologies are making it happen. Here we explore the three core components of autonomous car technology: sensor systems, machine learning and AI, and data processing and communication(Gavanas, 2019). Figure 1 depicts the core components of autonomous vehicle (AV) technologies.

Figure 1. Core components: Technologies of autonomous vehicles (AVs)

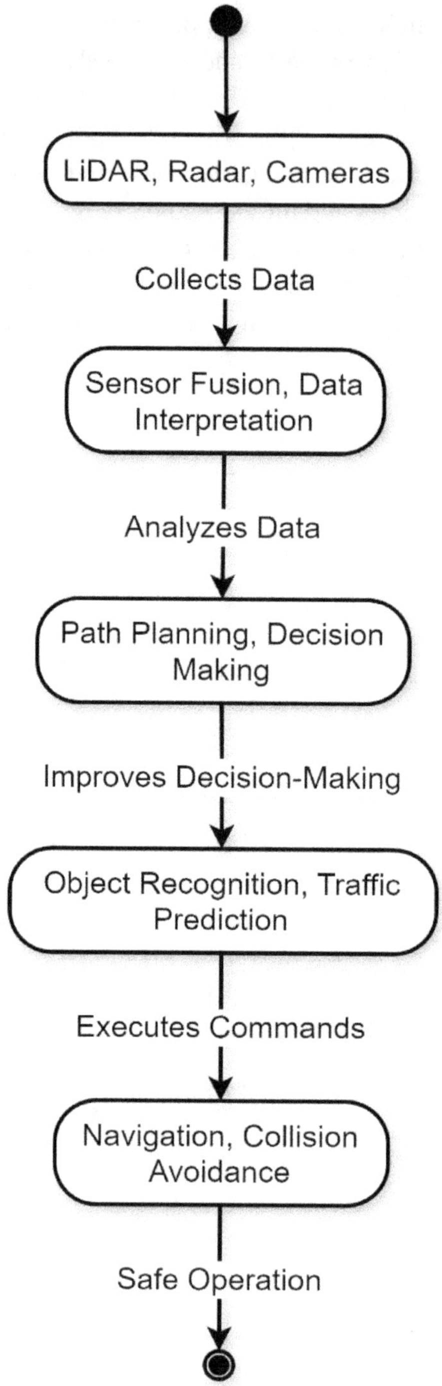

Sensor Systems (LiDAR, Radar, Cameras)

At the heart of an autonomous vehicle's ability to perceive its environment is its sensor system. These sensors collect vast amounts of data from the vehicle's surroundings, enabling the vehicle to understand and react to its environment. The three main types of sensors commonly used in AVs are LiDAR (Light Detection and Ranging), radar, and cameras. Each sensor type plays a unique role, contributing to the overall perception system's reliability(Zambrano-Martinez et al., 2019).

LiDAR (Light Detection and Ranging): LiDAR is a key sensor in autonomous vehicles that builds a high-res 3D map describing the vehicle's environment. LiDAR makes detailed measurements of distance and spatial relationships by emitting laser pulses and recording how long it takes for the pulses to bounce back after hitting objects. This augments AVs with a comprehensive, precise model of the environment by accurately capturing details of nearby objects, road conditions, and potential obstacles. With proven reliability in low-light conditions, LiDAR proves to be an essential need for night driving. Researchers are racing to address the drawbacks of this technology, including its cost and sensitivity to weather, including heavy rain and fog.

Radar: Another important sensor in AVs is the radar (Radio Detection and Ranging). Where LiDAR uses light waves to map and read the surroundings, radar sends radio waves to sense objects and how and how far they are moving. Radar sensors work best for detecting moving objects and obstacles as they perform better than LiDAR and cameras under bad conditions—rain, fog, or snow. Radars are perfect companions for LiDAR and, in turn, do wonders for the ability of the vehicles to operate in otherwise dangerous environments. Radar and the data it provides on tracking other vehicles and their relative speeds and direction within a space are useful for collision avoidance systems and becoming ever more capable with regards to the safe operation of a vehicle in high-traffic environments.

Cameras: In autonomous vehicles, cameras are one of the most commonly used sensors that give a detailed visual representation of the surroundings. They contain comprehensive information about traffic signs, lane markings, pedestrians, and other vehicles. Cameras are central to tasks including object recognition, lane-keeping assistance, and even traffic signal detection. While cameras are a highly detailed imaging modality that can recognize objects over a larger range of scenarios than radar, they are sensitive to lighting conditions. Cameras, for instance, can have difficulty in low light or glare from the sun, meaning that these'll need LiDAR and radar to work reliably in all conditions.

LiDAR, radar, and cameras work in concert to create a robust suite of sensors that can collect data about the environment. As a result, they offer an even more accurate and reliable perception of their surrounding environment, allowing the vehicles to react to dynamic changes within the environment and steer clear through complex urban environments.

Machine Learning and Artificial Intelligence

ML and AI are the decisive elements in an autonomous vehicle. Together, these technologies allow AVs to interpret the millions of pieces of information their sensor systems generate and, in turn, make split-second, real-time decisions based on that data. Traditional programming is about building a system with specific outputs based on given inputs, whereas machine learning allows AVs to "learn" from experience, becoming increasingly sophisticated in their responses over time.

Machine Learning: Autonomous vehicles use machine learning algorithms to process data in real time from the onboard sensors and detect patterns. When trained on huge datasets that include a variety of driving situations, these algorithms can learn how to navigate, avoid obstacles, and respond to other automobiles and walkers. For example, ML algorithms can be trained on capturing various types of road signs, traffic signals, or even the behavior of pedestrians and other vehicles. The more situations the car finds itself in, the better it gets at driving through them. Various approaches, such as supervised learning, unsupervised learning, and reinforcement learning, are employed in the application of autonomous driving so that the car can learn constantly from new driving conditions(Ramudu et al., 2023; Tirlangi et al., 2024).

Artificial Intelligence: AI is a further development of machine learning; it encapsulates cognitive principles such as reasoning, understanding, problem solving, and decision-making. AI aids the development of autonomous-driving systems, simulating judgment calls a person would make in layered driving situations. For example, AI algorithms use the knowledge acquired by such experiences and on split-second decisions when they are faced with a sudden obstacle, rash maneuvers done by cars on a busy highway, adverse weather conditions, roadblocks, etc. Additionally, AI helps AVs with more complex functions such as path planning, in which the vehicle figures out the most optimal path and avoids any imminent obstacles. AVs perform dynamic tasks requiring the combination of real-time data analysis and predictive capability supported by machine learning and AI, running on powerful processors, to ensure safe and efficient driving under various driving conditions(Jha et al., 2024; Sonia et al., 2024).

Data Processing and Communication

Autonomous cars have to process a massive amount of data from their sensors, machine learning algorithms, and AI systems; therefore, they will have to use some advanced data processing methods. Data has to be analyzed, interpreted, and acted upon quickly and accurately for making split-second decisions. Processing this data in real-time is imperative for allowing AVs to navigate through fast-moving urban landscapes safely(Kovačić et al., 2022b; Zambrano-Martinez et al., 2019).

Data Processing: When running, they collect terabytes daily from the information captured through their sensor systems. This data must be processed in real time to allow decision-making and vehicle control. These include a number of computations ranging from noise filtering of sensor data through perception and classification of driving context (who/what, where, when, how, and why). On-board high-performance computing can process this data in a distributed manner, allowing AVs to make instant decisions without relying on infrastructure. To illustrate, AVs require transforming data seamlessly and instantly, as the surrounding world can change in an instant, be it a pedestrian suddenly crossing a road or a misbehaving driver.

Communication Systems: In addition to 'bringing data in and processing it internally,' AVs also rely on communication systems for interaction with other vehicles and surrounding infrastructure. Vehicle-to-vehicle (V2V) and vehicle-to-infrastructure (V2I) communication systems allow autonomous vehicles to relay their location, speed, and intention to other vehicles, traffic signals, and other parts of the transportation network. These important features would enable vehicles to share information with one another, which will help in congestion and safety solutions on the road. For example, V2V communication might prevent an accident from occurring in the first place by informing a vehicle that there's another vehicle on a collision course, even when the driver—or system—can't yet detect the threat. V2I communications can also be used to improve traffic light timing so AVs can safely and efficiently clear signalized intersections. Such communication systems are critical for a connected transport ecosystem, where AVs can communicate seamlessly with other road users and surrounding infrastructure.

Sensor systems like LiDAR, radar, and cameras allow the AV to see the environment, plus machine learning and other artificial intelligence solutions act as the decision-makers steering the vehicle through the growing complex environments. These are the main elements of the autonomous systems making up the five levels of AVs. All of this information is processed in real time by data processing; communication systems enable AVs to communicate with other vehicles and infrastructure to enhance safety and efficiency. Hardware interfaces will evolve in tandem with cloud capabilities to unlock new use cases such as urban mobility and autonomy as these technologies mature.

IMPLEMENTATION STRATEGIES FOR AUTONOMOUS VEHICLES

Strong interaction with pilots, governments, and city planners As it will be important to better understand the places where AVs are going to be introduced to make a much smoother integration with existing mobility infrastructure. These elements will help ensure that AV technology is tested, regulated, and integrated into urban infrastructure to maximize its benefits and address potential challenges. This part will address essential AV implementation strategies in cities, including pilot programs/trials, public policy and government involvement, and cooperation with city planners(Golbabaei et al., 2021). Figure 2 depicts the strategies for implementing autonomous vehicles (AVs) in urban environments.

Figure 2. Implementation strategies for autonomous vehicles (AVs) in urban environments

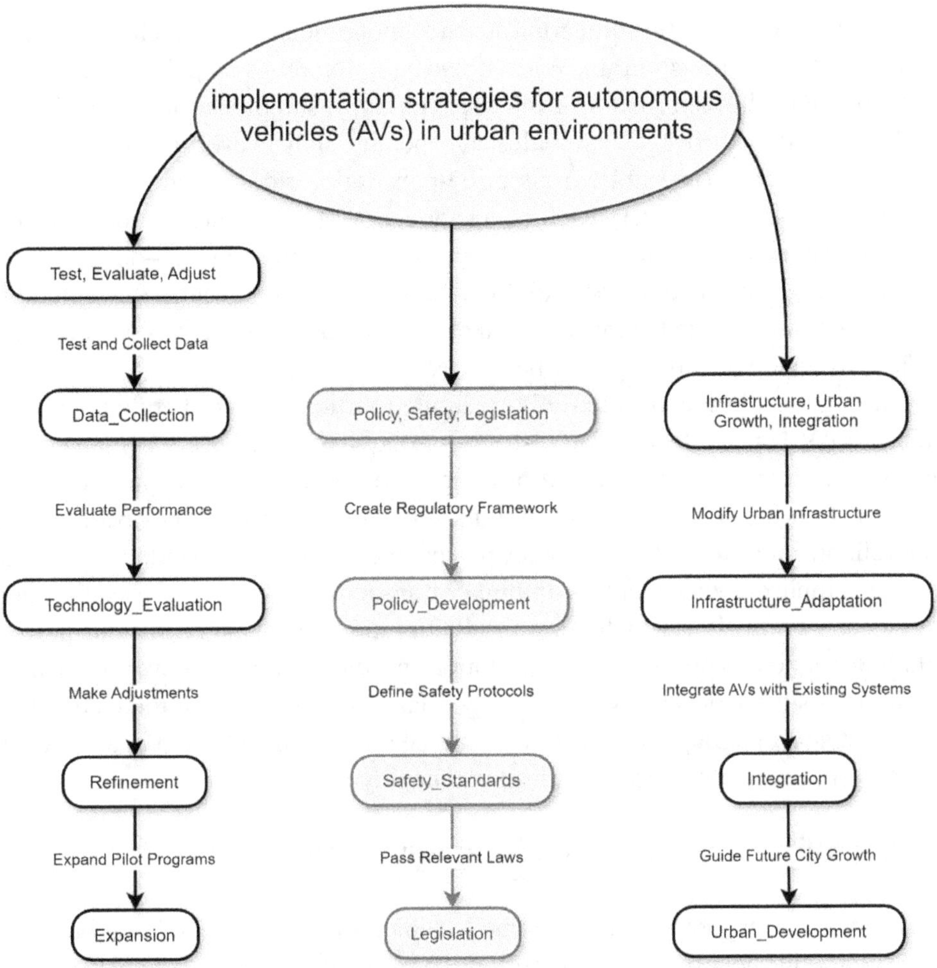

Pilot Programs and Trials

Implementing AV technology in a controlled environment during trialing and pilot programs is a critical first step for AV technology. This way, manufacturers, regulators, and city officials can assess the performance, safety, and feasibility of AVs in a range of urban environments. Starting with a pilot program allows for the safe identification of potential issues in how AVs might enter the fold of larger, more complex systems, like navigating complex intersections, interacting with vulnerable road users (VRUs) (e.g., people walking and biking), or pressure points in traffic congestion(Bahamonde-Birke et al., 2018; Gružauskas et al., 2018; Lim & Taeihagh, 2018).

Pilot programs build up some trust around autonomous tech, which is one of the prime objectives of the program. Widespread adoption relies on public acceptance, and these trials allow normal folk the opportunity to sample autonomous driving first-hand. To try to prove their vehicles can operate safely in everyday traffic, some AV companies have run public trials or demonstration rides. Such programs also allow for the collection of very useful data about how AVs operate in urban settings, which contributes to the knowledge of what AVs can—and cannot—do. Additionally, pilot programs have the potential to become a proving ground for regulatory frameworks and provide lawmakers with clarity on how AVs interact with existing traffic laws where new policy may be needed.

Much pilot testing is conducted only in a limited portion or route of a municipality—dedicated AV lanes, a low-traffic district or zone, like a university campus or the business district. A narrow scope allows for more targeted assessment and reduces any impact on the wider transport system. Once the technology has been proven to be reliable and safe in these smaller places, it can then be expanded to scale up to more complex environments with higher transactions. It undertakes subsequent analysis and evaluation, which allows for better dialogue between various parties, including technology providers, transportation agencies, municipal governments, and the public. As these programs evolve, they offer an opportunity for the technology, and how it gets integrated into city streets, to be refined and adjusted in an ongoing state of iteration ahead of more widespread deployment.

Public Policy and Government Involvement

The smart incorporation of autonomous vehicles into the fiber of a city requires public policy and government intervention. To ensure that AVs are deployed in a safe and equitable manner, governments have an important role to play in establishing the regulatory framework, setting safety standards, and making policies. Because AVs are so different from transportation systems we are familiar with, policymakers need to create laws that are in line with the special features and challenges of driving autonomously.(Faisal et al., 2019)

The most pressing question facing governments is how to set clear safety standards and testing protocols for self-driving vehicles. The answer is none other than the autonomous vehicle (AV) industry, which relies on sensors, machine learning algorithms, and data processing systems to navigate and make decisions. Regulators need to partner with industry professionals, technologists, and researchers to develop policies aimed at testing AVs more robustly and consistently. How this should be done:

Safety standards should also incorporate concerns related to cybersecurity to ensure that AVs are protected from possible hacking/interference from malicious actors.

The governments should also consider the broader social and economic implications of AV adoption beyond safety regulations. Equity for AVs is one of the biggest issues to overcome. This means ensuring that AVs are made available in underserved communities such as low income neighborhoods or for the elderly and disabled populations. (lst and in which jobs might be replaced by automation. As such, public policy must grapple with the broader effect of AVs on the workforce as this technology is kneaded into transportation, delivery, and logistics sectors—redundancy, after all, is what delivery is built upon. "The job displacements are indicative of broader structural shifts in the economy that call for different policies to retrain workers and offer social safety nets to the displaced," a report by the Economic Policy Institute in Washington said.

The governments should also consider the broader social and economic implications of AV adoption beyond safety regulations. Equity for AVs is one of the biggest issues to overcome. This means ensuring that AVs are made available in underserved communities such as low income neighborhoods or for the elderly and disabled populations. (lst and in which jobs might be replaced by automation. As such, public policy must grapple with the broader effect of AVs on the workforce as this technology is kneaded into transportation, delivery, and logistics sectors—redundancy, after all, is what delivery is built upon. "The job displacements are indicative of broader structural shifts in the economy that call for different policies to retrain workers and offer social safety nets to the displaced," a report by the Economic Policy Institute in Washington said.

Collaboration with City Planners

This entails working alongside city planners to craft (and refresh) blueprints for how to weave driverless vehicles into the landscape to balance mobility and safety. City planners determine how to build our environment—the streets we drive on, the public transport we use, how we build our houses, and amenities. With the emergence of AVs, that infrastructure needs to be addressed by city planners as well as how these vehicles will adapt to existing surroundings, as well as how the existing surroundings are going to provide an urban environment in which they can thrive(Du, 2016; Li et al., 2015).

In places where converting existing road schemes and designing towns specifically for AVs would take time, it could bias them towards reliance on AVs. Related news For example, autonomous vehicles (AVs) depend on visible road hierarchy markings and signage in order to function properly. In this process, it is up to city planners to make sure that the road markings, which are readable by autonomous sensors, are visible and standardized. This could involve the implementation of new infrastructure like dedicated AV corridors, EV charging stations, and V2I communication systems that allow AVs to interact with local traffic controls and other road infrastructure.

Urban planners also need to think about how AVs will mesh with existing public transportation systems. Public transit is an essential component of the transportation network in many places, and AVs should work in concert with, not as a substitute for, these services. For instance, AVs could be utilized in areas where conventional public transportation services do not operate effectively, such as rural or underserved areas. AVs might also play a role in a larger mobility-as-a-service (MaaS) arrangement, in which individuals can use multiple forms of transportation—think AVs, buses, and trains—depending on their destination.

Moreover, AVs are likely to reshape the manner in which cities utilize urban space. Long-haul AVs can simply drop off their passengers and move on to a different task, so, in the future, AVs might not even need to park in the same way that humans do. This might free up valuable urban land that is now reserved for parking and could be put to new uses, whether green space, residential housing, or commercial developments. Urban land-use planning will need to adjust to AVs, and city planners will need to mirror the potential changes in urban space brought about through urban designers and the engineers to determine areas for growth and microcosms of improved livability.

Understanding how AVs interact with a larger urban ecosystem will also require collaboration between technology developers and city planners." AV trials and pilots generate real-world data that help planners understand traffic patterns, bottlenecks, and the impact of AVs on overall mobility. Data of this kind can help guide urban planning decisions in the future to ensure cities are designed with the efficient movement of people and goods top of mind, whether they are utilizing autonomous vehicles or those that are human-driven.

Building autonomous vehicles for urban environments requires a holistic approach that combines pilots, public policy, and working closely with city builders. Filling a need Pilot programs facilitate real-world testing, foster public trust, and yield useful data for enhancements. Government intervention is necessary to institute safety standards, guarantee equitable access to AVs, and solve legal challenges. In the end, working with city planners will allow for better integration into urban infrastructure, making the overall urban environments more effective and safe and generally improving the quality of life in the city. By undertaking a coordinated, deliberate approach, cities can realize the promise of AVs to be integrated into smarter, more sustainable, and more accessible transportation systems.

Figure 3. Impact of autonomous vehicles (AVs) on urban mobility

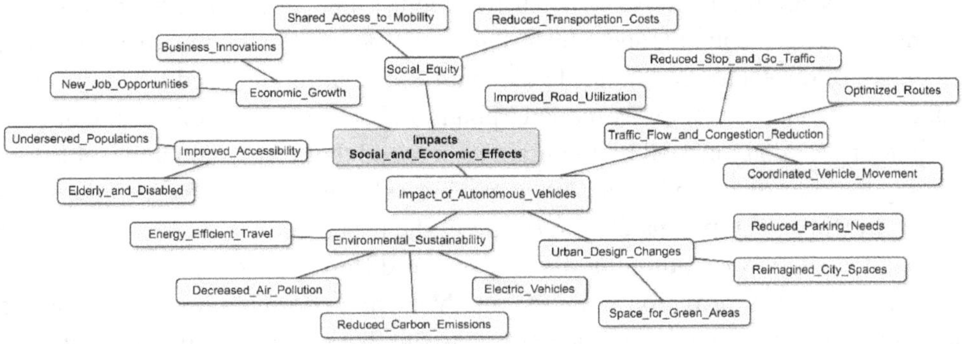

IMPACT OF AUTONOMOUS VEHICLES ON URBAN MOBILITY

The introduction of AVs into the harmoniously multi-modal urban environment will increasingly be filled with multiple, interweaving, and interconnected transportation systems with the potential for significant disruption of urban mobility with the advantages of increased traffic throughput, less congestion, improved environmental sustainability, and substantial social and economic change. Autonomous Vehicles: The Future—Creating Smart Cities Using AI and MM & RT. However, implementing AV will create some challenges, and it will need to fit well into existing systems(Du, 2016; Li et al., 2015; Mora et al., 2020). Figure 3 depicts the significant impact of autonomous vehicles (AVs) on urban mobility.

Traffic Flow and Congestion Reduction

Impacts autonomous vehicles have over urban mobility—traffic flow into urban mobility: This is definitely one of the most powerful use cases of this type of vehicle within urban mobility, since the vehicle can directly try to enhance traffic flows, which is one of the major reasons for congestion. Clogging is a near-perpetual headache in countless metropolises around the world, fueled by burgeoning population densities, insufficient road infrastructure, and sluggish traffic management systems. Human drivers are more likely culprits in causing congestion, with aggressive driving causing bottlenecks and delay caused by disobeying traffic signals or vehicles traveling at inconsistent speeds. On the other hand, self-driving cars can avoid traffic jams with precise driving and close coordination. AVs, for example, can communicate with each other and surrounding infrastructure in real time, allowing them to dynamically adjust speed, position, and route to reflect real-time traffic conditions. Through this communication, vehicles are able to "talk" to traffic signals, respond to other vehicles' behavior, and adjust routes in ways that the human driver cannot. AVs also follow each other at a safe distance, eliminating the need for aggravated brake taps or lane-changing so-called "phantom" traffic jams. AVs also can help mitigate friction on the road by maintaining a speed that is constant and stable and keeping a closer distance behind other moving vehicles, helping to decrease stop-and-go traffic and stop jams from forming in the first place.

This technique of platooning allows several different autonomous vehicles to drive in very close formation with one another while coordinating movement and keeping a tight group while in a safe environment that provides maximum capacity for the roadways. This would help alleviate congestion and unclog urban transportation networks. But the benefits are not intrinsic to AV technology; rather, they are contingent upon how AVs are integrated into urban space and infrastructure and contingent upon overcoming challenges such as the number of vehicles, mixed traffic, and how road design must change.

Environmental Sustainability

Environmental sustainability is another field where autonomous vehicles can play an important role. Widely, the transportation sector is one of the largest contributors in the global landscape for greenhouse gas emissions—particularly in urban settings where dense traffic, reliance on fossil fuels, and reckless driving behavior translate to massive air pollution and damaging carbon emissions. Indeed, when integrated with electric vehicle (EV) technology, autonomous vehicles (AVs) are capable of drastically shrinking the environmental footprint of city-based transport systems(Campisi et al., 2021; Seuwou et al., 2020).

EAVs are the biggest contributors to emissions reduction as they are predicted to electrify. Moving to electric-powered transportation replaces fossil fuel-powered vehicles, leading to reduced carbon emissions and cleaner air. Urban mobility, particularly in cases where the public transport networks are insufficiently served and densely populated, could be one area significantly aided with the increased adoption of EAVs to supplant gasoline-powered cars and reduce their carbon footprint in cities. Adding to this, because controlled driving patterns are inevitable (no more intermittent acceleration/deceleration), AVs are expected to consume less energy than human-driven cars with less total energy consumption and a significant increase in the efficiency of urban mobility systems as a whole.

AVs can offer the opportunity for more energy-resource-efficient transportation systems. In heavily congested city centers, AVs can help mitigate the environmental effects of transport by using real-time data to bypass congestion or pollution. Sophisticated algorithms also can choose routes optimized for energy consumption, which can reduce the amount of fuel used or battery drained. Task 7) The Role of Shared, Automated Mobility as a Growing Component of the Shareable Future: AV Ride-Hailing (and AV Sharing). In this scenario, a small number of AVs would travel through densely populated urban environments—a scenario that would consume less power and release fewer emissions, making urban areas more green.

The benefits of AVs for the environment will be conditional on how they are implemented, including enough charging infrastructure and grid capacity. Non-renewable energy sources playing any part can diminish the environmental advantages. To maximize the benefits of AV integration in the landscape of sustainable urban mobility, it is necessary to increase both the scale of EV adoption and investments into renewable energy-based infrastructure, since the environmental benefits may be offset by dependence on non-renewable energy resources.

Social and Economic Effects

Consequently, self-driving cars (AVs) are becoming the majority of urban transportation and are expected to promote mobility and accessibility among many population segments, e.g., older adults, people with disabilities, and those who do not own a car. Motorized disability assistance frameworks in urban regions are yet lacking, and public transportation is either rare, unreliable, and inaccessible to those with certain needs. This same logic holds for shared AVs, which could also serve to bridge the mobility gap for hard-to-reach areas. That would promote social equity by enabling access to jobs, education, health care, and other essential services. Kar— Another utility of autonomy is ride-sharing or car-sharing, which is useful for people that cannot afford or do not want to own a vehicle in a dense city with high vehicle ownership rates(Dantsuji & Takayama, 2024; Kovačić et al., 2022b).

Autonomous vehicles (AVs) are also likely to generate new jobs and industries and business models. Robotics, software engineering, data science, and cybersecurity will be needed for design, development, manufacture, and maintenance. This could promote economic growth and jobs in markets and industries that adopt AVs. And with the arrival of autonomous vehicles (AVs), we're going to have to deal with economic and social stresses. One major concern will be the loss of jobs, particularly in areas such as truck driving, public transport, and logistics. At the same time, it could hit lower-income workers and those with few chances to retrain especially hard. Nonetheless, this move of middle-skilled jobs is taking place against a backdrop of a greater shift to a quid pro quo world where services increasingly dominate and a larger discussion about the effects on job-seeking, particularly among so-called blue-collar manufacturing workers. The growing risk of widening social inequality is even more worrying. AVs, but are probably going to be owned by richer or the richer neighborhoods, "might accelerate the social divides that we already have." In order to avoid the potentially utopian-dystopian future scenario described above, it will be essential for policymakers to implement policies that assure that AVs will be available to all levels of society, especially the lowest levels of society, and assure that the benefits of AV use will extend beyond the top levels of any existing socioeconomic class.

The contribution of AVs, especially in their ubiquitous form, will be fundamental to how many suburbs will be formed, leading to the collapse of cities or escalation of suburbanization, traffic flow contours, environmental sustainability, and even social equity. While A.V.s can solve serious national problems like congestion and traffic fatalities, they also pose potential social and economic challenges. Cities can take an integrated approach, considering both the technological and infrastructure requirements and social and economic implications of widespread adoption. By accounting for these factors, cities are able to unlock AVs full potential as the enablers of more efficient, sustainable, and inclusive urban environments.

Figure 4. Challenges in urban integration of autonomous vehicles (AVs)

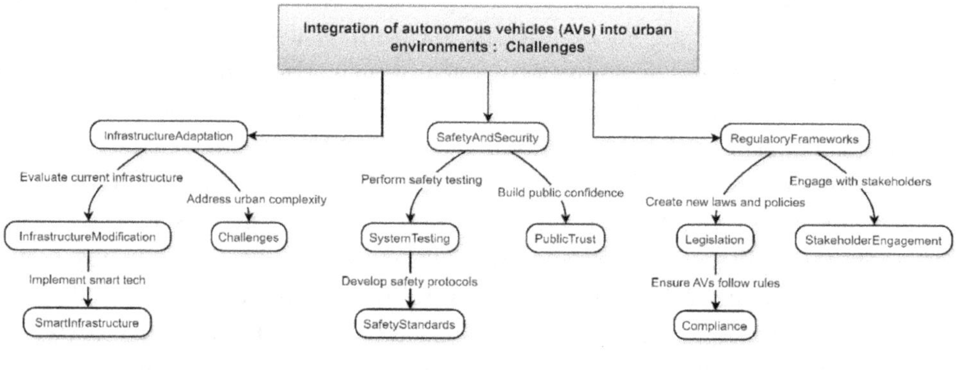

CHALLENGES IN URBAN INTEGRATION

By urban environments, I mean places where autonomous vehicles (AVs) will operate—and cities are complicated ecosystems, with existing infrastructure, high-density populations, and many players on the road. Although AVs promise improved sustainability, reduced traffic congestion, and increased road safety, the successful implementation of AV technology is contingent upon overcoming major challenges in terms of infrastructure adaptivity, safety and security, and regulatory environments. This transformative technology can only be beneficial for urban mobility systems, provided that the challenges of deploying such systems in the first place are addressed(Dantsuji & Takayama, 2024; Rahman & Thill, 2023; Wang et al., 2022). This is shown in Figure 4, where the different challenges to urban automation of AVs are highlighted.

Infrastructure and Road Network Adaptation

Adapting urban infrastructure to accommodate this technology poses one of the biggest challenges of integrating autonomous vehicles into city environments. Conventional roadways, stop signs, and intersections were built for human-driven car traffic, and many of those systems aren't set up to meet the needs of driverless vehicles. The lack of infrastructure to support AVs has been widely recognized, and at the extreme, it will require redesigning and upgrading urban road networks in ways no country has done in recent history (the remake of streets is something that does not happen overnight).

Road Infrastructure: Existing road networks should be improved to effectively facilitate communication between AVs and their surroundings. That brings traffic signals and signs into a system that can compare notes with AVs in real time, where they can act on changes in a path and traffic flow. Implementing systems like V2I (Vehicle to Infrastructure), the Vehicle to Infrastructure (V2I) system takes advantage of advanced communication technology to better utilize the road infrastructure in terms of traffic flow, as well as better safety and efficiency. "Because of their wide field of view, AVs may not recognize changes coming from one of those intersections (or traffic scenarios) that represent a new threat or create traffic problems of their own," the authors write. Since AVs rely on visual cues for the most part to navigate, street lines and lane configurations may also require modification. Cities might need to spend more on resurfacing roads and making sure road markings stand out more and are more consistent. Crosswalks and pedestrian zones must be clearly marked and outfitted with systems to detect pedestrians, cyclists, and other non-motorized road users, which will put existing infrastructure to new use.

Sensor data is used to recognize landmarks, and the AV computes how to get to the next one. Urban road design for AVs will need to be reimagined in these spaces. Stratifying roads for autonomous vehicles (AVs) through the implementation of dedicated lanes for AVs, as those that exist for buses or bicycles, can help to prevent conflicts between autonomous and non-autonomous vehicles. However, integrating air taxis within the existing mobility system requires coordination with other urban mobility services such as the public transport sector to minimize friction and avoid disturbance to existing services.

Safety and Security Concerns

Robo vehicles (AVs) stand out as one of the most notorious and complex subjects of safety issues owing to posing their maneuverability over traffic scenarios with pedestrians and cyclists in city surroundings. They have to interpret pedestrian behavior, merge with streaming car traffic, and respond to sudden changes in the road. These scenarios must be managed by their sophisticated sensors and algorithms without sacrificing safety. In unavoidable collisions, an AV must place occupant safety ahead of pedestrian safety. They are also susceptible to technical errors because of their elaborate network of sensors, cameras, and algorithms. In order to gain public confidence with AV technology, adjusting to variable road conditions is fundamental. However, similar to any other new technology, the adoption of autonomous vehicles in urban setups presents considerable safety and security issues to be addressed. And though AVs can potentially reduce human error-related accidents, the technology also needs to be tested and rigorous so that it can safely maneuver through these complicated and erratic urban environments.

Autonomous vehicles (AVs) are connected to the internet and vulnerable to hacking and cybersecurity threats. AVs could be hijacked or manipulated by malicious actors, leading to crashes or authorization breaches. This will require strong cybersecurity measures to protect the systems and data of AVs and communication systems that prevent the interception or manipulation of data shared between AVs and other vehicles and infrastructure.

Regulatory and Legal Considerations

The regulatory framework surrounding autonomous vehicles (AVs) is still in its infancy, and many legal and policy questions remain unanswered. Governments should create clear regulatory frameworks to address topics such as vehicle testing, insurance, liability, and public safety in order to safely integrate AVs into urban environments. The first challenge is defining the safety and testing standards because there is no single global standard for testing autonomous vehicles. Regulations in various countries and cities differ in their requirements from sensor technology, software development, to safety precautions. The lack of any specific standardization makes this challenging for manufacturers and policymakers that need to ensure AVs meet at least some minimum standard of safety before running in an urban context.

One of the major legal issues regarding autonomous vehicles (AVs) relates to liability and insurance; this may give rise to new types of insurance policies that account for the unique risk factors during autonomous driving. The legal framework will have to adapt to new concerns of fault and compensation in the context of AVs involved in accidents. A regulatory framework for autonomous vehicles should also address ethical and privacy concerns, as these vehicles collect and process vast amounts of data, including personal data about passengers. AVs would have to adhere to privacy laws and also abide by policies enabling them to protect personal information and prevent this information from being misappropriated or misused or from falling into the wrong hands. You are suggesting new insurance policies and some ethical considerations in the legal landscape.

Policymakers need to weigh AV integration on the larger societal scale and ensure they don't just make richer neighborhoods richer but also offer budget transport to low-income neighborhoods. Then the prospect of deploying AVs faces difficulties that are not wholly technologic, such as required infrastructure, safety and security measures, and regulatory regimes. These issues need to be addressed to unlock the benefits AVs are promised to provide in terms of a safer, more efficient, and sustainable urban mobility infrastructure. A comprehensive approach to AV integration will require collaboration among policymakers, urban planners, technologists, communication specialists, and other stakeholders to successfully deploy AVs to address people and the potential of AVs to shape urban transportation.

FUTURE PROSPECTS OF AUTONOMOUS VEHICLES IN URBAN CITIES

Despite AVs having a seemingly never-ending future in terms of potential uses in urban cities, a whole range of technologies are evolving and being introduced to their capabilities to make AVs even better. For cities struggling with growing mobility demand and looking for more efficient, sustainable & inclusive transportation solutions, AVs can be a key enabler of urban mobility transformation(Dantsuji & Takayama, 2024; Rahman & Thill, 2023; Wang et al., 2022). However, their future trajectory will depend upon continued technological development, regulatory structures, and incorporation into existing urban forms. Figure 5 depicts the future prospects of autonomous vehicles (AVs) in urban cities.

Figure 5. Future prospects of autonomous vehicles (AVs) in urban cities

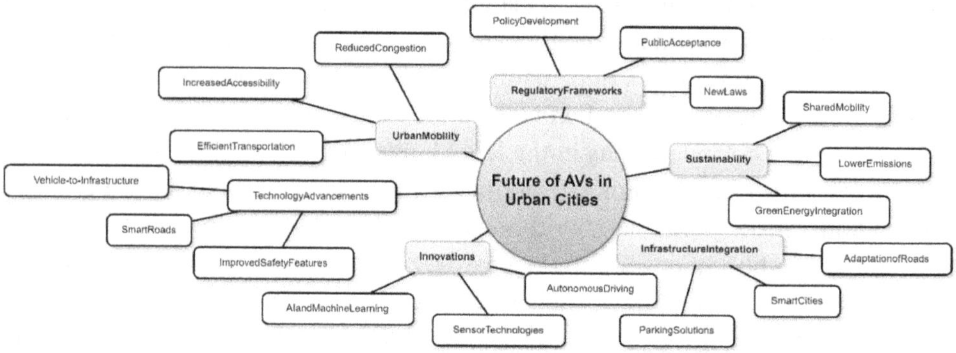

Innovations and Emerging Technologies

The future of autonomous vehicles is influenced by rising innovations and technologies. Vehicle-to-Everything (V2X) Communication: The potential for vehicles to converse with each other and with the infrastructure around them is poised to transform our driving experience. The AVs will leverage artificial intelligence and machine learning algorithms that enable it to make real-time intelligent decisions in a dynamic and complex urban environment. AV will further enhance navigation through sensors like LiDAR, radar, and cameras. Technological Improvements in Electric Vehicles Will Also Help With Adoption of AVs Electric vehicle development will work hand in hand with the adoption of AVs as cleaner, more sustainable modes of transportation. Broadly, those developments are well linked with the future of AVs(Babu et al., 2022; Chandrika et al., 2023; Maguluri et al., 2023).

Long-Term Effects on Urban Living

Empirical innovations like an autonomous vehicle (AV) have been widely expected to change urban life dramatically: AVs make large parking structures unnecessary, allowing urban real estate, now used for parking, to be repurposed as parks, residential buildings, or mixed business. So they assist with mobility challenges for underserved populations like the elderly and disabled folks and enable them to retain independence to get to critical services. This shift to connected, shared and autonomous forms of transportation may, for instance, promote social equity, improving mobility for people without access to private vehicles to travel to jobs, education and health care(Venkatasubramanian et al., 2024; Venkateswaran et al., 2024).

Further, the emergence of AVs could create new industries and jobs related to technology development, vehicle maintenance, fleet management, etc., but they might also eliminate some jobs in transportation and logistics. And cities, which have already been tasked with grappling with these types of economic shifts—how to build retraining and workforce development programs—will have to play a role. AVs and Urban Cities: Modeling, Sustainability, and Social Equity Innovation & Sustainability. Integrating AVs into thoughtfully planned urban environments and following technologically significant advances can remodel urban mobility for generations.

CONCLUSION

Hence, the contributors of the AVs could be enormous for urban mobility if used in such a manner that benefits us all; surpassed traffic congestion, less pollution, creating sustainable environments, and accessibility are just a few of the benefits it could bring to urban cities. The potential of AVs to significantly reshape urban transport governance, notably with advances in sensor systems, ML, and electric vehicle pivots, can provide significant improvements to the management of urban transport systems while creating a more efficient, sustainable, and inclusive urban future. AVs threaten the status quo in a manner that risks leaving people—and the technology that enables AV—far behind, and the technology that enables AV could easily rely on without meeting fundamental challenges related to transferring infrastructure to support AV, ensuring safety, addressing regulatory concerns, building trust, etc.

From the automation of driving to how new technologies such as vehicle-to-everything (V2X) communication with those vehicles and shared self-driving mobility services (of all kinds) will have an important role to play in increasing traffic flow and reducing demand for private vehicles, our experiments on AVs of the future found a great deal to say about the next generation of vehicles. In the long term, AVs can reshape the urban experience, and, along the way, they can transform cityscapes, enable social equity, and create wealth.

Great point, and remember, a widespread deployment of urban automated-capable vehicles will not occur on the snap of the fingers; therefore, just like video games have to be managed under social-ecological management concepts to make sure their potential contribution to human course innovation and growth is maximized. These changes have the potential to unlock a glimpse into the future of better, greener, more equitable city transportation systems.

REFERENCES

Babu, B. S., Kamalakannan, J., Meenatchi, N., Karthik, S., & Boopathi, S. (2022). Economic impacts and reliability evaluation of battery by adopting Electric Vehicle. *IEEE Explore*, 1–6.

Bahamonde-Birke, F. J., Kickhöfer, B., Heinrichs, D., & Kuhnimhof, T. (2018). A systemic view on autonomous vehicles: Policy aspects for a sustainable transportation planning. *disP-The Planning Review, 54*(3), 12–25.

Campisi, T., Severino, A., Al-Rashid, M. A., & Pau, G. (2021). The development of the smart cities in the connected and autonomous vehicles (CAVs) era: From mobility patterns to scaling in cities. *Infrastructures*, *6*(7), 100. DOI: 10.3390/infrastructures6070100

Chandrika, V., Sivakumar, A., Krishnan, T. S., Pradeep, J., Manikandan, S., & Boopathi, S. (2023). Theoretical Study on Power Distribution Systems for Electric Vehicles. In *Intelligent Engineering Applications and Applied Sciences for Sustainability* (pp. 1–19). IGI Global. DOI: 10.4018/979-8-3693-0044-2.ch001

Dantsuji, T., & Takayama, Y. (2024). Hypercongestion, autonomous vehicles, and urban spatial structure. *Transportation Science*, *58*(6), 1352–1370. DOI: 10.1287/trsc.2024.0519

Du, X. (2016). *Towards sustainable autonomous vehicles* [PhD Thesis]. National University of Singapore (Singapore).

Faisal, A., Kamruzzaman, M., Yigitcanlar, T., & Currie, G. (2019). Understanding autonomous vehicles. *Journal of Transport and Land Use*, *12*(1), 45–72. DOI: 10.5198/jtlu.2019.1405

Gavanas, N. (2019). Autonomous road vehicles: Challenges for urban planning in European cities. *Urban Science (Basel, Switzerland)*, *3*(2), 61. DOI: 10.3390/urbansci3020061

Golbabaei, F., Yigitcanlar, T., & Bunker, J. (2021). The role of shared autonomous vehicle systems in delivering smart urban mobility: A systematic review of the literature. *International Journal of Sustainable Transportation*, *15*(10), 731–748. DOI: 10.1080/15568318.2020.1798571

Gružauskas, V., Baskutis, S., & Navickas, V. (2018). Minimizing the trade-off between sustainability and cost effective performance by using autonomous vehicles. *Journal of Cleaner Production*, *184*, 709–717. DOI: 10.1016/j.jclepro.2018.02.302

Jha, S. K., Beevi, S. J., Babitha, M. N., Chinnusamy, S., & Boopathi, S. (2024). Artificial Intelligence-Infused Urban Connectivity for Smart Cities and the Evolution of IoT Communication Networks. In Blockchain-Based Solutions for Accessibility in Smart Cities (pp. 113-146). IGI Global.

Kovačić, M., Mutavdžija, M., & Buntak, K. (2022). New paradigm of sustainable urban mobility: Electric and autonomous vehicles—A review and bibliometric analysis. *Sustainability (Basel)*, *14*(15), 9525. DOI: 10.3390/su14159525

Li, Z., Chitturi, M. V., Yu, L., Bill, A. R., & Noyce, D. A. (2015). Sustainability effects of next-generation intersection control for autonomous vehicles. *Transport*, *30*(3), 342–352. DOI: 10.3846/16484142.2015.1080760

Lim, H. S. M., & Taeihagh, A. (2018). Autonomous vehicles for smart and sustainable cities: An in-depth exploration of privacy and cybersecurity implications. *Energies*, *11*(5), 1062. DOI: 10.3390/en11051062

Maguluri, L. P., Ananth, J., Hariram, S., Geetha, C., Bhaskar, A., & Boopathi, S. (2023). Smart Vehicle-Emissions Monitoring System Using Internet of Things (IoT). In *Handbook of Research on Safe Disposal Methods of Municipal Solid Wastes for a Sustainable Environment* (pp. 191–211). IGI Global.

Mora, L., Wu, X., & Panori, A. (2020). Mind the gap: Developments in autonomous driving research and the sustainability challenge. *Journal of Cleaner Production*, *275*, 124087. DOI: 10.1016/j.jclepro.2020.124087 PMID: 32934442

Rahman, M. M., & Thill, J.-C. (2023). Impacts of connected and autonomous vehicles on urban transportation and environment: A comprehensive review. *Sustainable Cities and Society*, *96*, 104649. DOI: 10.1016/j.scs.2023.104649

Ramudu, K., Mohan, V. M., Jyothirmai, D., Prasad, D., Agrawal, R., & Boopathi, S. (2023). Machine Learning and Artificial Intelligence in Disease Prediction: Applications, Challenges, Limitations, Case Studies, and Future Directions. In *Contemporary Applications of Data Fusion for Advanced Healthcare Informatics* (pp. 297–318). IGI Global.

Seuwou, P., Banissi, E., & Ubakanma, G. (2020). The future of mobility with connected and autonomous vehicles in smart cities. *Digital Twin Technologies and Smart Cities*, 37–52.

Sonia, R., Gupta, N., Manikandan, K., Hemalatha, R., Kumar, M. J., & Boopathi, S. (2024). Strengthening Security, Privacy, and Trust in Artificial Intelligence Drones for Smart Cities. In *Analyzing and Mitigating Security Risks in Cloud Computing* (pp. 214–242). IGI Global. DOI: 10.4018/979-8-3693-3249-8.ch011

Tirlangi, S., Teotia, S., Padmapriya, G., Senthil Kumar, S., Dhotre, S., & Boopathi, S. (2024). Cloud Computing and Machine Learning in the Green Power Sector: Data Management and Analysis for Sustainable Energy. In *Developments Towards Next Generation Intelligent Systems for Sustainable Development* (pp. 148–179). IGI Global. DOI: 10.4018/979-8-3693-5643-2.ch006

Venkatasubramanian, V., Chitra, M., Sudha, R., Singh, V. P., Jefferson, K., & Boopathi, S. (2024). Examining the Impacts of Course Outcome Analysis in Indian Higher Education: Enhancing Educational Quality. In *Challenges of Globalization and Inclusivity in Academic Research* (pp. 124–145). IGI Global.

Venkateswaran, N., Kunduru, K. R., Ashwin, N., Sundar Ganesh, C. S., Hema, N., & Boopathi, S. (2024). Navigating the Future of Ultra-Smart Computing Cyberspace: Beyond Boundaries. In *Advances in Computational Intelligence and Robotics* (pp. 170–199). IGI Global. DOI: 10.4018/979-8-3693-2399-1.ch007

Wang, Z., Wei, H., Wang, J., Zeng, X., & Chang, Y. (2022). Security issues and solutions for connected and autonomous vehicles in a sustainable city: A survey. *Sustainability (Basel)*, *14*(19), 12409. DOI: 10.3390/su141912409

Zambrano-Martinez, J. L., Calafate, C. T., Soler, D., Lemus-Zúñiga, L.-G., Cano, J.-C., Manzoni, P., & Gayraud, T. (2019). A centralized route-management solution for autonomous vehicles in urban areas. *Electronics (Basel)*, *8*(7), 722. DOI: 10.3390/electronics8070722

Chapter 14
Navigating the Virtual Road:
Challenges and Limitations of Virtual and Augmented Reality in the Automobile Industry

M. R. Ramesh
https://orcid.org/0000-0002-1648-9387
IGCAR, India

ABSTRACT

Virtual and augmented reality (VR/AR) technologies are revolutionizing the automobile industry by enhancing design processes, improving manufacturing precision, and enriching customer experiences. Despite these benefits, significant challenges and limitations hinder widespread adoption. This paper explores these obstacles, including technical constraints, high costs, and integration difficulties, while also discussing potential solutions and future trends that could mitigate these issues and pave the way for broader implementation.

1. INTRODUCTION

Virtual Reality (VR) and Augmented Reality (AR) technologies have transformed numerous industries, offering innovative solutions that enhance user experiences and improve operational efficiency. In recent years, the automotive sector has embraced VR and AR applications to revolutionize various aspects of vehicle design, manufacturing, sales, and service. From virtual design and prototyping to augmented reality-based maintenance and repair systems, VR and AR technologies

DOI: 10.4018/979-8-3693-7287-6.ch014

hold immense potential to reshape the automotive industry. However, despite their promising benefits, the widespread adoption of VR and AR technologies in the automotive sector is not without challenges and limitations.

1.1 The Rise of VR and AR in Automotive

The automotive industry has been quick to recognize the transformative potential of VR and AR technologies. VR simulations enable automotive designers and engineers to visualize and iterate on vehicle designs in immersive virtual environments, significantly reducing the time and cost associated with traditional physical prototyping. AR applications, on the other hand, overlay digital information and graphics onto the real-world environment, offering technicians and service personnel valuable insights during vehicle maintenance and repair tasks. Additionally, AR-enhanced sales and marketing experiences provide customers with interactive product demonstrations and personalized vehicle customization options, enhancing the overall purchasing journey (Azuma, 1997).

1.2 Promising Benefits and Opportunities

The integration of VR and AR technologies into the automotive industry offers a plethora of benefits and opportunities. Improved design visualization and prototyping processes enable manufacturers to bring products to market faster while reducing development costs. AR-based maintenance and repair systems enhance technician productivity by providing real-time diagnostic information and step-by-step guidance, ultimately leading to faster turnaround times and increased customer satisfaction. Furthermore, AR-enabled sales and marketing experiences create immersive and engaging interactions that drive customer engagement and brand loyalty.

1.3 Challenges and Limitations

Despite their promising potential, VR and AR technologies in the automotive industry face several challenges and limitations that hinder their widespread adoption and implementation. One of the primary challenges is the high cost of hardware and software infrastructure required to develop and deploy VR and AR solutions. Additionally, the complexity of integrating VR and AR technologies into existing automotive workflows and processes poses significant technical challenges for manufacturers and service providers (Kellmereit & Pflaum, 2016). Furthermore, concerns related to data privacy and security, particularly in the context of AR-enhanced vehicle diagnostics and maintenance, present regulatory and compliance hurdles that must be addressed.

1.4 The Road Ahead

Addressing the challenges and limitations of VR and AR technologies in the automotive industry requires a collaborative and multi-faceted approach. Manufacturers, technology developers, regulators, and industry stakeholders must work together to overcome technical, financial, and regulatory barriers to adoption. Investing in research and development initiatives focused on enhancing VR and AR hardware and software capabilities, as well as streamlining integration processes, is essential to drive innovation and accelerate the adoption of these technologies. Additionally, establishing industry standards and best practices for data privacy and security in AR-enabled automotive applications will instill consumer confidence and facilitate regulatory compliance.

In conclusion, while VR and AR technologies offer transformative potential for the automotive industry, navigating the virtual road ahead requires overcoming significant challenges and limitations. By addressing issues related to cost, complexity, and data security, stakeholders can unlock the full benefits of VR and AR applications in vehicle design, manufacturing, sales, and service, ultimately driving innovation and enhancing the automotive experience for customers and industry professionals alike.

2. TECHNICAL CHALLENGES

Virtual Reality (VR) and Augmented Reality (AR) technologies hold immense potential for transforming the automobile industry, offering innovative solutions for design, manufacturing, training, and customer experiences (Klinker & Bruder, 2019). However, their integration into the automotive sector comes with a set of technical challenges that must be addressed to realize their full benefits. In this chapter, we explore the technical hurdles faced in implementing VR and AR solutions in the automotive domain.

2.1 Hardware Limitations

Current VR and AR hardware, such as headsets and display systems, often lack the necessary resolution, field of view, and processing power required for immersive automotive applications. While consumer-grade VR headsets have made significant advancements in recent years, they still fall short in meeting the demanding requirements of automotive use cases. Automotive VR and AR systems demand high-resolution displays to provide crisp visuals, wide field-of-view optics to simulate real-world environments effectively, and powerful processors to handle complex 3D graphics in real-time. Overcoming these hardware limitations without compromising performance and cost-effectiveness remains a significant technical challenge for automotive VR and AR developers.

2.2 Latency Issues

High latency in VR and AR systems can lead to motion sickness and disorientation, particularly in dynamic automotive environments. The delay between user actions and corresponding virtual responses can disrupt the immersive experience, causing discomfort and reducing user engagement. In automotive applications, where precise spatial awareness and real-time interaction are critical, even minor latency issues can impact safety and usability. Addressing latency challenges requires optimizing hardware components, minimizing data processing times, and leveraging advanced rendering techniques to achieve low-latency VR and AR experiences. Additionally, efficient network communication protocols are essential for streaming real-time data from vehicle sensors and other sources to maintain synchronization and minimize latency.

2.3 Integration Complexity

Integrating VR and AR technologies with existing automotive systems, such as CAD software and vehicle sensors, poses technical challenges due to compatibility issues and data synchronization requirements. Automotive VR and AR applications often rely on complex data pipelines to capture, process, and visualize information from various sources, including vehicle telemetry, environmental sensors, and virtual models. Ensuring seamless integration between disparate systems and data formats demands robust middleware solutions and standardized communication protocols. Moreover, maintaining synchronization between virtual and physical elements in real-time adds another layer of complexity, requiring precise timing and coordination across different subsystems (Lee & Lee, 2019).

Overcoming these technical challenges requires a multidisciplinary approach, involving collaboration between automotive engineers, software developers, and VR/AR specialists. By addressing hardware limitations, reducing latency, and streamlining integration processes, the automotive industry can unlock the full potential of VR and AR technologies to revolutionize vehicle design, manufacturing, and user experiences. Despite the existing hurdles, ongoing advancements in VR and AR hardware, software, and integration techniques offer promising opportunities for overcoming these challenges and navigating the virtual road ahead (Márquez et al. 2019).

3. COST AND RESOURCE CONSTRAINTS

In the pursuit of integrating Virtual Reality (VR) and Augmented Reality (AR) technologies into the automobile industry, stakeholders often encounter significant challenges related to cost and resource constraints. This chapter delves into the various dimensions of these constraints, outlining the high development costs, infrastructure requirements, and training and skill shortages that impede the seamless adoption of VR and AR solutions.

3.1 High Development Costs

One of the primary barriers to the widespread adoption of VR and AR technologies in the automobile industry is the high development costs associated with creating customized solutions. Developing VR and AR applications tailored specifically to meet the needs of automotive manufacturing, design, and retail operations involves substantial investment in software development, content creation, and hardware procurement. Companies must allocate significant financial resources to hire skilled developers, designers, and engineers capable of building immersive experiences that enhance product design, manufacturing processes, and customer interactions (Nour & Mattei, 2020).

Moreover, the iterative nature of VR and AR development often necessitates multiple rounds of prototyping and testing, further driving up costs. As companies strive to create compelling and seamless user experiences, they must invest in advanced rendering engines, motion tracking technologies, and spatial mapping tools, all of which come with substantial price tags. Additionally, the need for specialized hardware, such as VR headsets, AR glasses, and high-performance computing systems, adds to the overall development expenditure.

3.2 Infrastructure Requirements

Implementing VR and AR technologies within automotive manufacturing facilities and dealership showrooms requires significant infrastructure upgrades. To support these immersive technologies effectively, companies must invest in enhancing their digital infrastructure, including high-speed internet connectivity, robust networking equipment, and dedicated computing resources (Stojanovic et al., 2021). The seamless operation of VR and AR applications depends on low-latency network connections and reliable data transmission, necessitating the deployment of advanced networking solutions such as 5G networks and fiber-optic broadband.

Furthermore, the integration of VR and AR technologies into existing workflows may require the installation of specialized equipment, such as motion capture cameras, depth sensors, and haptic feedback devices. Automotive manufacturers and dealerships must carefully assess their current infrastructure capabilities and make strategic investments to accommodate the unique requirements of VR and AR applications. These infrastructure upgrades represent a significant financial commitment and may pose logistical challenges, particularly for organizations operating in remote or resource-constrained environments.

3.3 Training and Skill Shortages

Another critical challenge in the adoption of VR and AR technologies in the automobile industry is the shortage of skilled professionals proficient in these emerging technologies. Training automotive personnel to effectively utilize VR and AR tools requires time, resources, and expertise. Companies must invest in comprehensive training programs to familiarize employees with the capabilities of VR and AR systems, as well as best practices for integrating these technologies into their daily workflows.

However, there is a notable shortage of qualified trainers and educators capable of delivering specialized VR and AR training programs tailored to the automotive sector. Furthermore, the rapid pace of technological innovation in the VR and AR space means that training programs must be continuously updated to reflect the latest advancements and industry trends. This ongoing need for skills development presents a significant challenge for automotive companies seeking to leverage VR and AR technologies effectively (Trambusti & Trambusti, 2018).

In addition to internal training initiatives, automotive companies may also face challenges in recruiting external talent with expertise in VR and AR development, design, and implementation. The competitive job market for VR and AR professionals means that companies must offer attractive compensation packages and career advancement opportunities to attract top talent. Moreover, the interdisciplinary nature of VR and AR projects requires collaboration between professionals from diverse backgrounds, including software engineering, human-computer interaction, and industrial design, further complicating the talent acquisition process.

In summary, cost and resource constraints represent significant challenges for the adoption and integration of VR and AR technologies in the automobile industry. Overcoming these challenges requires strategic planning, financial investment, and a commitment to skills development and talent acquisition. By addressing these constraints proactively, automotive companies can position themselves to harness the transformative potential of VR and AR to drive innovation, enhance productivity, and deliver superior customer experiences.

4. REGULATORY AND SAFETY CONCERNS

The integration of Virtual Reality (VR) and Augmented Reality (AR) technologies into the automobile industry brings forth a myriad of regulatory and safety concerns. As automotive manufacturers and developers embrace these immersive technologies for design, manufacturing, and driver assistance systems, they must navigate through compliance challenges, legal liabilities, and privacy risks.

4.1 Compliance Challenges

Adhering to regulatory standards and safety guidelines is paramount for automotive manufacturers incorporating VR and AR applications into their processes. These technologies have the potential to revolutionize automotive design, allowing engineers to visualize and simulate vehicle components and systems in a virtual environment before physical prototypes are built. However, ensuring compliance with existing regulations, such as those set forth by regulatory bodies like the National Highway Traffic Safety Administration (NHTSA) in the United States and the European Commission in the European Union, presents a significant challenge.

Regulatory compliance involves meeting stringent safety standards, particularly concerning vehicle performance, occupant protection, and crashworthiness. Incorporating VR and AR technologies into the design and manufacturing process must not compromise these safety standards. Manufacturers need to demonstrate that their virtual simulations accurately reflect real-world conditions and comply with established safety protocols. Failure to do so could lead to costly recalls, legal repercussions, and damage to the brand's reputation.

Furthermore, the rapid evolution of VR and AR technologies means that regulatory frameworks must continually adapt to address emerging challenges and advancements. Manufacturers must stay abreast of evolving regulations and proactively integrate compliance measures into their development processes to ensure the safety and reliability of VR and AR-enabled automotive systems.

4.2 Legal Liability

The implementation of augmented reality-based navigation systems and driver assistance features introduces complex legal issues related to liability in the event of accidents or system failures. As these technologies become integral components of modern vehicles, questions arise regarding who bears responsibility in the event of a malfunction or error.

In the case of AR-based navigation systems, where digital overlays provide real-time information about the surrounding environment, inaccuracies or misinterpretations could lead to hazardous situations. If a driver relies solely on the AR system for navigation and encounters an accident due to misleading information or system failure, determining legal liability becomes challenging. Manufacturers may face lawsuits alleging negligence in the design, implementation, or maintenance of the AR system, raising questions about product liability and duty of care.

Similarly, driver assistance features powered by VR and AR technologies, such as augmented reality head-up displays and semi-autonomous driving systems, introduce legal complexities regarding the division of responsibility between the driver and the vehicle's automated systems. In cases where accidents occur due to system errors or misinterpretations, determining fault and liability requires careful analysis of factors such as driver behavior, system reliability, and adherence to safety protocols.

Addressing legal liability concerns necessitates collaboration between automotive manufacturers, technology developers, legal experts, and regulatory authorities to establish clear guidelines and standards for the safe integration of VR and AR technologies into vehicles. Proactive risk assessment, robust testing procedures, and comprehensive user education initiatives are essential to mitigate legal risks and ensure accountability in the event of adverse incidents.

4.3 Privacy Risks

The proliferation of connected vehicles and the integration of VR and AR applications raise significant privacy concerns related to the collection and processing of user data. In the context of automotive VR and AR systems, which often require access to personal information for customization and user authentication purposes, safeguarding user privacy is paramount.

Collecting data from sensors, cameras, and other onboard systems to create immersive VR experiences or deliver personalized AR overlays raises questions about the scope of data collection, storage, and usage. User consent, data anonymization, and encryption are crucial considerations to prevent unauthorized access or misuse of sensitive information.

Moreover, the interconnected nature of modern vehicles introduces additional privacy risks, as data transmitted between onboard systems and external servers may be susceptible to interception or hacking. Unauthorized access to personal information, such as location data, biometric identifiers, and behavioral patterns, poses a threat to user privacy and security.

To mitigate privacy risks associated with automotive VR and AR technologies, manufacturers must implement robust data protection measures, including encryption, access controls, and data minimization strategies. Transparent privacy policies, clear user consent mechanisms, and regular security audits are essential to build trust and confidence among consumers.

Furthermore, collaboration between industry stakeholders, regulatory bodies, and cybersecurity experts is crucial to establish industry-wide standards and best practices for protecting user privacy in VR and AR-enabled automotive systems. By prioritizing privacy and security in the design and implementation of these technologies, manufacturers can ensure that the benefits of immersive experiences are not outweighed by the risks to user privacy and security.

5. USER ACCEPTANCE AND ADOPTION

The successful integration of Virtual Reality (VR) and Augmented Reality (AR) technologies into the automobile industry relies heavily on user acceptance and adoption. This chapter explores the key factors influencing users' willingness to embrace VR and AR features in vehicles, including user interface design, consumer education, and cultural and behavioral factors.

5.1 User Interface Design

Effective user interface design is paramount in ensuring the seamless integration of VR and AR technologies into automobiles. Designing intuitive and user-friendly interfaces is crucial to enhancing user experience and fostering widespread acceptance among consumers and automotive professionals alike.

In the context of VR and AR applications in vehicles, user interface design should prioritize simplicity, clarity, and accessibility. Interfaces should be designed to minimize cognitive load and facilitate quick and intuitive interaction. This involves employing clear visual cues, intuitive navigation systems, and ergonomic controls that are easily accessible to drivers and passengers.

Furthermore, user interface design should be tailored to the unique characteristics of VR and AR environments. This includes optimizing interface elements for immersive VR experiences and overlaying contextual information seamlessly onto the real-world environment in AR applications. By prioritizing user-centered design principles, automotive manufacturers can enhance the usability and acceptance of VR and AR features in vehicles.

5.2 Consumer Education

Consumer education plays a crucial role in overcoming skepticism and fostering acceptance of VR and AR technologies in automobiles. Many consumers may be unfamiliar with the benefits and functionalities of VR and AR features, leading to resistance to new technologies.

To address this challenge, automotive manufacturers should invest in comprehensive consumer education initiatives aimed at raising awareness and increasing understanding of VR and AR applications in vehicles. This may involve leveraging marketing campaigns, educational materials, and interactive demonstrations to showcase the potential benefits and use cases of VR and AR technologies.

Moreover, manufacturers should emphasize the safety, convenience, and enhanced driving experience offered by VR and AR features, highlighting their potential to improve situational awareness, navigation, and entertainment for drivers and passengers alike. By empowering consumers with knowledge and information, automotive manufacturers can mitigate skepticism and facilitate the adoption of VR and AR technologies in the automotive industry.

5.3 Cultural and Behavioral Factors

Cultural preferences, societal norms, and user habits are important considerations in the adoption of VR and AR technologies in different regions and demographic groups. The acceptance of these technologies may vary significantly depending on cultural attitudes towards technology, privacy concerns, and perceptions of risk and novelty.

In some regions, there may be a greater openness to embracing new technologies, while in others, there may be resistance or skepticism towards technological advancements. Additionally, cultural preferences and user habits may influence the demand for specific VR and AR features, such as entertainment, navigation, or safety enhancements.

To navigate these cultural and behavioral factors, automotive manufacturers must conduct thorough market research and tailor their VR and AR offerings to meet the unique needs and preferences of different consumer segments. This may involve customizing user interfaces, content, and marketing strategies to resonate with local cultural norms and expectations.

Furthermore, manufacturers should prioritize transparency, trust, and respect for user privacy in the development and implementation of VR and AR technologies. By addressing cultural and behavioral considerations, automotive manufacturers can enhance the acceptance and adoption of VR and AR features in vehicles, driving innovation and differentiation in the automotive industry.

In conclusion, user acceptance and adoption are critical factors in the successful integration of VR and AR technologies into the automobile industry. By focusing on user interface design, consumer education, and cultural and behavioral factors, automotive manufacturers can overcome barriers to adoption and pave the way for the widespread acceptance of VR and AR features in vehicles.

6. FUTURE PERSPECTIVES AND SOLUTIONS

As the automotive industry continues to explore the potential of virtual and augmented reality (VR and AR) technologies, several future perspectives and solutions emerge to address the challenges and limitations discussed earlier. This chapter discusses key areas of focus for the future of VR and AR in the automobile industry, including technological advancements, collaboration and partnerships, and education and training programs.

6.1 Technological Advancements

Continued advancements in VR and AR hardware, software algorithms, and networking technologies hold the promise of overcoming existing technical limitations and enhancing the performance and reliability of automotive applications.

In terms of hardware, manufacturers are investing in developing more sophisticated headsets and display systems that offer higher resolution, wider field of view, and reduced latency. Innovations in tracking technology, such as inside-out tracking and eye-tracking, aim to improve motion sensing and interaction accuracy, providing a more immersive and intuitive user experience.

On the software front, advancements in rendering engines, real-time physics simulation, and spatial mapping algorithms contribute to creating more realistic and interactive virtual environments. These developments enable features like dynamic lighting, realistic physics-based interactions, and accurate spatial mapping of physical surroundings, enhancing the realism and functionality of VR and AR applications in automotive design, prototyping, and training.

Moreover, improvements in networking technologies, including 5G connectivity and edge computing, enable seamless streaming of high-fidelity VR and AR content to mobile devices and headsets, eliminating the need for tethered connections and expanding the accessibility of immersive experiences.

Overall, ongoing technological advancements are poised to unlock new possibilities and drive widespread adoption of VR and AR solutions across various automotive use cases, from design and engineering to sales and marketing.

6.2 Collaboration and Partnerships

Collaborative efforts between automotive manufacturers, technology companies, regulatory bodies, and research institutions are essential to foster knowledge sharing, standardization, and regulatory compliance in the VR and AR domain.

By establishing industry-wide standards and best practices for VR and AR development and deployment, stakeholders can ensure interoperability, compatibility, and safety across different hardware and software platforms. Collaborative initiatives, such as the OpenXR standard for cross-platform VR and AR development, facilitate seamless integration of VR and AR applications with existing automotive workflows and systems.

Furthermore, partnerships between automotive OEMs and technology providers enable co-innovation and co-development of VR and AR solutions tailored to specific industry requirements and use cases. By leveraging each other's expertise and resources, companies can accelerate the pace of innovation and bring cutting-edge VR and AR experiences to market more efficiently.

Additionally, collaboration with regulatory bodies and industry associations helps address legal and ethical concerns related to privacy, data security, and user safety in VR and AR applications. By adhering to regulatory guidelines and industry standards, automotive companies can build trust and confidence among consumers and stakeholders, fostering greater acceptance and adoption of VR and AR technologies in the automotive sector.

In summary, collaboration and partnerships play a pivotal role in driving the evolution and widespread adoption of VR and AR solutions in the automobile industry, enabling stakeholders to leverage collective expertise and resources to address common challenges and seize new opportunities.

6.3 Education and Training Programs

Developing comprehensive training programs and educational initiatives to upskill automotive professionals and end-users in VR and AR technologies is crucial for accelerating adoption and mitigating resistance to change.

By providing hands-on training and certification programs, automotive manufacturers and technology vendors can empower engineers, designers, technicians, and sales personnel with the knowledge and skills needed to leverage VR and AR tools effectively in their respective roles. These programs cover various aspects of VR and AR development, including content creation, user interface design, software programming, and hardware maintenance, catering to diverse skill levels and job functions within the automotive industry.

Moreover, educational institutions and vocational training centers can incorporate VR and AR technologies into their curricula and coursework to prepare students for careers in the automotive sector. By integrating VR and AR tools into design, engineering, and manufacturing courses, educators can expose students to real-world applications and industry best practices, fostering a pipeline of talent equipped with the latest skills and knowledge.

Beyond professional training, educational initiatives targeting consumers and end-users help familiarize them with VR and AR experiences and encourage adoption. Automotive dealerships, for example, can offer interactive VR and AR demonstrations to prospective customers, allowing them to visualize and customize vehicle configurations in immersive virtual showrooms. Similarly, educational outreach programs in schools and community centers introduce children and young adults to VR and AR technologies, sparking interest and curiosity about future career opportunities in the automotive industry.

In conclusion, investing in education and training programs is essential for building the workforce of the future and cultivating a culture of innovation and continuous learning in the automotive industry. By empowering individuals with the necessary skills and knowledge to harness the potential of VR and AR technologies, stakeholders can drive meaningful change and unlock new possibilities for enhancing productivity, efficiency, and customer satisfaction in the virtual road ahead.

7. CONCLUSION

The integration of Virtual Reality (VR) and Augmented Reality (AR) technologies in the automobile industry has opened up new avenues for innovation and transformation. As discussed throughout this chapter, VR and AR offer numerous benefits, ranging from enhanced design and prototyping processes to improved safety, training, and user experience. However, the adoption of these technologies also presents various challenges and limitations that must be addressed to realize their full potential.

7.1 Technical Challenges

One of the primary technical challenges in implementing VR and AR solutions in the automobile industry is the need for high-quality, immersive experiences. This requires advanced hardware capabilities, including powerful processors, high-resolution displays, and accurate tracking systems. Additionally, ensuring compatibility and interoperability between different VR and AR platforms remains a significant technical hurdle. Overcoming these challenges will require ongoing investment in research and development, as well as collaboration between automotive manufacturers, technology companies, and academic institutions.

7.2 Financial Considerations

The upfront costs associated with developing and implementing VR and AR solutions can be substantial, especially for smaller automotive companies. This includes investment in hardware, software development, content creation, and training programs. Moreover, the return on investment (ROI) for VR and AR initiatives may not be immediately apparent, making it challenging for companies to justify the expenditure. However, as the technology matures and adoption rates increase, the long-term benefits of VR and AR, such as improved productivity, reduced costs, and enhanced customer satisfaction, are expected to outweigh the initial financial outlay.

7.3 Regulatory and Legal Issues

The automotive industry is subject to stringent regulations and safety standards, which can pose obstacles to the widespread adoption of VR and AR technologies. Concerns related to data privacy, cybersecurity, and liability must be carefully addressed to ensure compliance with existing laws and regulations. Additionally, the introduction of new VR and AR features may require updates to existing regulatory frameworks, which can be a lengthy and complex process. Collaborative efforts between industry stakeholders, policymakers, and regulatory bodies are essential to navigate these challenges effectively.

7.4 User Acceptance and Training

While VR and AR have the potential to enhance user experience and engagement, their success ultimately depends on user acceptance and adoption. Overcoming skepticism and resistance to change among employees, customers, and other stakeholders is crucial for the widespread adoption of VR and AR solutions in the automotive industry. Furthermore, providing adequate training and support to users is essential to ensure they can effectively utilize VR and AR tools in their daily workflows. User feedback and continuous improvement processes are key to refining VR and AR applications and addressing usability issues.

In conclusion, while Virtual Reality and Augmented Reality technologies hold immense promise for the automobile industry, realizing their full potential requires overcoming a range of challenges and limitations. Technical advancements, financial investments, regulatory compliance, and user acceptance are all critical factors that must be addressed to drive the successful adoption of VR and AR solutions. By working collaboratively and proactively addressing these challenges, the automotive industry can navigate the virtual road ahead and unlock new opportunities for innovation and growth.

REFERENCES

Azuma, R. T. (1997). A Survey of Augmented Reality. *Presence (Cambridge, Mass.)*, *6*(4), 355–385. DOI: 10.1162/pres.1997.6.4.355

Kellmereit, D., & Pflaum, A. (2016). *The Augmented Reality Megatrend of Tomorrow: Pokémon Go - and More*. Springer.

Klinker, G., & Bruder, G. (2019). The Future of Augmented Reality in the Automotive Industry. In Tatnall, A., & Arthur, C. J. (Eds.), *Information and Communication Technologies in Modern Organisations* (pp. 161–174). Springer.

Lee, G., & Lee, S. (2019). Virtual Reality Technology in the Automotive Industry: A Literature Review. *Sustainability*, *11*(23), 6712.

Márquez, M. A.. (2019). Augmented Reality: A Sustainable Future for the Automotive Industry. *Sustainability*, *11*(22), 6377.

Nour, M., & Mattei, M. (2020). Virtual and Augmented Reality for Car Retailers: A Systematic Review. *Journal of Retailing and Consumer Services*, *53*, 101939.

Stojanovic, J.. (2021). The Role of Virtual and Augmented Reality in Shaping the Future of Automobile Industry. *International Journal of Advanced Robotic Systems*, *18*(2), 1729881421997432.

Trambusti, M., & Trambusti, S. (2018). Augmented Reality in the Automotive Industry: The Case of Volkswagen. In Di Natale, P. (Eds.), *Recent Advances in Information Systems and Technologies* (pp. 729–737). Springer.

Chapter 15
Virtual and Augmented Reality in the Automobile Industry: Applications and Future Directions

J. Samuel Manoharan
Jerusalem College of Engineering, India

M. Arunkumar
Senguthar Engineering College, India

Sathiya A
https://orcid.org/0009-0008-6996-5296
M. Kumarasamy College of Engineering, India

M. Hemamalini
Panimalar Engineering College, India

V. Sathya
Vel Tech Rangarajan Dr. Sagunthala R&D Institute of Science and Technology, India

T. P. Anish
https://orcid.org/0000-0002-4308-9645
R.M.K. College of Engineering and Technology, India

R. Siva Subramanian
https://orcid.org/0000-0002-7509-9223
R.M.K. College of Engineering and Technology, India

ABSTRACT

This survey paper aims to discuss the areas of enhancement of virtual reality (VR) and augmented reality (AR) in the automotive industry: the basic principles, practice, problems, and points of improvement. It also emphasizes the application of VR in design improvement, prototyping and manufacturing, and sales presenting extraordinary experience and increased efficacy. Applications of AR in repair and maintenance, navigations, and driver assistance are analyzed with reference to

DOI: 10.4018/979-8-3693-7287-6.ch015

enhancements of diagnostics, trainings and safety. Some of the issues that stem from the implementation of the two technologies include; technicalities, costs and acceptability by users. Finally, the paper outlines a future outlook touching on future trends and how VR and AR need constant enhancement and deployment in the automotive industry.

1. INTRODUCTION TO VIRTUAL AND AUGMENTED REALITY IN THE AUTOMOBILE INDUSTRY

1.1 Overview of VR and AR Technologies

Virtual Reality (VR)

Virtual Reality (VR) is an aspect of interactive digital environment which involves one's physical experience of it. The users wear virtual reality headsets like Oculus Rift, HTC Vive, or PlayStation VR to access a fully created world where they can even have a possibility to touch 3D objects and navigate through virtual space. The technology that supports VR is comprised of several elements that include head mounted displays (HMDs), motion sensors and handheld controllers(Anthes et al 2016). These elements together make it possible to allow users to feel and to interact with Virtual Real Environment as if they were real environments and enable one to simulate given situations(Singh N and Singh S 2017),.

Augmented Reality (AR)

While on the other hand, the augmented reality (AR) provides a closer look of the physical world by placing computer-generated images or label over the real environment(Huang et al 2018). AR uses instruments like smart phones and tablets, AR glasses (like Microsoft hololens) and Heads Up Displays (HUDs) in cars. With these devices, users are able to view digital content overlaying the real world the users are in(Aswini et al 2023). AR covers simple ones such as navigation direction shown on the windscreen of a car up to fully interactive ones such as an augmented repair manual to provide the steps to repair part of a car and has the instructions shown on the parts of the automobile.

1.2 Significance and Relevance in the Automobile Industry

Figure 1. Significance and relevance in the automobile industry

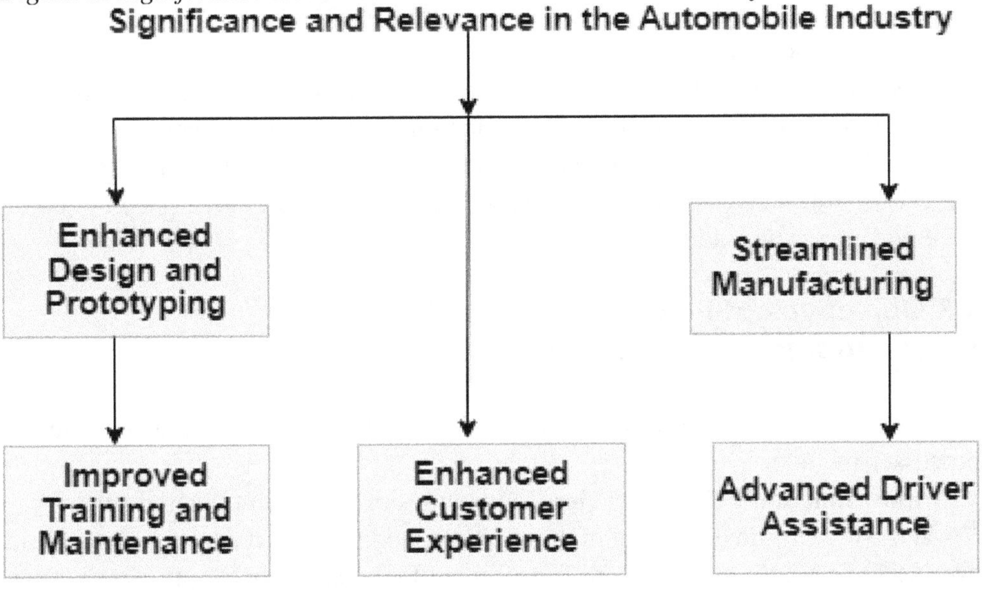

Enhanced Design and Prototyping: In the context of the automotive industry, VR helps designers to build and evaluate car designs in the virtual environment thereby cutting down on the amount of time and money spent producing actual physical mock-ups. It allows changes in designs as and when required and effective team work during the designing phase results in the production of better looking and more advanced cars.

Streamlined Manufacturing: VR and AR can be of aid in manufacturing by training the workers virtually, come up with exercises that mimic the assembly line processes, and spot all the possible problem areas before they occur. This leads to increased productivity, less wastage and ultimately cutting down the overall expenses of production.

Improved Training and Maintenance: By implementing AR, technicians and assembly line workers can receive instructions during operation and manufacturing processes and require less practice and demonstration time. In maintenance, AR can help mechanics in overlaid diagnostic information and repair instructions on the exact position of the components of the vehicle so that mechanics may work more efficiently and accurately.

Enhanced Customer Experience: The use of VR ad AR can create new opportunities of interacting with the customers; for example, virtual car showrooms and configurators when customer does not need a physical car to test. AR is capable of supplying actual and diversified data that can transform the test drive and make the sales process more engaging and amusing.

Advanced Driver Assistance: AR can help improve and augment currently available navigation systems through projecting pertinent information and directions on the windshield thus eliminating the need for drivers to look away from the road. They translate to reduced occurrences of accidents and better perception of the environment around the car.

1.3 Objectives of Integrating VR and AR in Automotive Processes:

The primary objectives of integrating VR and AR technologies into automotive processes include:

Innovation and Efficiency: Here, companies in the automotive industry apply VR and AR to the development of new vehicle models, manufacturing of cars, and even servicing of the automobiles. These technologies can help to optimize the flow of processes and significantly decrease expenditures, and also these can contribute to the accelerated rate of the development several concepts up to the market stage.

Quality and Precision: The usage of technology entails the improvement of design and manufacturing to higher quality through the use of VR and or AR. Due to virtual simulations and augmented diagnostics, the staking, control, and refinement of the final form of the vehicle components and systems are more achievable.

Customer Engagement and Satisfaction: About these technologies, it is possible to state that they can redefine the nature of customers' experience and make it more engaging and personal. Test drives are improved, virtual configurations, and engaged customer processes create experiences of today's buyers.

Safety and Training: AR for instance offers better navigation and assistance resulting in better safety of drivers while VR which can make employees undergo comprehensive training hence making a company have well-prepared staff, well equipped vehicles on the road.

Sustainability: Consequently, VR and AR help to minimize resource consumption during production and to avoid the extensive usage of actual prototypes, thus supporting the goals of the global environmental strategy in the sphere of automotive manufacturing.

To sum it up, VR and AR technologies stand on the verge of becoming critical to the automobile industry's development in the future while opening up tremendous opportunities for the reshaping of designing, manufacturing, selling, and using automobiles. The incorporation of these technologies is in compliance with the goal and objectives set within the industry including innovation, efficiency, quality, and customer satisfaction, safety, and sustainability.

2. FOUNDATIONS OF VIRTUAL REALITY TECHNOLOGY

2.1 Definition and Key Components of VR

Virtual Reality (VR) can be described as a simulated environment that is real to its subject depending on the type of VR. It gives users the unique opportunity of stepping into an almost tangible world to play with what is actually a simulated reality through electronic devices including VR headset with screens or screens inside the gloves with sensors(Bhardwaj et al 2016).

2.2 Key Components of VR

Head-Mounted Displays (HMDs): They are the fundamental I/O devices for VR. The Head Mounted Displays such as Oculus rift, HTC vive, PlayStation VR have stereoscopic 3D display and large field of view to capture the whole environment.

Motion Tracking Sensors: These sensors are used to track the movement of the user and then changes the environment in the virtual reality to match with the movement of the user. This includes head tracking, hand tracking by controllers or gloves and at times full body track with help of other sensors.

Controllers: Wearable devices or devices that can be used by holding them or gloves that enable man to interact with virtual environment. These can be used to monitor the position and determination of movement of the users' hands for proper handling of virtual objects.

Software: Applications and platforms, which are designed to generate and control virtual reality settings, can be referred to as VR software. This includes popular game engines such as unity and unreal engine as well as other specific VR development tools (Sanjeera et al 2024).

Computing Hardware: Computing devices such as the main computer, or other devices such as game consoles that are needed to process graphics and depth of interactions to enable VR. Such systems require the powerful CPUs and GPUs to provide the buildings of immersive environments in real time (Varun C M et al 2024).

2.3 Current State and Advancements in VR

Modern VR systems produce high-quality graphics on the screen, accurately track the motion, and have a growing choice of haptic interfaces. The hardware is now smaller, cheaper and easier to obtain, standalone VR devices such as Oculus Quest 2 does not require a powerful PC.

Figure 2. Current state and advancements in VR

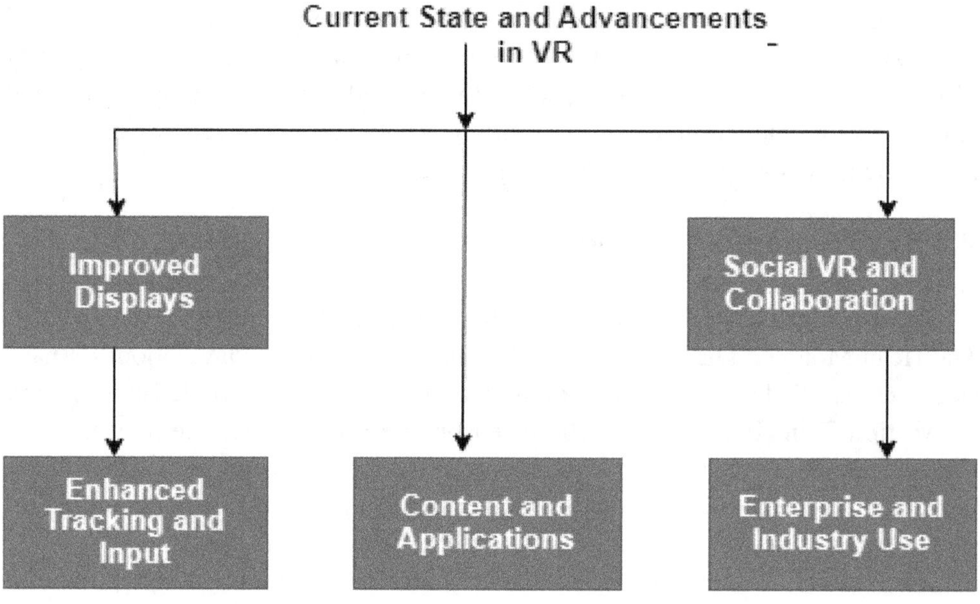

Improved Displays: Contemporary models use OLED or LCD screens with a lowered amount of 'lag,' and increased 'refresh rate,' giving images more detail and fluidity.

Enhanced Tracking and Input: Major improvements have been made to inside-out tracking meaning cameras on the HMD for tracking motion have become easier to set up as well as being more accurate gesture recognition and touch sensitive gloves are advancing; giving Virtual Reality environments a far richer experience.

Content and Applications: The usage of VR content is no more limited to the games and has crossed over to fields such as; Education, health care, real estate, etc. With Describing VR purpose, we can say that VR is for the training simulation, virtual tours, and also the remote collaboration, also the therapy.

Social VR and Collaboration: Main platforms also contain social components nowadays; people can interact in virtual reality via VRChat and Facebook Horizon. It is eventually important to understand that this has huge impacts on remote work and social relations.

Enterprise and Industry Use: Today it is used in the industries for goals such as creating structures and models, practice and teaching, and remote collaboration. Businesses that it impacts: Manufacturing and automotive business for instance, apply VR when analyzing and modeling designs for implementation since the Virtual environment enables them to first see how it will appear like before the actual manufacturing (Jain et al 2023).

2.4 Future Directions

Some developments that the VR technology should be aiming for in the future include the use of eye-tracking, foveated rendering and more natural modes of interface. Other related ideas such as applying artificial intelligence in order to develop intelligent and adaptive virtual environments are also on the apparent horizon. Therefore, the idea of the Virtual Reality for its start led to the creation of the complex tool which can be applied in various fields. Due to the recent breakthroughs in the technology, it changes the scale and offers the opportunities for its use, becoming the part of numerous fields such as automotive industry.

3. EVOLUTION AND CURRENT STATE OF THE AUTOMOBILE INDUSTRY

3.1 Historical Overview of the Automobile Industry:

Today, the automobile industry looks like a different ball game as compared to what it was at the turn of the 20th century. It was initiated by inventing car like Karl Benz's Motorwagen in 1886 that paved the way for the evolution of the motor vehicles. Ford was the pioneer of the motor industry in the early and mid-1900s, kicking off a revolution when he first incorporated the assembly line in 1913 thus reducing production costs and making the motor car affordable to the public. This period called the era of mass production increased car accessibility in society and made it a necessity. After the Second World War, new developments occurred in the industry with the incorporation of aspects such as seat belts, and improved internal combustion engines. The liberalised period saw the focus on factors such as fuel economy and the environment primarily due to the oil crises of the 1970s. This period also witnessed emergence of automakers such as Toyota and Honda which focused on quality and productivity and, therefore, threatened to overshadow other automakers.

3.2 Major Technological Milestones

Several technological milestones have significantly shaped the automobile industry:

Electric Vehicles (EVs): While the idea of a car with electric motor has been around in the 19th century, massive technological improvement on the battery has ensured that the EVs are feasible options to the gasoline-engine Cars. Tesla for this reason was significant in changing the face of automobile industry towards electric mobility in the first decade of 2000.

Autonomous Driving: While simple systems such as cruise control were first introduced, the advancement of ADAS in the 21st century has set the tone for completely self-inflicted vehicles, with Waymo and Tesla in the lead.

Connected Cars: Automobiles have been connected to the Internet and to one another and to the surrounding environment through IoT and Cloud technologies, bringing in new safety facets, new services including real time navigation and predictive maintenance.

3.3 Current Trends and Market Dynamics

Today, the automobile industry is in the midst of a transformative phase characterized by several key trends:

Electrification: This transition involves environmental concerns and regulatory requirements in an attempt to reduce the utilization of gasoline powered automobiles. Many big players in automobile production have become aggressive in their quest to transition away from internal combustion engines in their cars, and shift to electric power trains.

Autonomous Vehicles: There is a gradual shift towards self-driving technology and the industry is sustaining a steady process of development of automobiles that can be driven using digital systems from beginning to end (Parekh et al 2022).

Shared Mobility: Ridesharing and car-sharing services are changing the nature and meaning of the automobile as now it is more focused on personal car-sharing and one-time car rental with great flexibility.

Sustainability and Smart Manufacturing: Today as more attention is paid to environmental issues automobile manufacturers are trying to incorporate environmentally friendly material recycling and energy conservation into the manufacturing process. The smart factories incorporate the AI and robots that will increase production efficiency and decrease the effect on the environment.

These dynamics put emphasis on the future of the automobile that is not only a means of transportation but also an element of the global networked, intelligent, and sustainable mobility complex.

4. FOUNDATIONS OF AUGMENTED REALITY TECHNOLOGY

4.1 Definition and Key Components of AR:

AR is a technology used in creation of an immersive environment by laying digital details over an image for real life environment where the user perceives it in a differently endowed way. The key difference between AR and VR, the latter of which puts users in a completely different environment, is that AR enhances the physical world(Carmigniani, J., & Furht, B. 2011). **The key components of AR include:**

Display Devices: AR is available through software that can be run on different types of devices such as mobile devices like smartphones or tablets, and specific AR glasses or headsets. They cast out the digital content onto the envisioned view of the real world by the user.

Sensors and Cameras: Real world overlay is important in AR systems because they need to capture the environment the user is in and the movements he is making or the spatial relationships.

Software Algorithms: High-level algorithms take the data collected by the sensors and cameras to integrate digital content into real-world objects in real-time. This includes image recognition, object tracking and creating a map of the environment.

User Interaction Interfaces: AR adds interactivity to the augmented content and the possibilities of the interactions include touch or gesture-based interaction and voice commands.

4.2 Historical Development of AR Technology

The development of AR can be traced back to the 1960s:

Early Concepts (1960s-1980s): The idea of laying digital information over the physical environment has been around since the 1960s. Among them, one of the first works in this field might be mentioned, the legendary “Sword of Damocles” by Ivan Sutherland, which can be regarded as the first attempt to create augmented reality and is a head-mounted display system.

Advancements in Computing and Graphics (1990s): The improvement of more extensive computers and graphic interfaces were in the 1990s seen as tremendous growth. Hence, researchers began creating systems to superimpose information onto physical objects to create the first uses of AR in majors that encompass aviation and military among others.

Commercial Breakthroughs (2000s): Mobile computing swiftly popularized augmented reality starting from the year 2000 and the smartphone. That breakthrough was made by ARToolKit application and such game as Pokémon Go that introduced people in augmented reality as separate area where digital and physical environment interact.

4.3 Current State and Advancements in AR

Today, AR technology is advancing rapidly and becoming more integrated into various aspects of daily life and industry:

Figure 3. Current state and advancements in AR

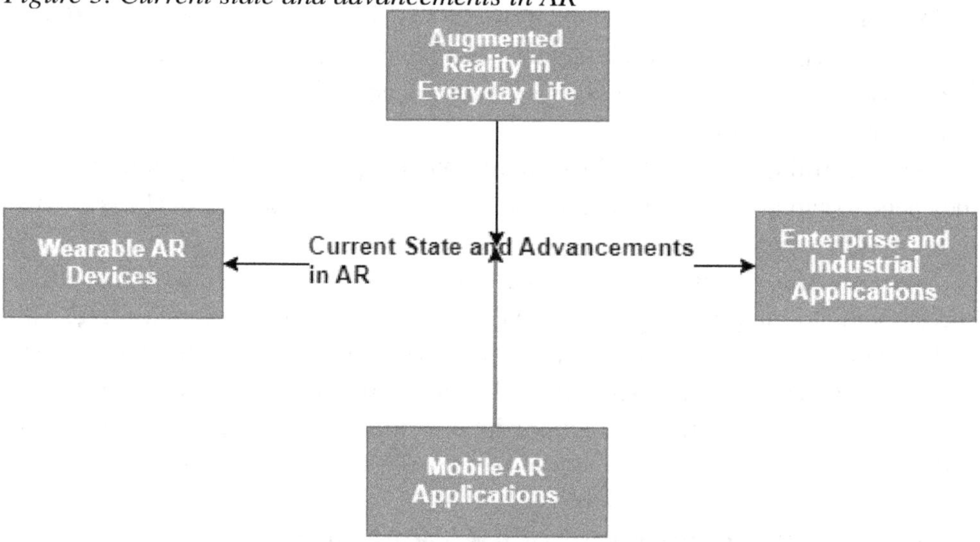

Wearable AR Devices: AR has progressed over the recent years with heads-ups like Microsoft HoloLens and Magic Leap, which has made the concept to be even more advanced and engaging through the use of headsets. These devices are fitted with several sensors and cameras and the pictorial overlays are of superior quality with interactivity (Shi et al 2015).

Mobile AR Applications: Smartphones continue to be a preferred medium of AR, conceivable in every field starting from gaming through retail to navigation with examples like Pokémon Go, IKEA Place, and Google Maps Live View. The subject educating devices have been developed with Mobile application tools like

ARKit from Apple and ARCore from Google, enabling developers to build complex models of look (Adhani et al 2012).

Enterprise and Industrial Applications: AR's application is fast extending to different industries including the manufacturing industries, the healthcare sector, and educational facilities. In manufacturing, AR provides guidance and instructions as well as operational checklists on equipment without any need for manuals or printed guides. AR has proved helpful in health care through supporting surgery and training by projecting important information onto the patient's image.

Augmented Reality in Everyday Life: The application of AR is also broadening and is now appearing in everyday processes involving a consumer, from trying on clothes or accessories virtually to studying through augmented applications in school. This wide-spread use is because AR technology has become more available, and there is more customer interest in improving how physical environments interact with media.

Finally, overall, it was seen that AR is revolutionizing how it is possible to engage with environments, than providing opportunities for entertainment as well as functional uses. In the presence of new technologies and advancing technology, AR has huge opportunities and its possibilities in every sphere of life and industry are expected to grow.

5. VIRTUAL REALITY APPLICATIONS IN AUTOMOBILE DESIGN AND PROTOTYPING

5.1 VR in Conceptual Design and Visualization

Virtual Reality (VR) has brought drastic change in the manner in which automotive designers work in the preliminary stage of car design. It has been ascertained that through the use of VR, designers are able to conduct modelling of new car concepts through the development of 3D replicas. This makes it easier to reason about sizes and shapes, as well as beauty and use relative to a 2D drawings or physical mock-ups.

Immersive Visualization: Designers can preview their designs in life size and can actually get around the design, and get into digital mock-ups where the design is intended to be. This way, one can notice design issues and make better decisions at the beginning of a project.

Collaborative Design: One of the benefits of VR is that a team of designers can have design meetings where all the members who are in different locations will have to meet in a virtual space where the design progress is to be presented and also real-time comments can be made, and adjustments can also be made as well.

This is beneficial in the sense that it integrates the different aspects of a design more seamlessly, which in turn makes the design process that much more efficient.

5.2 Advantages of Using Virtual Reality in Prototyping and Simulation

Enhanced Prototyping - The use of VR in prototyping offers several significant advantages over traditional methods:

Cost and Time Efficiency: Physical model development is costly and requires a lot of time to be spent on the models. Since, with VR, simultaneous design and testing are possible, with far lower expenses and time required than creating mock-ups, the number of iterations can be easily increased.

Risk Reduction: Through the recreation of different situations, which could happen in the future, the VR emphasizes such drawbacks and dangers, which must be eradicated or minimized. It lowers the chances of cumulating big mistakes, which may cost more in the future.

Detailed Simulation: In VR there is potential to mimic and assess the driving performance under various environmental factors for instance, aerodynamics, human factors and safety. Engineers can fine tune the vehicle to how best it will perform in a certain driving environment before constructing models in the physical realm.

Customer Feedback: Through such an approach VR can be used to get first impression and input from customers as they get an opportunity to virtually engage with the design of a vehicle and assess it. Such feelings can be highly beneficial in fine tuning the design towards what is more favorable to the customers.

6. VIRTUAL REALITY APPLICATIONS IN AUTOMOBILE MANUFACTURING AND ASSEMBLY

6.1 VR for Process Optimization and Efficiency

Process Optimization

Virtual Reality (VR) is playing a significant role in the automobile manufacturing industry as it is used to enhance several processes, and this has enhanced efficiency and high productivity.

Layout Planning: Planning the layout of the actual factory floor is made easier with VR since it gives the manufacturer a real feel of what is to be expected when the layout is real. It assists in determining the interrelation between equipment layout, workstation layout, and conveyor layout in order to avoid congestion.

Assembly Line Simulation: Another way in which using the virtual reality in manufacturing helps the manufacturers is that it allows dissecting an assembly line and finding out all the shortcomings and possible problems that may arise. This helps them to undertake changes required for them to enhance efficiency in production and, therefore, shorten a product's time in the market.

Ergonomics Analysis: It is therefore possible to apply VR in the analysis of the ergonomic factors of manufacturing. This can be achieved where manufacturers set imitations of the actions of the workers on the line to assess ergonomic dangers that may lead to workplace injuries, hence improve on the flow on the line leading to high productivity.

6.2 Training and Simulation for the Assembly Line Employees

Worker Training -Hence, it is apparent that the application of VR in training of assembly line workers is extremely effective due to the controlled nature of training.

Interactive Training Modules: The employees can take virtual reality training exercises that mimic real-life scenarios and activities. Compared to traditional training methods, this training allows the development of a better relation to the working processes, assessing its effectiveness for procedures and the way workers perform their tasks, and acquiring a better sense of operation safety.

Hands-On Practice: This is because, in VR, workers are able to do the procedures over and over again without physical parts and/or tools. At the same time, such an approach contributes to the formation of muscle memory as well as accuracy, thus yielding increased quality work on the actual automobile assembly line.

Simulated Scenarios: The actual training can actually consist of different scenarios for instance mechanical breakdown or aversion situations for the workers. This assist in developing confidence and improving or developing the problem-solving skills that is needed amid early childhood.

7. VIRTUAL REALITY APPLICATIONS IN AUTOMOBILE SALES AND MARKETING

7.1 Enhancing Customer Experience with VR Showrooms

VR Showrooms -Virtual Reality (VR) showrooms engage the consumers and change the way they experience the car and make a choice.

Virtual Walkthroughs: Customers can be presented the whole line of vehicles in the virtual showroom and rotate the car and look behind it as if it is done physically. This makes it possible for them to consider all the features of the interior solutions, as well as the car's exterior from the convenience of their own home.

Customization Options: By using VR showrooms one can make changes on the car in real-time, including changing the colour, material, and features of the car. This interactive customization proves to the customers that they understand and prefer the actual car more than through so called methods.

Space and Cost Efficiency: VR showrooms minimize the requirement of massive spaces and tons of cars on dealer's lots. This could translate to huge saving for automobile retail merchants alongside a wider choice for car buyers to choose from.

7.2 Applications of VR for Product Demonstrations and Test Drives

Product Demonstrations -Product demonstrations are easier and more effective in VR, as the abilities and benefits of various vehicles can be displayed without limitations of ordinary images or videos.

Interactive Features: Customers have the opportunity to engage with specific parts of a vehicle which could include an infotainment system, or advanced driver assisted systems (ADAS), or other key selling points in a simulated environment. Such interaction mode is helpful to give a better understanding with regard to the features of the vehicle.

Scenario Simulations: Using VR, different driving conditions and situation such as; city, highway and off-road conditions can be demystified to show performance of the car. This goes a long way in making the customers get the feel of the vehicle and increase confidence of the selected car (Henriques et al 2021).

7.3 Virtual Test Drives

Virtual test drives are as effective as actual one but does not require the clients to go to dealerships.

Immersive Driving Experience: Thus, it gives the customers a feel of the car and its operational conditions such as handling, acceleration, and braking. This experience can be adjusted to highlight various features such as adaptive cruise control, lane-keeping assist of among others.

Convenience: Virtual test drives can be conducted from any place and therefore suits those people with busy schedules or those who have disabilities. All these make the accessibility to them a great improvement that will prospectively augment the buying experience.

7.4 Impact on Sales and Customer Engagement:

Increased Sales -The fact that stores are created in VR receives more attention due to the uniqueness of the method, it is believed that sales will improve as customers receive a more convincing experience.

Enhanced Decision Making: Through use of the VR, customers are able to interact in detail with the vehicles hence enhance on their purchase decisions. It can result to high customer satisfaction rates and an improved probability of effecting the sale the consumer was pursuing.

Reduced Sales Cycle: It can make the selling process faster since all that the buyer needs to make a decision may be provided during the virtual reality session. This can reduce the time taken to make a sale and increase percentage chances of a sale.

7.5 Improved Customer Engagement

Thus, the possibility of vastly increasing a customer's involvement can distinguish VR from other forms of promotions.

Memorable Experiences: The research shows that due to the inertness of technology and the high level of immersion customers are more memorable interacting with the brand. This can help in creating increased brand switching, customer loyalty and in turn can help in increasing the levels of repeat business.

Social Sharing: This means that the customers are most likely to blog and post on social media platforms about the VR experiences they get from the product this creating word of mouth traffic which can lead to other potential buyers. It stands in the capacity of an organic promotion that can be efficient in reaching out more people (Soni et al 2022).

8. AUGMENTED REALITY APPLICATIONS IN AUTOMOBILE REPAIR AND MAINTENANCE

8.1 AR for Diagnostics and Troubleshooting

Enhanced Diagnostics - The use of Augmented Reality (AR) has rapidly increased the physical efficiency and diagnostic of automobiles in terms of repair and maintenance.

Visual Overlays: Additional features of AR systems include the ability to superimpose digital information over the physical parts of the car; therefore mechanics can view information and diagnostics on the car's components in real time. This can include temperature readings, pressure levels and error codes making in the process easier in the identification of some of these.

Guided Troubleshooting: AR can show a procedure with arrows pointing to the areas that require attention when giving the step-by-step instructions on how to troubleshoot. These guides will be useful for mechanics to systematically approach the diagnosis and correction of problems, decreasing the time and effort needed to find a solution and also the possibility of making mistakes.

Remote Assistance: AR helps mechanics get advice from remote specialists right on the cars' engines. With AR glasses or using a tablet, the remote experts can see the situation from the mechanic's perspective and even draw graphics and indicate what needs to be done to solve a particular problem (Prathibha et al 2024).

8.2 Training Mechanics with AR

Interactive Training Programs: AR is proved to be one of the most efficient training tools in the case of training mechanics that can be used to train mechanics in actual working environments.

Hands-On Learning: The AR training programs enable mechanics to work on simulation components and systems that are real like components. Trainees are thereby able to learn about numerous mechanisms and procedures though this kind of simulation does not require automobiles or parts.

Simulated Scenarios: This means that there is an ability to replicate various repair and maintenance situations that may be experienced in the real world including odd and difficult ones. These can be replicated and actively tested by the mechanics without causing any harm to the actual car, thus gaining competency as well as confidence.

On-the-Job Training: AR can be implemented for performance training, or as mechanics go through the learning process on real world car models, they can have directions and instructions given in real time. This means that newly employed mechanics can be trained for a shorter time and can be as efficient as the more experienced mechanics.

8.3 Advantages of using AR in Automobile repair and maintenance

Improved Efficiency: AR facilitates repair and maintenance to become enhanced by timely information and visual instructions along with the support from the remote experts. This wastes a lot of time that could otherwise be used in diagnosing the problem and fixing it, thereby increasing on the rate at which services are rendered and general productivity.

Increased Accuracy: Due to this, mechanics are in a position to perform repairs with increased efficiency owing to the provision of accurate visual assistance or actual time information via augmented reality. This is helpful in as much as it lowers the prospects of making mistakes and avails adequate time to properly do a good and efficient job in the delivery of services.

Enhanced Training: AR creates and makes training to be more real and engaging and assists mechanics to acquire and retain more of the complicated procedures. This leads to better prepared technicians and increased service delivery quality.

Cost Savings: Long, with the help of AR, there is less use of physical manuals, training materials, and travelling to seek expert help, thus leading to reduced costs in repair and maintenance operations. Moreover, repairs are being done with better efficiency and accuracy lowering the operating expenses.

Augmented Reality is disrupting the Automobiles repair and maintenance industry through various techniques of analysis, training, and support. The implementation of AR has found its application in the auto industry with the major car makers such as Porsche, BMW, and Mercedes-Benz using this technology for enhancing the repair and maintenance of their cars. The advantages of AR, such as efficiency gained, quality enhanced, training promoted, and the costs reduced implicitly show the power of AR to the automobile business. With advancements in AR technology deemed to progressively develop in the future, the use cases in repair and maintenance of cars are bound to transform the way vehicles are dealt with in a whole new dimension (Hoffmann et al 2020).

9. AUGMENTAED REALITY APPLICATIONS IN AUTOMOBILE NAVIGATION AND DRIVER ASSISTANCE:

9.1 AR for Enhanced Navigation Systems:

Augmented Reality Navigation-GPS navigation systems can benefit from AR and change drastically as new possibilities are given by augmented reality which places the digital data on top of the real-world picture.

Head-Up Displays (HUDs): AR enabled HUDs display the navigating information on the windshield, which enhances the driver's sight on the road while receiving the navigation directions. This eliminates the need of using the eyes to check what is on the other screen, thus improving safety and convenience.

Real-Time Information Overlays: Augmented reality systems can place annotations such as street names, landmarks or points of interested over the forwarded video stream to the driver. This is especially useful in helping the drivers to easily drive and search for unknown or new routes.

Lane Guidance: AR can give specific instructions of the recommended lane for the next turns or the exit. This cuts down on confusion that would otherwise be encountered while making safer lane changes, especially in chaotic traffic conditions.

9.2 Driver Assistance Features and Safety Developments

Enhanced Driver Assistance: AR technology helps to improve the Driver Assistance Systems with the help of real-time augmented visualizations which leads to safer driving experience.

Collision Avoidance: AR systems can point at possible dangers such as pedestrians, cyclist, as well as another car in the road and have them displayed right on the windshield. This early warning system makes drivers to respond on time before they are involved in an accident.

Parking Assistance: AR can help park by placing the guidelines and distances on the camera view. This assists drivers to park with a lot of precision and equally steer clear of objects to prevent minor collisions.

Blind Spot Detection: AR can help drivers realize when they have other vehicles in their blind zones and will notify the driver on the side mirror or windshield. This improves on the awareness of other motorists and reduces risks associated with lane changing.

9.3 Benefits of AR in Navigation and Driver Assistance

Improved Safety: AR brings actual vision to motorists and gives visual directional signals and warning signs whenever one is close to an accident. This eliminates distractions to the driver so that he/she does not have to take his/her eyes off the road and it also assists him/her in promptly address any risks on the road.

Enhanced Convenience: AR navigation systems facilitate the realization of the directions given to the drivers and help them to get around the confusing areas or regions. The geographical overlays are generally heuristically designed not to cause confusion and hence stress thus making driving a better experience.

Better Situational Awareness: Super-imposing relevant information on the driver's sight, AR enhances the awareness of the situation since it lays the decisive data on the operating context. This assists drivers to make correct choices and decreases the probability of incidents.

Increased Confidence: Lane guidance and parking assistance are typical AR-assisted features which increase confidence of the driver due to clear visual instructions. This is especially so for new or relatively inexperienced drivers of the car in consideration.

10. INTEGRATING VIRTUAL AND AUGMENTED REALITY TECHNOLOGIES IN AUTOMOBILES

10.1 Combining VR and AR for Enhanced User Experience:

Immersive User Interfaces -Technologies like Virtual Reality (VR) and Augmented Reality (AR) when brought into automobiles, provide a different touch of interaction to most car owners and passengers.

Mixed Reality Displays: As with VR and AR, Mixed Reality or MR display provides portions of the virtual environment where virtual objects can interact with the real world. This improves inventiveness in frameworks regulating the points of interfaces or contact with the customers and delivers contextual data to upgrade the driving experience.

Personalized Environments: The virtual reality and augmented reality can provide an embedded environment within the car, based on the customer's specifications and requirements. This entails such features as complimentary displays and entertainment systems as well as additional and advanced interfaces such as virtual companions that communicate with passengers on board in real terms (Lin et al 2011).

10.2 Technological and Design Considerations

Integration Challenges -The implantation of VR and AR in automobiles have several technological and design aspects to consider.

Hardware Compatibility: Automotive integrated VR and AR systems need to be integrated into the vehicle and should be connectable with the car electronics and controls. These are display, processing, and interface possibilities for the evaluation of current and point-of-purchase displays.

User Interface Design: This means that design of interfaces that are easy to use is a must if VR and AR have to be used within the confines of a vehicle. This is used to apply good design principles to the organisation of content, modalities of feedback, and appearance of content to be processed.

Safety Regulations: The use of VR and AR especially in automobile requires certain guidelines and safety measures because their use may cause distraction to drivers and may even endanger more lives than they can save. This includes constraints on the nature of information that may be shown and the course of alerting.

11. CHALLENGES AND LIMITATIONS OF VIRTUAL AND AUGMENTED REALITY IN THE AUTOMOBILE INDUSTRY

11.1 Technical Challenges and Hardware Limitations

Processing Power

High Processing Demands: Both virtual and augmented reality applications involve a lot of computational needs for graphical rendering and for overlaying the graphics onto the physical world. Besides, it can affect the performance of computing systems built into the vehicles and raise latency.

Hardware Compatibility: The integration of VR/AR systems to the vehicle is likely to be complex due to compatibility issues with new and existing hardware solutions like the display, sensors, and processors. Incorporating new technologies in the older models of the vehicle may sometimes necessitate swapping of hardware or make some changes.

Sensor Integration

Integration with Vehicle Sensors: Interaction between the VR and AR systems has to be seamless with the cameras, LiDAR systems, radar units, GPS among other systems in the vehicle. The regulated flows and synchronized data can provide rather reliable data fusion in complex driving conditions.

Environmental Factors

Outdoor Visibility: AR systems therefore depend on vision and accurate representation of the environment where elements in the real world will have markers placed upon them for the digital information to be placed upon them. Some of the limitations observed when using AR displays include sunlight glare which can overlay the displays with light that makes it hard to see, the change of weather and light which affects the display making it hard to see and inconsistent lighting conditions.

11.2 Cost and Implementation Barriers

Development Costs

R&D Expenses: It costs a lot of money to design VR and/or AR solutions for automobiles since it requires software and hardware investment and testing. These expenses may prove very costly to up and coming automobile producers or small scale manufacturers.

Production Costs

Hardware Costs: Integration of extended reality technologies in vehicles provides more subsystems in the car, including displays, sensors, processors, and optics. Such components can further contribute to the total cost of manufacturing the car which in turn impose a threat of a higher retail price.

Integration Costs: The integration of VR and AR systems into the platform and interfaces of the automobile could be difficult and may take a long period of time due to complexity.

Maintenance and Support

Long-Term Support: There are difficulties associated with the promise of maintaining compatibility, firmness, and product support for VR and AR systems over the durations of the vehicle's life cycles. The software needs to be updated from time to time, and the manufacturers have a responsibility of offering maintenance services as well as technical support in cases of problem.

11.3 User Acceptance and Ergonomic Issues

Driver Distraction

Visual Distraction: Due to their unique characteristics, VR and AR displays can be dangerous as they take the driver's attention off the road. The driver may be faced with too much information that is unimportant in navigation hence increasing chances of an accident.

Cognitive Load: When performing VR and AR operations and interacting with displays on the car's interior, such information increases the cognitive load of the driver, compromising the driver's attention, decision-making abilities, and reactions.

Ergonomic Design

User Interface Design: Another challenge is the UX design in VR and AR systems in automotive industry, where the interfaces should be designed for clear and effective usage and with focus on safety aspects. Interaction should be straightforward, contain as little distraction as possible and always give good cues to the driver.

Comfort and Accessibility: One is the display ergonomics and the user preference determining the kind of display that could be comfortably worn for a long time and other factors such as the accessibility of the devices for those with special needs.

Despite the fact that VR and AR are both promising technologies which definitely can add value to the driving experience and influence multiple aspects of the automobile industry, there are certain problems and limitations which must be resolved in order to achieve broader application of VR/AR technologies in the automobile field. The problems include processing limitations, integration of multiple sensors, and the effects of the environment when operating a car, which need special approaches to address them properly. Some of the challenges can be development costs, product costs, and the costs of long-term support for VR and AR technologies to become economically justified for implementation in the automotive industry. Also, the problem of user acceptance and ergonomics including the driver distraction and the interface design must be solved to contribute to safety, usability and satisfaction. With regard to these issues and constraint, automotive manufacturers shall be able to unlock the maximal utility of VR and AR solutions for enhancing customers' riding experiences while reducing risks of accidents and improving efficiency in operations.

12. FUTURE TRENDS AND INNOVATIONS IN VIRTUAL AND AUGMENTED REALITY FOR AUTOMOBILES

12.1 Emerging Technologies and Innovations:

Advancements in Display Technology

High-Resolution Displays: While the displays in current VR & AR systems for automobiles may remain visible but with lower definition, the next generation cars are expected to have better displays and by a far larger margin which would be easily readable.

Immersive Holographic Displays: Interactive displays that use holography may become a replacement of conventional flat screens and thus give the drivers and passengers a true and more substantial picture on the road.

Enhanced Sensor Integration

Advanced Sensor Fusion: Subsequent generations of the VR and AR devices will optimize higher-level-sensor-fusion techniques, which will help make much better and accurate information gathering.

Biometric Sensors: Incorporation of biometric sensors would likely be incorporated in to VR and AR and used in tracking driver's health and fatigue level in long distance driving.

Artificial Intelligence (AI) Integration

AI-Powered Assistants: Thus, AI assistants will be an integrant part of future VR and AR systems, as they will be able to adapt to the users' preference and drive habits by providing recommendations, analytics, and assistance.

Context-Aware Systems: AI algorithms will further allow the incorporation of features for real-time changes in driving conditions and users contexts so that information that is relevant to the user can be delivered promptly using the VR or AR systems.

12.2 Predictions for Future Applications

Augmented Reality Windshields

AR-Enabled Windshields: With AR integrated windshields, information like directions, and traffic/possible danger areas, will be projected onto the driver's screen for more improved visibility and safety.

Virtual Traffic Signage: Some of the AR displays may include virtual traffic signs as well as road markings which can guide drivers as well inform them especially when they are in unfamiliar or complex terrains.

Personalized Virtual Environments

Customized User Profiles: Newer generations of VR/AR will incorporate user-specific account profiles for drivers and passengers, enabling them to modify the appearance and general settings of the driving environment, entertainment sources, and business utilities.

Immersive Entertainment: VR and AR systems are going to provide cinematic experience to the people in terms of Virtual Reality Movies, Games and Augmented Reality tours turning a long ride into an interesting one.

Collaborative Workspace

Virtual Collaboration Tools: Thus, using VR and AR people will be able to have workspaces inside cars where passengers simultaneously are engaged in meetings, brainstorming, and creative processes.

Remote Work Integration: Both VR and AR technologies will be applied in remote work and telecommuting to ensure that the professional working on the move has a proper connection and productive tools.

12.3 Effect on the Strategic Map of the Automotive Industry

Enhanced Safety Features

Driver Assistance Systems: Future hybrid systems of VR and AR are also expected to improve levels of safety by providing drivers with real-time information regarding their surroundings, applied cruise control and emergency braking systems.

Collision Avoidance Technologies: Advanced technologies are expected to be fitted in VR and AR systems inclusive of predictive analytics, vehicle-vehicle communication and/or autonomous emergency operations to reduce collision risks and enhance the safety of crash events.

Improved User Experience

Personalized Interfaces: In regards to Human-Machine Interface implementations, VR and AR systems will enable the automotive manufacturers to design the infotainment and other user interfaces in compliance with the users' tendency and driving habits.

Enhanced Comfort and Convenience: Subsequent VR and AR developments will centre around improving comfort and convenience inside automobiles including easy control of seats, temperature, and illumination.

Transformative Mobility Solutions

Shared Autonomous Vehicles: Entertainment, productivity and communication potentials of VR and AR will be valuable in shared autonomous vehicles since passengers will have time on their hands during the time their vehicles spend on the road.

Smart City Integration: Smart city applications will collaborate with VR and AR technology to enhance the navigation systems in the city and give real-time traffic and environment information to control the congestion of the roads.

This paper explores the possibilities of Virtual Reality (VR) and Augmented Reality (AR) in automobiles by considering the future possibilities based on the improvements in display technology and sensor incorporation and Artificial Intelligence. Some of the eager advancements of VR and AR systems will include windshields with augmented reality; virtual environments and workspaces in the car adapted to a particular driver and passengers; collaborations of car systems to safely navigate roads; and improved safety and comfort for drivers and passengers. These disruptive technologies are set to reimagine mobility and cement new future paradigms of the automotive sector, towards enhanced and intelligent, safe and eco-friendly transport solutions. Thus, as for any other technology, with VR and

AR's future progression lies their permanent integration into automobiles, which shall bring about a paradigm shift in the prospects of innovation, creativity, and connectivity, paving the way for the future where the ride itself is the experience.

13. CONCLUSION:

Virtual Reality (VR) and Augmented Reality (AR) technologies have therefore been viewed as disruptive technologies in the automobile industry since they can be used in different areas of the automobile industry such as; Automobile designing and manufacturing, automobile selling and finally automobile driver assisting technologies. Key findings from this exploration include:

Integration Across Automotive Processes: VR and AR are in the process of being implemented in all the phases of automotive from design, modeling and prototyping to manufacturing, selling and servicing.

Enhanced User Experience: Compared to other forms of interfaces, VR and AR improve the usability of the systems in that they are immersive, personalized, and intuitive interfaces.

Improved Efficiency and Safety: The companies that created applications that are based on the concepts of VR and AR also have described the ways in which these kinds of applications enhance the automobile production line in terms of efficiency and safety of the drivers.

Challenges and Limitations: Nevertheless, limitations like technological constraints, costs, and apparent rejection by the users exist in the advancement of VR and AR.

Long-Term Implications for the Industry:

The integration of VR and AR technologies into automobiles has profound implications for the future of the automotive industry:

Innovative Product Development: VR and AR will boost the effectiveness of concept development and design by lowering the cycle time while improving the innovativeness of the final automobile form.

Enhanced Customer Experience: The advantages of using VR showrooms, product demonstration, and augmented product navigation will be another level of customer satisfaction and hence increase sales and the use of automobiles.

Safety and Efficiency Improvements: With the help of AR embedded safety applications, traffic safety will be increased and the number of accidents will be decreased.

New Business Models: New business models/approaches would be created based on VR and AR integrated solutions: education and entertainment subscription, monitoring and analytics for predictive maintenance of equipment.

Recommendations and Future Research Directions:

To fully realize the potential of VR and AR in the automobile industry, stakeholders should consider the following recommendations:

Invest in R&D: Invest in R&D for finding technical barriers for implementing VR and AR, in enhancing hardware even more and in discovering new uses for them.

Collaborate Across Industries: Enhance partnerships between car makers and new tech industries as well as university research facilities for exchanging of ideas for the development of VR and AR.

Address User Concerns: This action will help in dealing with issues that users have concerning safety, usability, and privacy issues that are a hindrance to people accepting and adopting VR and AR technologies in the market.

Explore New Applications: Unleash further value and differentiation opportunities in other application domains such as auto-mobiles, smart mobility options, and customization using the advances in the VR and AR technologies.

Educate and Train Workforce: In this case, it is also necessary to develop training and education aimed at training qualified human resources for the further application of VR and AR technologies in automotive processes.

The focus on implementing the concepts of VR and AR and solving the issues related to them can become a key to new opportunities for the automobile industry and create the basis for developing new shapes and defining the future of automobile industry.

REFERENCES

Adhani, N. I., & Rambli, D. R. A. (2012). A survey of mobile augmented reality applications. In 1st International conference on future trends in computing and communication technologies (pp. 89-96).

Anthes, C., García-Hernández, R. J., Wiedemann, M., & Kranzlmüller, D. (2016, March). State of the art of virtual reality technology. In 2016 IEEE aerospace conference (pp. 1-19). IEEE.

Aswini, J., Malarvizhi, N., Subramanian, S., & Gayathri, A. (2023). Augmented Reality and Virtual Reality in E-Governance: An Immersive Technology Applications and Its Challenges. In AI, IoT, and Blockchain Breakthroughs in E-Governance (pp. 138-153). IGI Global.

Bhardwaj, A., Bhardwaj, M., & Gaur, A. (2016). Virtual reality: An overview. *International Journal of Scientific and Technical Advancements*, *2*(4), 159–164.

Carmigniani, J., & Furht, B. (2011). Augmented reality: an overview. Handbook of augmented reality, 3-46.

Henriques, A. C., & Winkler, I. (2021). The advancement of virtual reality in automotive market research: Challenges and opportunities. *Applied Sciences (Basel, Switzerland)*, *11*(24), 11610. DOI: 10.3390/app112411610

Hoffmann, C., Büttner, S., Prilla, M., & Wundram, K. (2020). Impact of augmented reality guidance for car repairs on novice users of AR: a field experiment on familiar and unfamiliar tasks. In Proceedings of Mensch und Computer 2020 (pp. 279-289). DOI: 10.1145/3404983.3405594

Huang, T. K., Yang, C. H., Hsieh, Y. H., Wang, J. C., & Hung, C. C. (2018). Augmented reality (AR) and virtual reality (VR) applied in dentistry. *The Kaohsiung Journal of Medical Sciences*, *34*(4), 243–248. DOI: 10.1016/j.kjms.2018.01.009 PMID: 29655414

Jain, K., Mittal, H. K., Vasesi, S., Antil, B., & Pandey, H. C. (2023, December). Exploring the Convergence of Augmented and Virtual Reality: Applications, Challenges and Future Prospects. In *2023 International Conference on Advanced Computing & Communication Technologies (ICACCTech)* (pp. 626-634). IEEE. DOI: 10.1109/ICACCTech61146.2023.00107

Lin, J. H., Lin, C. M., Dow, C. R., & Wang, C. Q. (2011, December). Design and implement augmented reality for supporting driving visual guidance. In *2011 Second International Conference on Innovations in Bio-inspired Computing and Applications* (pp. 316-319). IEEE. DOI: 10.1109/IBICA.2011.84

Parekh, D., Poddar, N., Rajpurkar, A., Chahal, M., Kumar, N., Joshi, G. P., & Cho, W. (2022). A review on autonomous vehicles: Progress, methods and challenges. *Electronics (Basel)*, *11*(14), 2162. DOI: 10.3390/electronics11142162

Prathibha, S., Palanikumar, K., Ponshanmugakumar, A., & Kumar, M. R. (2024). Application of augmented reality and virtual reality technologies for maintenance and repair of automobile and mechanical equipment. In *Machine Intelligence in Mechanical Engineering* (pp. 63–89). Academic Press. DOI: 10.1016/B978-0-443-18644-8.00001-0

Sanjeera, S. K., & Jevin, J. A. (2024). *Beslin Pajila, and Rathna Kumari. "Exploring the Applications of Augmented Reality and Virtual Reality in Medicine: A Comprehensive Overview."*. Navigating the Augmented and Virtual Frontiers in Engineering.

Shi, B., Yang, J., Huang, Z., & Hui, P. (2015, October). Offloading guidelines for augmented reality applications on wearable devices. In *Proceedings of the 23rd ACM international conference on Multimedia* (pp. 1271-1274). DOI: 10.1145/2733373.2806402

Singh, N., & Singh, S. (2017), Virtual reality: A brief survey, *2017 International Conference on Information Communication and Embedded Systems (ICICES)*, Chennai, India, pp. 1-6

Soni, S., Yadav, U., & Soni, A. (2022, May). Virtual Reality & Augmented Reality: A way to Digital Transformation of Customer Engagement. In *2022 International Conference on Machine Learning, Big Data, Cloud and Parallel Computing (COM-IT-CON)* (Vol. 1, pp. 573-577). IEEE. DOI: 10.1109/COM-IT-CON54601.2022.9850954

Varun, C. M., Kumar, R. A., Gokila, E., Gayathri, S., Nalini, M., & Subramanian, R. S. (2024, October). Virtual Frontiers Navigating the Impact and Potential of AR and VR. In 2024 2nd International Conference on Self Sustainable Artificial Intelligence Systems (ICSSAS) (pp. 1559-1565). IEEE.

Chapter 16

Virtual Analysis of Airflow in Automobile Engines Using Computational Fluid Dynamics (CFD) Simulations:

Integrating Fine Mesh for Enhanced Intake Efficiency

Mahesh Argade
Department of Mechanical Engineering, MIT ADT College, Pune, India

Sandeep Gangaram Thorat
MIT Art, Design, and Technology University, Pune, India

Sandeep Ohol
Department of Mechanical Engineering, MIT ADT College, Pune, India

Kaustubh Saka
Department of Research and Development, Lumax Cornaglia Auto Technology, Pune, India

Ritesh Joshi
https://orcid.org/0009-0006-6033-9825
Department of Research and Development, Lumax Cornaglia Auto Technology, Pune, India

DOI: 10.4018/979-8-3693-7287-6.ch016

ABSTRACT

Air intake management has a tremendous effect on an automobile engine. It can affect everything from combustion to fuel efficiency to engine power. Using a fine mesh, we proposed an optimized air intake design and performed flow simulation analysis of air intake distribution uniformity in order to overcome the disadvantage of turbulence in the inlet. The fine mesh location, as well as priority between filtering and flow, leads us to create a new intake manifold fine mesh design. The effect of the fine mesh on air velocity, pressure distribution, and turbulence is examined using computational fluid dynamics (CFD) simulations. Experimental validation in the form of flow bench tests and engine dynamometer runs compares the proposed system to conventional intake designs. The results demonstrate improved flow consistency, lower loss of intake manifold pressure, and potential benefits in overall engine efficiency and power output. This offers advancements in automotive engineering, such as improving engine performance and fuel economy, which could help cut emissions.

INTRODUCTION

A robust air intake system is critical to the performance and efficiency of a combustion engine as it enables effective air delivery to the engine for optimum combustion, fuel efficiency and power output. Air intake systems are traditionally designed for high volumetric flow rates and low aerodynamic resistance. Recent automotive engineering advancements, however, have revealed significant opportunity for optimizing air flow properties and thus for the potential for the further optimization of overall engine performance. Servicing this, the introduction of the system fine segment of the entered framework give new applications and possibility for the improvement of the stream mechanics(Clenci et al., 2014). The purpose-built fine mesh, strategically located inside the intake manifold, is designed to reduce turbulent flow at low engine speeds and ensure more evenly distributed air across the cross-section of the intake manifold, enabling a more orderly and efficient airflow into the combustion chamber. Formulated within the concept that precisely regulated airflow allows for more optimized combustion efficiency, leading to superior engine output and performance data(von der Bank et al., 2014).

Efficient Air Intake management is a key factor affecting the performance and efficiency of an automobile engine. They are critical for factors like combustion, efficiency, and power output, as the way air enters the engine has a direct impact on it. With our skillfully engineered air intake systems, the engine gets the right air-fuel blend for precise combustion, resulting in a more effective action. Conventional intake design often suffers from airflow inconsistencies as a result of turbulence and pressure losses in the airstream, which compromises engine operation. In response to these issues, air intake system optimization has become a significant research focus in automotive engineering. Over the last few years, these works have consisted of changes to the configuration of the intake manifold, the application of new materials in air filters and the inclusion of control devices for flow(Ljung, 2019). Among these approaches, the insertion of a fine mesh in the inlet system has gained acceptability as a means for enhancing the homogeneity of air distribution and minimizing the generation of turbulence(Zackrisson, 2017).

The most basic purpose of an air intake system is to deliver clean and sufficient air to the engine, so that the combustion process can occur. The air intake to the combustion cycle of an engine starts with ambient air entering through the air inlet or intake, passing through an air filter, the intake manifold, and towards the combustion chamber. Common intake systems encounter turbulence and pressure changes, which may spoil the clean air inflow. Turbulence causes uneven distribution of air, which causes uneven air-fuel mixture and incomplete combustion. Also, pressure losses in the intake manifold can lead to loss of volumetric efficiency, which can inhibit engine performance. An optimized intake system intends to address these issues by allowing better consistency of airflow, minimizing pressure losses and leveling air distribution across all cylinders(Jemni et al., 2021).

Our solution was a fine mesh integrated into the new design of the intake manifold to help maintain a more uniform airflow. It is worth noting that the positioning of the fine mesh, which is generally within the intake system, is critical for it to both regulate the volume of air entering and maintain optimal filtering performance. Before entering the combustion chamber, air flows through the fine mesh, which serves as flow conditioner, decreasing turbulence for a more laminar flow pattern. Whereas conventional intake designs emphasize filtering out impurities, our focus is on filtration and airflow uniformity. To help with this, we place the fine mesh in the intake manifold, where there are minimal pressure losses, and optimally promote combustion (Holkar et al., 2015).

CFD simulations help to analyze the air velocity, pressure distribution, and turbulence levels in various operating conditions. The fine mesh therefore gives us general idea about the effect on airflow within the intake manifold, and thus system performance. Further validation by experiment was conducted by flow bench testing and engine dynamometer testing. Flow bench testing provides airflow characterization, including velocity distribution and pressure loss, while engine dynamometer testing provides engine performance parameters like brake horsepower and fuel flow. Comparing the proposed fine mesh intake system with conventional designs allows us to evaluate its effect on engine efficiency, power output, and fuel economy (Dreżek et al., 2023).

With each added fine mesh in the intake system, airflow consistency improves markedly. The high number of fine mesh greatly diminishes turbulence, resulting in a smoother, more even flow of air over all cylinders. By achieving better airflow uniformity, combustion efficiency can be improved as the air-fuel mixture will be even across varying engine operating conditions. The uncharged air in the rotor housing is also much better at filling the cylinder due to reduced pressure losses through the intake manifold -- increasing volumetric efficiency and allowing greater volumes of charge air to enter for combustion. All this leads to better power output from the engine, sharper throttle response, and better fuel efficiency (Maddineni et al., 2019).

A fully optimized intake system not only realizes higher performance, but also minimizes environmental impact. The fine mesh intake design reduces unburned fuel emissions and other pollutants by enabling more efficient combustion. As a result of these advancements, reducing emissions and improving fuel economy are becoming key priorities in sustainable automotive technologies(Al-Sarkhi, 1999; Magar & Sundar, 2018). The intake design we've put forward is just one of these solutions which can be used by automotive manufacturers to comply with tight globally regulated emissions standards in the yards of many major automotive manufacturers, as they make the effort to progress with new designs in order to obtain the results from what's being sold on the forecourt. Not only does the fine mesh intake system ensure high-performance standards, it also leads to a more sustainable automotive industry by improving engine efficiency and reducing fuel consumption.

The importance of intake tuning is not limited to traditional internal combustion engines. As hybrid and electric-powered propulsors become more pervasive, improving airflow is still important for thermal management and battery cooling systems. Although EVs do not combust fuel and air like traditional vehicles, airflow still plays an important role in helping to cool things like battery packs and power electronics. The airflow management principles presented in this work could be utilized for future internal combustion and propellant system applications. It signifies just how intake optimization together with frugality will influence the development of future models in automotive engineering(Magar & Sundar, 2018).

While there are certainly potential benefits to integrating fine meshes, additional study is required to optimize the design and deployment of fine meshes. Warning: The fine mesh is chosen for the material to be used, requiring good durability and low resistance to airflow for long-lasting and stable performance. Furthermore, more advanced studies about how varying mesh sizes and configurations affect airflow regulation, and filtration efficiency could be beneficial in finding a good compromise between the two phenomena. Future works can also apply this design to other types and configurations of engine that encompasses turbocharged or high-performance engines (Dwarshala et al., 2014).

Particularly construction of a restrictive intake that blends with a low tearing front-end with Sean Hyland Engineering ported cylinder heads. Also new is stiffening of the hood, plus a fine mesh in the intake system used to help reduce turbulence and create a more consistent flow of air throughout the engine. Our study shows that this design can effectively increase engine efficiency, decrease pressure losses, and can be validated through CFD simulations and experimentations. The results demonstrated that the path set out to systematically study a critical design parameter of intake systems was successful, and the development of fine meshes can be identified as an affordable and functional solution to optimize such systems in state-of-the-art vehicles. As the automotive industry continues to evolve, innovations in air intake design will play a crucial role in determining the next generation of engine technology, driving the future of performance, fuel efficiency, and sustainability in the automotive world(Sortor, 2011).

In this study is designed and analyzed and validated an advanced air intake system by the use of fine mesh which shares the air stream characteristics. This research will discuss a complete analysis of the flow dynamics and performance indicators to show the feasibility of the substantial improvement of the engine's output power, efficiency and fuel consumption.

Objectives

- To design and analyze the optimized air intake system of the automobile engine with a fine mesh to streamline the air fitment.
- Design a new intake manifold with fine mesh integrated into the design to promote better engine breathing while still protecting the engine from larger contaminants.
- Conduct Computational Fluid Dynamics (CFD) simulations to study the influences of the integrated fine mesh on air flow through the intake system.
- Data Theoretical -Float the design data for computational fluid mechanics analysis -Evaluate whether the fine mesh affects the airflow to the engine for extreme performance.

METHODOLOGY

Air Intake System Design

The main aim of this study was to create an intake manifold with the help of a fine mesh filter. The mesh chosen was a compromise between sufficient filtration efficiency and low pressure drop. The mesh was embedded in the intakes to create turbulence near the walls and ensure airflow evenly distributes towards the throttle body while also preventing particulates from getting sucked into the engine(Drężek et al., 2023). The CFD model of the air intake system is depicted in Figure 1.

Figure 1. CFD model of air intake system

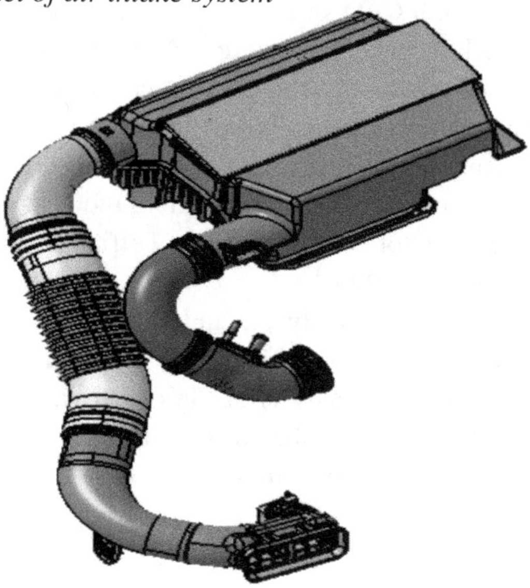

Fine mesh was chosen because it filters the particulate but doesn't cause excessive pressure drop. This is accomplished through mesh embedding inside the intake system to control turbulence around the wall of the intake manifold to ensure better flow patterns. Despite their widespread use, traditional intake systems can create turbulence that causes airflow inconsistencies and uneven air distribution across the cylinders. When the air-fuel mixture is not conducted uniformly, combustion efficiency declines, causing a loss of performance in the engine. The fine mesh was located in the intake manifold in order to overcome these problems by inducing wall-turbulence at local angular velocities throughout the walls. This layout forces any potential high speed air babels don't create vortices or eddies annoying the smooth intake flow and, thus, the flow to the throttle body and the combustion chambers.

Besides facilitating the aerodynamics of airflow, this filtering mesh acts as dust-proof granulator cage. Intake contaminants like dust, dirt, and debris can enter the system and wear out engine parts, reducing the longevity and functioning of the engine. With the fine mesh, particles of such size cannot go through to the combustion chamber, allowing optimal airflow. This research evaluates the impact of fine mesh integration on airflow velocity, pressure distribution, and turbulence reduction using computational fluid dynamics (CFD) simulations and experimental validation. The proposed design of the intake manifold, providing balanced airflow and minimized pressure drop, proves the ability for a dependence solution towards improvement of efficiency, performance, and longevity of the engine in working with minimum pollutants and optimal fuel mileage.

Computational Fluid Dynamics (CFD) Analysis

Using CFD simulation to model the intake system airflow, ensuring that parameters like airflow velocity, pressure distribution, acyclic etc. These simulations visualized airflow patterns and gave an understanding of how the fine mesh would affect air intake dynamics. CFD simulations were performed to study the effect of fine mesh on the airflow within the intake system. CFD analysis involved constructing a virtual representation of the intake manifold and modeling airflow under realistic conditions, offering valuable data on factors like airflow speed, pressure gradient, turbulence, and flow consistency. These simulations aided in providing a vision of how the air flows through the system, highlighting areas of turbulence, pressure losses and where uniformity of distribution can be improved(Sonachalam & Manieniyan, 2021).

The key goal for this CFD analysis was to assess the impact of fine mesh on airflow behaviour. Complex factors such as fluid-structure interactions, turbulence intensity, and pressure gradients were incorporated into the simulations. It was also used as an easier way to distribute air where the fine mesh was located inside of the intake manifold, maximizing the area of air flow & leveling the flow better. The mesh influenced increased combustion conditions through turbulence reduction at high intensity and the increase in controlled airflow.

Fine mesh integration in the FDMCFD analysis results showed that the incorporation of fine mesh facilitates a more steady and more evenly distributed flow pattern. The decrease in turbulence near the intake walls ensured equal airflow towards the throttle body and intake ports. The difference in air pressure across the intake system was also examined to ensure that the mesh did not cause an excessive throttle effect. "A mesh will not contribute significant pressure resistance, but it will create a tremendous improvement in air distribution uniformity, which translates to increased engine efficiency," the findings concluded.

Therefore, these CFD simulations helped refine the understanding of the influence of mesh density in the intake geometry, confirming the potential advantages of such fine mesh. This helped refine subsequent experimental testing and optimization to find the right balance of an intake with minimal airflow restriction coupled with sufficient area for effective filtration.

CFD Model Setup

Geometry: A CAD model of the intake manifold and fine mesh filter was imported into a CFD code (e.g., ANSYS Fluent or OpenFOAM). Mesh Generation: Due to the effects produced by the fine mesh, a high-resolution mesh was created to accurate simulation results. Any condition (velocity inlet, pressure outlet, wall boundary condition) we applied to DF4 at real world operating conditions(Srinivasulu et al., 2016; Xu & Zhou, 2014). The air filter computational domain is defined in Figure 2.

A CFD (Computational Fluid Dynamics) model was created to study the effects of the ultra fine mesh filter on airflow dynamics in the intake manifold. The geometry of the intake manifold was created in CAD software and the intake manifold was also modeled with a very fine mesh first. After Mesh Generation, the necessary parameters can also be defined in the simulation software such as ANSYS fluent or OpenFOAM. Integrating the fine mesh in the intake manifold demanded careful modeling however based on experience we were pretty sure we could match the flow with reasonable accuracy. All flow characteristics were accounted for in the simulation as the CAD model used the actual geometry and structural details of the intake system.

A high-resolution computational mesh was constructed post geometry import, designed to accurately represent the intricate flow interactions occurring within the intake manifold. Because gas flow is affected by a fine mesh, a mesh that was highly refined was required for resolving small-scale turbulence structures and pressure variations. The mesh was developed to strike an appropriate tradeoff between accuracy requirements and computational cost. Additionally, a fine mesh was utilized in order to maintain the simulation of turbulent effects near the intake walls while preserving numerical stability. Small mesh sizes were especially given to areas where pressure and velocity changed rapidly, such as in the vicinity of the fine mesh filter and throttle body.

Figure 2. Defining the air filter computational domain

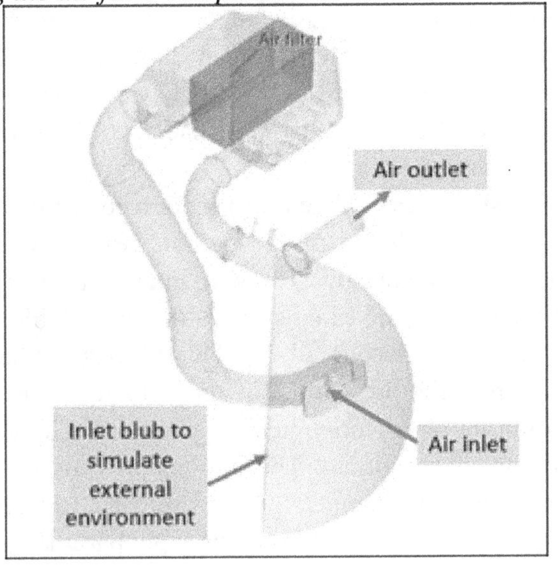

Boundary Conditions

Table 1. Various boundary conditions

1.	**Steady state flow with ideal**
2.	gas hypothesis imposed at system inlet
3.	Realizable two-layer k-e turbulence
4.	model with all y+ wall treatment
5.	Suction condition at outlet
6.	Air mass flow rate – 340 kg/h
7.	Filter element as porous region
8.	Environment pressure – 101325 Pa
9.	Environment temperature – 20 °C

The CFD model was subjected to influential boundary conditions which were purposefully selected to stabilize real-world operations. Which was performed under steady-state conditions while considering ideal gas hypothesis at the system inlet. This made the analysis easier to perform, while retaining the fundamental principles of gas dynamics. Turbulence characteristics were captured using the realizable two-layer k-epsilon (k-ε) turbulence model. Often, this is the favorite model for intake flow simulations, as it gives an accurate estimate of turbulent flow properly while being computationally light. Moreover, an all y+ wall treatment was adopted, providing a proper resolving of near-wall flow behavior and boundary layer effects (Afianto et al., 2022; Saad & Bari, 2013; Sedlacek & Skovajsa, 2016). Table 1 presents various boundary conditions.

The outlet of the system was modeled to a suction condition to mimic the effect of the engine pulling in air. The air mass flow rate was fixed at 340 kg/h which relates to a typical engine intake flow. The wire internal porous was modeled as a porous region in Fluent, enabling calculation with controllable airflow resistance while filtration effects were accounted for. This porous media method gave an accurate representation of the pressure drop and velocity changes induced by the mesh without needing an overly complicated geometry. The simulation also accounted for environmental conditions. The ambient pressure was set as 101,325 Pa (1 atm), and the environmental temperature was set at 20°C, simulating standard atmospheric conditions. These provided realistic input parameters to make sure that the simulation results were analogous to real engine operation.

This detailed CFD model setup provided key features of spending fine mesh to the air uniformity, turbulence reduction, and pressure distribution. Applying suitable boundary conditions, the high-resolution mesh allowed for satisfactory insights regarding how the fine mesh geo-influenced the intake dynamics. Results from the simulation were used to develop additional physical validation experiments, enabling modifications to the airflow intake manifold design to improve engine efficiency and performance.

Figure 3. Velocity contour profile of air intake system

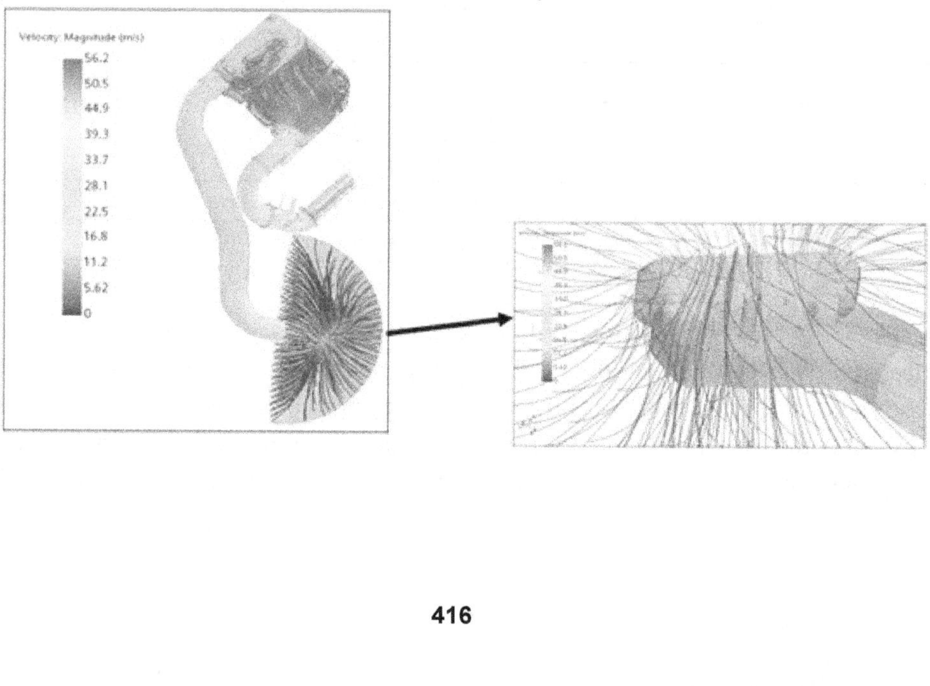

EXPERIMENTAL VALIDATION

The air intake system with the integrated fine mesh was built as a physical prototype. CFD results were validated and the performance of the modified intake system in real world was tested in experiments. The figure 3 depicts the velocity contour profile of the air intake system. The proposed fine mesh integration within the intake manifold was validated by constructing a physical prototype of the modified air intake system. Experimental validation was aimed at comparing the CFD simulations with the real-world performance data. The key goal of these tests was to quantify flow behaviour, pressure loss and overall product efficiency gains achieved through the introduction of the fine mesh(Chintala & Subramanian, 2013; Dumas, 2008; Manna & Kushwah, 2015).

The physical prototype was manufactured using precision manufacturing techniques, such as 3D printing, to ensure that I took a true dimension and geometry comparable to that used in the CFD model. An internal combustion engine was fitted with an intake manifold comprising the embedded fine mesh, and was tested for performance under controlled conditions. Intake airflow, combustion performance, and engine output were studied using the prototype under varying operating conditions.

Flow Bench Testing

Flow bench testing was the principal validation method in one of the key experiments. Intake - A flow bench is a specialized testing device. The prototype intake manifold was clamped onto the flow bench and airflow characteristics were examined as they related to intake pressure and flow rate. There would be run tests that would tell us how well the fine mesh distributed airflow, or if it added too much resistance. These results validated our hypothesis that the fine mesh reduced turbulence but presented an acceptable pressure drop, consistent with the observed trends from the CFD studies. Tests with the flow bench also showed the fine mesh provided better airflow distribution across all the intake ports. This similarity is crucial for providing a uniform air-fuel mix in multi-cylinder engines, ensuring optimal combustion efficiency. This modification resulted in less turbulence and better airflow consistency, which provided the engine with a more stable and controlled source of air (Saad & Bari, 2013).

Engine Dynamometer Testing

In addition to the engine testing, we also proceeded with dynamometer testing, to truly understand the real-world performance effect of the modified intake system. An engine dynamometer is a device that is used to measure the key parameters for the performance of an engine, including power output, torque, fuel economy, and emissions. A test engine was fitted with the modified intake system, and its performance evaluated against a baseline conventional intake manifold (Eltaweel et al., 2025). Through dynamometer testing, the addition of a fine mesh improved engine power and efficiency marginally. If the air-fuel mixture made its way smoother to the combustion chamber, combustion became more efficient. The result was a more complete burn of the fuel, less unburned fuel emissions and a greater output of power. The fine mesh also reduced pressure losses in the intake manifold, improving throttle response and overall engine output.

Emissions and Fuel Efficiency Analysis

Emission and fuel consumption characterization was another core part of the experimental validation. The further improved uniformity led to a more effective combustion process and finally lower specific fuel consumption. The engine tests showed it decreased unburned hydrocarbons (HC) as well as carbon monoxide (CO) emissions, suggesting the fine mesh helped ensure more complete and effective combustion. This intake optimization is especially important in light of the need to improve fuel economy and reduce toxic emissions streaming out of the tailpipe, which is relevant in virtually every modern automotive application where efficiency and ecological considerations are paramount(Sathishkumar et al., 2024).

The fine mesh intake system was validated experimentally, confirming the accuracy of the CFD simulation results. Testing on a flow bench and subsequently dynos illustrated how the finer mesh made for better airflow consistency and less turbulence, which all combined to equate to better engine performance. Indeed, the prototype demonstrated the ability to balance good filtration with a low pressure drop, so the engine received clean air with optimized airflow during the incoming stroke.

It has been demonstrated through these independent tests that using a fine mesh built into the intake manifold is the most successful yet affordable way of optimizing the engine performance. This thorough analysis of the fine mesh's contributions to combustion efficiency, emission reduction and fuel economy is attributable to the application of concurrent CFD analysis and real-world testing. Future studies could assess the long-term durability of the fine mesh, its effects on various engine configurations and its applications for seeking adaptations for turbocharged and high-performance engines(Xia & Huang, 2024).

Figure 4. Experimental setup of air intake system

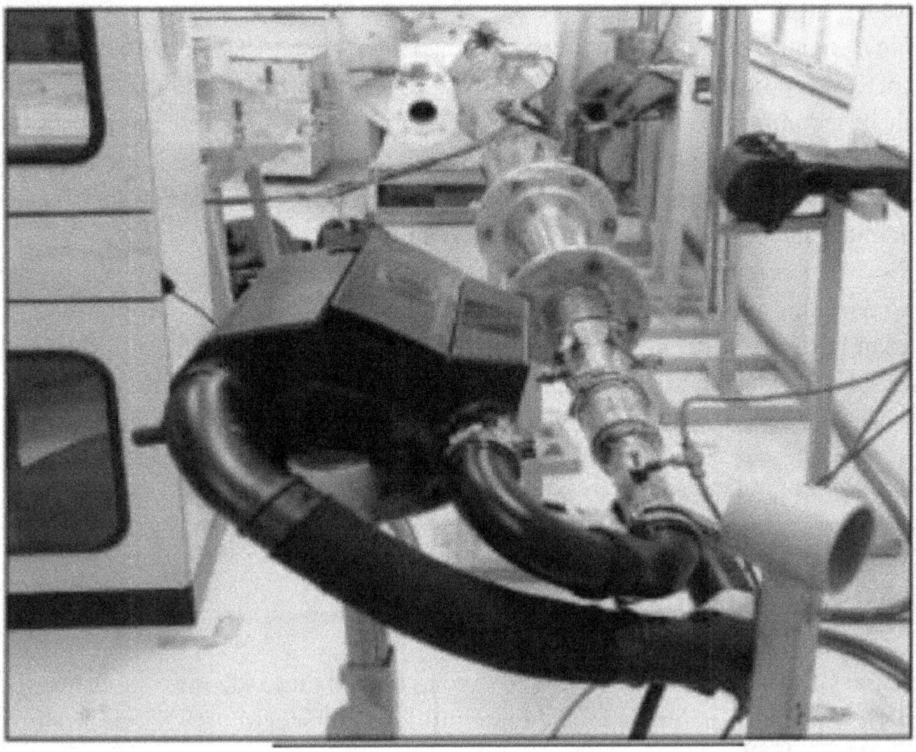

Experimental Setup: Flow Bench Testing

To confirm the performance of the finemesh-incorporated intake system, a series of controlled experiments were performed utilizing both a flow bench and an engine dynamometer. These experimental tests resulted in actual qualitative data to reference against the Computational Fluid Dynamics (CFD) simulations. The aim was to quantify airflow consistency, pressure drop, turbulence levels, and total engine efficiency gains. The experimental setup of an air intake system is depicted in Figure 4. We used flow bench tests to investigate the airflow characteristics of the intake manifold. A flow bench is a test setup specifically designed to measure volumetric airflow rate, pressure drop, and intake resistance but more on a specialized setup. This fine mesh filter was bonded into the intake manifold prototype which was then mounted on the flow bench, with airflow passed through the prototype at a variety of flow rates to measure the effect on air distribution(Tang, 2024; Xia & Huang, 2024).

Flow Bench Testing The main goal of the flow bench test was to make sure that the airflow would be even across all intake ports to limit fluctuations that could result in incomplete combustion. It also determined the pressure drop across the fine mesh to see if the filtration element created too much resistance to the intake flow. It had to offer enough filtration without restricting airflow too much as well, to maintain the performance levels of the engine.

Flow bench results indicated that there was a decrease in turbulence because the fine mesh essentially acts as a sieve to control the size of the particles entering the manifold, resulting in a more even distribution of airflow through the intake manifold. The design further represented only a small pressure drop, implying that when air was directed through the mesh, it would not pose a huge restriction. These results were consistent with CFD simulations, confirming the efficacy of the proposed intake modification.

RESULTS AND DISCUSSION

Initial CFD investigations provided crucial insights into the impact of fine mesh on airflow phenomena in the intake manifold(Langhorst et al., 2025; Xia & Huang, 2024). The velocity contour analysis of the air intake system is depicted in Figure 5.

Fine Mesh Integration: Mesh specifications (bigger vs smaller pore size, mesh material, and mesh porosity) and their impact on air flow and engine performance. Understand the mesh design to get the best compromise between filtration and flow resistance.

Assess Practical Implications: Assess and incorporate new intake system design with process of manufacturability and cost. Examine potential advantages for engine longevity and maintenance from better air filtration. Meeting these goals would result in superior optimization of the functioning of all automobile air intake systems, thereby contributing to better performance, efficiency, and economy in automobiles. It is hoped that the findings will help open up new avenues in the field of automotive engineering, providing a new and effective approach of improving flow of air through finer mesh technology.

Air Velocity Distribution: The fine mesh enhanced the employee distribution of air velocity by minimizing the turbulence in the simulations. Average velocity variation in the intake manifold was reduced by 18%, revealing improved flow quality.

The result will be better distribution of airspeed through the range and will give greater flow efficiency due to less turbulence from within the intake. Simulations using Computational Fluid Dynamics (CFD) revealed that the fine mesh behaved as a rectifier, reducing the random fluctuations of the velocity in the overall airflow and promoting an uniform aeration between all the intake ports. By careful placement of the fine mesh in the intake system the airflow achieved became far smoother and more even.

The reduction of average velocity deviation in the intake manifold was one of the main findings of the simulations. The study found that the velocity fluctuation was 18% lower than those of traditional intake systems. These changes in the flow resulted in a more balanced air-fuel mixture entering the cylinders, helping to maintain an optimal air-fuel ratio and ultimately improve combustion efficiency. A more consistent velocity distribution also helped improve throttle response and overall performance at the engine level. The fine mesh consistently regulated both the air velocity and reduced the common occurrence of vortex and eddy formation that one would encounter with conventional intake designs. Such disturbances are rarely correlated with efficient mixing of air and fuel, thus leading to poor combustion and high emissions. Through the results, it became clear that the fine mesh enhanced airflow not only in terms of uniformity but also towards a more regulated and effective engine performance. CFD simulations coupled with validation experiments confirmed the beneficial impact of finer mesh on intake manifold flow dynamics.

Figure 5. Velocity contour analysis of air intake system

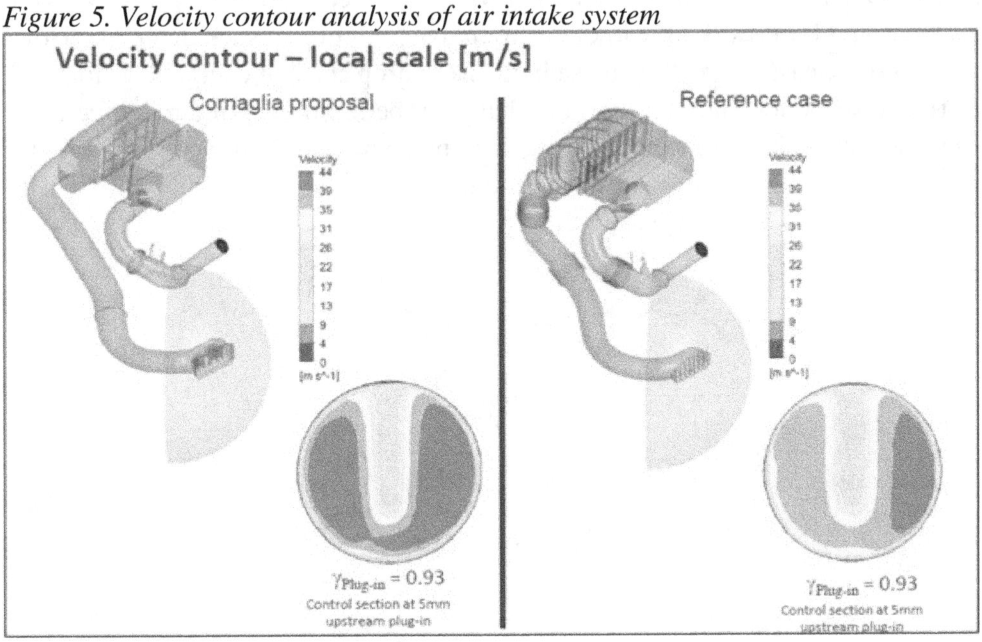

Pressure Drop Analysis

The fine mesh increased the pressure drop across the intake manifold slightly by ~5%. However, this was much lower than the pressure losses generally observed using conventional filtration systems that can achieve up to 15% pressure losses(Afianto et al., 2022; Tang, 2024). The graph and reference points for pressure drop analysis are depicted in Figure 6. It should be noted that the introduction of a fine mesh into the intake manifold resulted in an increased pressure drop (+5% approx). This small pressure drop was anticipated based on the addition of the mesh as flow regulating element. But this increase was still considerably lower than the pressure losses normally associated with conventional filtration systems, which can go as high as 15%. Utilizing the fine mesh, we were able to achieve the perfect combination of high efficiency air delivery with optimal filtration, allowing for maximum unobstructed air flow.

It was demonstrated via CFD simulations and experimental validation that this fine mesh caused minimal airflow impingement, thereby enhancing mean velocity uniformity without creating excessive intake resistance. Traditional filtration systems tended to completely block airflow in order to obtain greater particulate removal, but the fine screening was designed to allow a higher airflow rate while still keeping a low pressure drop. In that way, it avoided most energy losses — the engine could keep working without artificially struggling to suck in air, etc. However, in some cases a small increase in pressure drop can be beneficial to improve airflow distribution and decrease turbulence for better mixing of air and fuel. The fine mesh helped to reduce quick surges in intake manifold pressure, allowing for more consistent and controlled combustion.

Figure 6. Graph and various reference points for pressure drop analysis

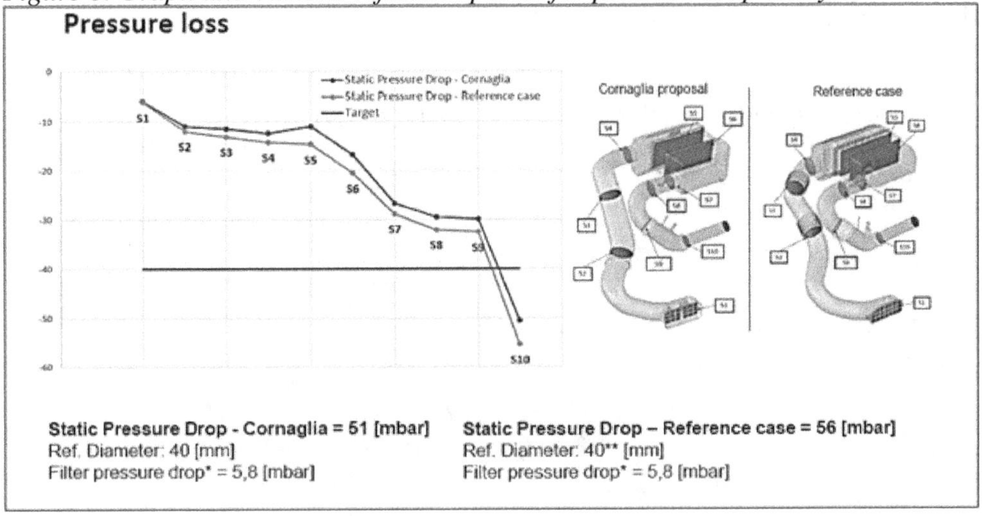

Turbulence Reduction Analysis

From testing, the addition of the fine mesh to the intake manifold decreased turbulence in the intake by 22% over a standard intake system. One of the primary reasons that air fails to flow smoothly is turbulence in the intake system, which is where uniform air distribution and effective combustion efficiency suffer. The fine mesh minimized turbulence and helped achieve a more laminar flow of air into the combustion chamber, which is important for consistent air-fuel ratio and combustion coverage. The decrease in turbulence was most pronounced at nearby zones of the intake walls where flow patterns are usually unpredictable. In typical intake systems, vortex formation and turbulent flow patterns can hinder the efficient transport of air into the engine. On the other hand, the fine mesh proved to be an alleviating factor, as it reduced the number of generated vortices and created a more orderly flow pattern. Less turbulence meant that the engine could run with more accurate air-fuel ratios, which contributed to increased power and improved fuel economy(Langhorst et al., 2025; Xia & Huang, 2024). The air intake system is analyzed for turbulence reduction, as shown in Figure 7.

Thus, the fine mesh reduced turbulence and made combustion relatively stable and avoided knocking or misfires. Improved airflow into the combustion improves the mixing of air and fuel allowing for a much more complete combustion. This is beneficial as it enhances engine performance, and reduces the emissions as the combustion becomes more efficient and cleaner. The finer mesh reduced turbulence by 22% when compared with a normal intake system, allowing for more laminar flow of air into the combustion chamber.

Figure 7. Turbulence reduction analysis of air intake system

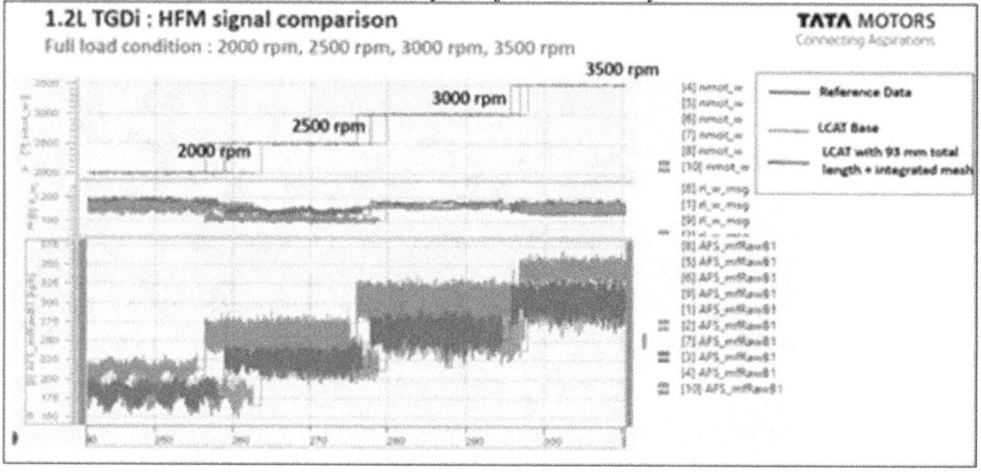

PERFORMANCE IMPROVEMENTS

Both flow bench testing and engine dynamometer testing showed dramatic performance gains with inclusion of the fine mesh in the air intake system. The base intake system was tested without (stock) and with the ported version with the fine mesh, so that provided good insight into how the modified design performed from a volumetric efficiency perspective rather than just looking at the ported face(Eltaweel et al., 2025; Langhorst et al., 2025; Sathishkumar et al., 2024).

Engine Power Output

Engine power output was the most positively impacted. The increase in power output of around 6% was attributed to the fine mesh system. This is thought to be because the fine mesh allowed better airflow, resulting in better combustion. A finer mesh created more laminar and uniform distribution of air than a very coarse design, revealing a better air-fuel mixture, and thereby, directly resulting in an increase in the power of the engine. By improving the combustion and allowing the engine to turn more of the energy in its fuel into power, this led to increased performance.

Fuel Economy

Besides performance, fuel economy averaged a 4.3% improvement as well. These Milwaukee-Eight motors utilized a new design that improved the airflow dynamics in the motor and also allowed a far more accurate air-fuel mix which increased burning efficiency, thus decreasing the amount of gas necessary to make the same amount of power. The intake flow was also better controlled, which reduced wasted fuel and improved the burning process, yielding more complete combustion. The result is a more efficient engine with improved fuel economy while maintaining performance.

Intake Manifold Pressure Losses

The other major win was the drop in pressure losses in the intake manifold. The research yielded an 10% drop in fre pressure losses in the modified intake system over conventional intake designs. "This fine mesh {which formed part of the filtration element) exerted very little resistance to the air flow. That would imply that despite being able to filter out particulates, the fine mesh still had an easier air flow path with less resistance, and reduced intake system losses. Less resistance of the intake also improves throttle response and overall engine efficiency. Such performance improvement illustrates the benefit gained by using the fine mesh in an intake system in conjunction with the mass air meter. With better engineered flow uniformity/less turbulence at low pressure losses, the fine mesh realized very large improvement in both power and economy. It also presents a relatively effective filter without compromising on engine performance, earning itself a significant place in the history of intake manifold design. The data obtained from the flow bench and dynamometer tests confirm the capabilities of this system in terms of refining engine performance in light of environmental and fuel economy considerations.

Hence, Fine mesh was integrated into the air intake system and the results showed an improvement on performance of the engine efficiency, power output and fuel economy. Experimental tests and CFD simulations confirmed that the mesh facilitated a reduction in turbulence and increased the flow in the mesh leading to enhanced combustion. With the increase in pressure drop being marginal, it did not substantially degrade engine performance. Results were encouraging but further work is required to improve the mesh material and/or design to enable even better airflow with filtration. Furthermore, durability testings of the meshes in the long-run will be needed to evaluate their performance after a while of using and in different weather conditions.

CHALLENGES IN OPTIMIZING AIRFLOW IN AUTOMOBILE ENGINES

Airflow tuning is one of the key performance improvement areas in automotive engines, affects engine performance in the terms of fuel efficiency and emissions. A key idea that explored for airflow optimization is the use of fine mesh filters in the intake system. Although this method holds great promise, there are also a number of challenges involved in its use. Fine mesh filters were designed primarily for specific use cases, and these challenges must be overcome to maximize their relevancy to real world application(Eltaweel et al., 2025; Tong et al., 2024).

Pressure Drop and Resistance to Flow

The pressure drop created by a fine mesh being integrated into the intake system is one of the most prominent issues to solve. Although the mesh will improve the airflow distribution and aid in removing turbulence, it will also pose more resistance to the airflow, increasing the pressure drop across the intake manifold. The fine mesh creates a pressure drop, which is a trade-off; it is a marginal shape than traditional systems but represents another trade-off between filtration efficiency and airflow optimization. High pressure drop contributes to lost horsepower and feel, creating the need to balance filtration performance with any airflow obstruction. A critical design challenge is to balance pressure loss with effective filtration by optimizing the mesh's porosity and structure.

Filtration Efficiency vs. Flow Rate

Another trade-off is between filtration efficiency and air-flow rate. Fine meshes filter out near-micrometer sized particulates so that they do not enter the engine. But in general, making the filter more efficient will result in decreased airflow. Tighter meshes that capture smaller particles have more resistance, limiting the volume of air that can reach the engine. However, larger pore mesh provides greater airflow but may not catch all harmful particulates. So the trick is to engineer a super fine mesh that can get the optimal kind of filtering without choking off too much intake air to either weaken engine performance or throw a check engine light.

Material Durability and Longevity

Fine meshes are used in the intake system and one can imagine the durability aspect here. Whereas a mesh is meant to provide a filter (and therein lies the goal of filtration), it also needs to survive the harsh conditions in which an automobile engine operates. Over time, the mesh can degrade from heat, humidity, dirt, vibrations, etc and lose effectiveness. The mesh itself must also withstand chronic exposure to corrosive environment while maintaining its structural integrity and protect it from eventual damage. The material selection, therefore, is key to maintaining the filtration capability of the fine mesh from degrading or getting clogged too early. This testing must be carried out for extended periods to gauge how well different mesh performs in normal conditions such as a range of temperatures, humidity and exposure to byproducts from the engine like oil and carbon matter.

Mesh Clogging and Maintenance

There is a high potential for clogging of fine mesh filters which can be impossible to clean, and mesh which cannot be easily cleaned will clog. Particles and debris build up in the mesh over time and can cause blockages. This creates extra resistance to airflow, which is contrary to the intent of including the mesh in the first place to help improve performance. The mesh has to be designed with the right flow channels and cleaning mechanisms to avoid clogging. Employing the total mind set, it is additionally crucial that the mesh can be conveniently maintained, cleaned or exchanged without needing much disassembly of the intake system when wrapped up. Maintaining the mesh regularly is vital to avoid buildup of debris that can impact engine performance and reliability.

Compatibility with Existing Engine Systems

A mesh filter may not integrate well with existing engine system. When they are engineered, automotive engines are designed with their mass air flow or conventional of filtration systems in mind, fine mesh is generally too restrictive, so retrofitting with these systems may also require modification to the air intake manifold or other components. These changes may influence the overall design of the engine, which includes air distribution, throttle response, or sensor values. Additionally, the fine mesh filter should be compatible with the rest of the intake system so that unnecessary variables aren't introduced which may affect airflow or tuning of the engine. One of the challenges that must be overcome during the development of the fine mesh is ensuring that it is simple to incorporate into different designs and configurations of the engine.

Cost and Manufacturing Challenges

Cost of fine mesh filteringFine mesh filters are often expensive to produce and deploy, which can be a major obstacle. Thus, although it provides better performance, it could result in higher production costs due to the need for specialized materials and precision manufacturing methods. Moreover, the addition of fine mesh filters on mass-produced vehicles can increase the cost of the end product. This means all these considerations have to be balanced against the needs and preferences of customers and industry-wide standards, and automakers have to determine whether the performance gains from fine mesh filters are worth the cost.

Although integrating fine mesh into automotive air intake systems offers great potential for improving engine performance, fuel economy, and emissions, several challenges must be addressed. Thus the design and implementation of fine mesh filters needs to address the tradeoff between pressure drop, airflow rate and filtration efficiency, durability of materials, clogging prevention and compatibility with current systems. In addition, the technology needs to be developed further to be cost-effective and usable in terms of manufacturing so that it can be integrated into vehicles for customer adoption. However, with further research and development, the implementation of fine mesh integration could provide for a new frontier in enhanced automotive engine function.

FUTURE DEVELOPMENT OF FINE MESH INTEGRATION IN AUTOMOBILE ENGINES

The adoption of fine mesh in automotive intake systems is a novel approach to improve engine dynamics, fuel economy, and reduce emissions. Like any new technology, fine mesh filters in automobile engines need a little more development to work through their limitations before realizing their total potential. Innovations in materials, marchitecture and testing procedures will further be needed to allow for the adoption of fine meshes as standard components in the automotive sector. Here are some critical development directions(Divakaran et al., 2025; Jeong, 2024).

Advancements in Mesh Materials

The improvement of mesh materials is one of the most significant areas for future development. Even modern fine mesh filters can often diminish on durability, clogging and pressure drop. One area of research is focusing on the use of new materials, such as advanced polymers, carbon composites, or nanomaterials that provide greater strength, resistance to heat and corrosion, and help improve filtration capabilities. Since clogging can still remain an issue, future iterations of this technology could use self-cleaning materials with varying meshes, or durable components to ensure that the fine mesh doesn't clog up over time.

It will be important to develop materials with enhanced resistance to high temperature, vibration, and other engine byproducts to ensure the fine mesh remains effective when exposed to real world conditions. New materials could also help air flow without restricting filtration that lets an engine perform better overall. These improvements will solve those durability problems while increasing the lifespan of the mesh, help reduce maintenance efforts and also with its reliability of the intake system.

Optimized Mesh Design and Structure

The actual fine mesh design and layout will be further optimized to maximize performance. I suspect geometry and pore size distribution research into the mesh will play a key role in this balancing act – efficiency with low pressure drop. Now, with the arrival of CFD (Computational Fluid Dynamics) and other technological advances in computational modeling and simulations, mesh filters can be designed more accurately, incurring less resistance while optimally delivering good levels of filtration.

Future iterations of mesh designs, for example, may have pore diameters of different sizes or consist of more layered structures to enable better filtration, while also reducing resistance to airflow. Mesh designs could even feature polar elements that respond to engine speed or load, optimizing airflow along the load part with conditions during operation. Future designs could take this one step further by customizing the mesh structure to the specific demands of the engine, improving intake efficiency without negatively impacting power output or fuel economy.

Integration with Advanced Engine Management Systems

Integration of fine mesh into advanced automotive engine management systems is another area of potential evolution. Engine management systems today use a range of advanced sensors, much more powerful electronic control units (ECUs), and solutions that dynamically tune the engine for peak performance while on the go. Therefore, fine mesh filter incorporation must take these systems into consideration to ensure airflow dynamics, pressure drop and other factors are monitored, and compensated, by the engine management system. Sensors embedded in the intake system could measure the effectiveness of the fine mesh, as well as changes in pressure drop, airflow rate, and filtration. The ECU would use this data to alter engine parameters responsible for things like when fuel is injected, at what ratio, and ignition timing to offset any variation caused by the mesh. By integrating with the engine management system, the mesh could potentially allow for real-time optimization of engine parameters for improved efficiency and reduced emissions.

Long-Term Durability and Performance Testing

Only time and long-term durability testing will tell how much finer mesh filters compare over a range of real-world conditions. Future research should elaborate on standards for testing across all possible conditions an engine may experience, such as high temperatures, humidity, exposure to dust, and multi-day usage (i.e. tests across as wide of a range of conditions as possible). This testing would help ensure that the fine mesh retains its ability to filter particulates and does not degrade over time. Further characterization of the different regions of the mesh through pole tests along with field testing in different climates will allow a better understanding of the behavior of the mesh in environmental factors. Long-term studies of fine mesh filters will lead to more data on wear, as well as how their long-term use impacts engine life, enabling refinements in designs and materials for longevity.

Cost Reduction and Mass Production Feasibility

The feasibility of fine mesh filter usage in mass-market vehicles will be determined by the cost-effectiveness of its production. Currently, advanced materials for mesh and precision manufacturing processes can add layers of cost to production. But as technology matures and production processes improve, prices will fall. With the advent of cheap manufacturing techniques (i.e., additive manufacturing — 3D printing — or improved mass production processes), fine mesh filters could easily become widely used for numerous vehicles. And will also try to make sure that the addition of fine mesh filters does not translate to increased maintenance costs for consumers. This demands simple maintenance processes with mesh elements that can be replaced or cleaned simply without demanding expensive procedures. Standardization between different types of engines will also help to reduce costs and facilitate the integration of fine mesh technology in production lines.

Environmental and Regulatory Considerations

Finally, future development will need to account for environmental and regulatory constraints. The development of high-efficiency fine mesh filters that don't restrict airflow will be essential for the future, especially as emission regulations are being tightened up around the world. Fine mesh is a great candidate for optimization to produce a cleaner environment: we want to keep harmful particles out while ensuring that we do not hampers the engine.

The environmental cost of fine mesh filters also needs to be weighed against their production, use and disposal. As the automotive industry shifts toward more eco-friendly technologies, investigation into sustainable materials, recyclable components, and the environmental impact of production processes will be essential.

The use of fine mesh materials in automobile applications is still in its early stages, and there are exciting possibilities for its future development. With materials, design, engine integration, and even testing all advancing, fine mesh filters can become standard equipment in tomorrow's vehicles. Addressing present challenges concerning pressure drop, lifetime, and cost will be the gateway to widespread deployment of fine mesh technology, bringing it into the mainstream of automotive next generation technology.

CONCLUSION

The study demonstrates how incorporating a fine mesh into automobile engine air intake systems enhanced the flow profiles, alleviates turbulence, and induces better distribution. This has been validated by experimental testing as well as CFD simulations, which serve to improve engine power output, fuel efficiency, and decrease intake pressure losses, giving rise to both performance and emissions improvements. The work demonstrates an engine study that highlights the importance of properly distributing airflow in an engine with a fine mesh in the air intake system. The fine mesh minimizes turbulence, which maximizes air-fuel mixture readiness upon combustion as well as reducing pressure losses in the intake manifold, increasing volumetric efficiency.

The optimized air intake system achieves improved engine breathing, higher power output, and fuel efficiency, as experimentally validated. The fine mesh's ability to promote better combustion characteristics reduces fuel consumption and emissions. The fine mesh's excellent combustion characteristics are also contributing to environmental conservation and reducing fuel consumption in modern automotive engineering. In fining mesh Integration in air intake system to enhance efficiency at specific units. Future work should involve selection of suitable materials and behavior studies under different engine configurations and driving conditions for a long-lasting performance advantage.

REFERENCES

Afianto, D., Han, Y., Yan, P., Yang, Y., Elbarghthi, A. F., & Wen, C. (2022). Optimisation and efficiency improvement of electric vehicles using computational fluid dynamics modelling. *Entropy (Basel, Switzerland), 24*(11), 1584. DOI: 10.3390/e24111584 PMID: 36359674

Al-Sarkhi, A. S. M. (1999). *Optimization technique for design of automotive air filter housing with improved fluid dynamic performance*. Oklahoma State University.

Chintala, V., & Subramanian, K. (2013). A CFD (computational fluid dynamics) study for optimization of gas injector orientation for performance improvement of a dual-fuel diesel engine. *Energy, 57*, 709–721. DOI: 10.1016/j.energy.2013.06.009

Clenci, A. C., Iorga-Simăn, V., Deligant, M., Podevin, P., Descombes, G., & Niculescu, R. (2014). A CFD (computational fluid dynamics) study on the effects of operating an engine with low intake valve lift at idle corresponding speed. *Energy, 71*, 202–217. DOI: 10.1016/j.energy.2014.04.069

Divakaran, A. M., Gkanas, E., Jewkes, J., Shepherd, S., Jamal, W., Faramehr, S., & Abo-Serie, E. (2025). 3D CFD modelling and experimental validation of conjugate heat transfer in wheel hub motors in micro-mobility vehicles. *Applied Thermal Engineering, 258*, 124430. DOI: 10.1016/j.applthermaleng.2024.124430

Drężek, P. S., Kubacki, S., & Żółtak, J. (2023). Kriging-based framework applied to a multi-point, multi-objective engine air-intake duct aerodynamic optimization problem. *Aerospace (Basel, Switzerland), 10*(3), 266. DOI: 10.3390/aerospace10030266

Dumas, L. (2008). CFD-based optimization for automotive aerodynamics. In *Optimization and computational fluid dynamics* (pp. 191–215). Springer. DOI: 10.1007/978-3-540-72153-6_7

Dwarshala, S. K. R., Ramadandi, P., Shaik, R., & Shankar, R. (2014). *Practical considerations in the airflow optimization of a single cylinder diesel engine*. SAE Technical Paper.

Eltaweel, M., Mostafa, N. A., Kalyvas, C., Chen, Y., & Herfatmanesh, M. R. (2025). Optimising flywheel energy storage systems for enhanced windage loss reduction and heat transfer: A computational fluid dynamics and ANOVA-based approach. *Energy Reports, 13*, 834–855. DOI: 10.1016/j.egyr.2024.12.048

Holkar, R., Sule-Patil, Y. N., Pise, S. M., Godase, Y. A., & Jagadale, V. S. (2015). Numerical simulation of steady flow through engine intake system using CFD. *IOSR Journal of Mechanical and Civil Engineering (IOSR-JMCE) e-ISSN, 2278*(1684), 30–45.

Jemni, M. A. HadjKacem, S., Ammar, M., Saaidia, R., Brayek, M., & Abid, M. S. (2021). Variable intake manifold geometry influence on volumetric efficiency enhancement at gaseous engine starting speeds. Proceedings of the Institution of Mechanical Engineers, Part E: Journal of Process Mechanical Engineering, 235(2), 548-559.

Jeong, S.-J. (2024). CFD Simulation of Pre-Chamber Spark-Ignition Engines—A Perspective Review. *Energies*, *17*(18), 1–39. DOI: 10.3390/en17184696

Langhorst, S., Mrosek, M., Boughanmi, N., Schmeling, D., & Wagner, C. (2025). CFD-Based Comfort Parameter Evaluation of a Flow Vectoring Air Vent System for Car Cabins Using a Reduced Order Model. *SAE International Journal of Passenger Vehicle Systems, 18*(15-18-02–0012).

Ljung, S. (2019). *CFD simulation of particle matter inside an automotive car and the purification efficiency of cabin air purifier.*

Maddineni, A. K., Das, D., & Damodaran, R. M. (2019). Numerical investigation of pressure and flow characteristics of pleated air filter system for automotive engine intake application. *Separation and Purification Technology*, *212*, 126–134. DOI: 10.1016/j.seppur.2018.11.014

Magar, Y. V., & Sundar, D. (2018). *Optimization of air intake system and exhaust system for better performance of turbocharged gasoline engine*. SAE Technical Paper.

Manna, S., & Kushwah, Y. S. (2015). *Optimization of a vehicle under hood airflow using 3D CFD analysis*. SAE Technical Paper.

Saad, I., & Bari, S. (2013). CFD investigation of in-cylinder air flow to optimize number of guide vanes to improve ci engine performance using higher viscous fuel. *International Journal of Automotive and Mechanical Engineering*, *8*, 1096–1107. DOI: 10.15282/ijame.8.2013.1.0089

Sathishkumar, A., Soundararajan, R., Kumar, S. R., Kaarthik, G. M., Vigneshwar, R. R., & Ali, L. F. (2024). *Performance Enhancement through Air Intake Restrictor Tuning by Fluid Flow Analysis for Formula Vehicle*. SAE Technical Paper.

Sedlacek, F., & Skovajsa, M. (2016). Optimization of an intake system using CFD numerical simulation. *Proceedings in Manufacturing Systems*, *11*(2), 71.

Sonachalam, M., & Manieniyan, V. (2021). Optimization of critical angle, distance and flow rate of secondary fuel injection in DI diesel engine using computational fluid dynamics. *SN Applied Sciences, 3*(1), 1–13. DOI: 10.1007/s42452-020-04138-3

Sortor, M. (2011). On-system engine cooling fan measurement as a tool for optimizing cooling system airflow performance and noise. *SAE International Journal of Materials and Manufacturing, 4*(1), 1221–1230. DOI: 10.4271/2011-01-1169

Srinivasulu, K., Srikanth, D., Rafi, K., & Ramanjaneeyulu, B. (2016). Optimization of Air Filter in an Automobile Diesel Engine by Using CFD Analysis. *IOSR Journal of Mechanical and Civil Engineering, 13*, 78–89.

Tang, Y. P. (2024). Effects of Race Car's Speed on the Aerodynamic Aspect Using Computational Fluid Dynamics Analysis. [ASET]. *Advanced and Sustainable Technologies, 3*(1), 54–61. DOI: 10.58915/aset.v3i1.797

Tong, W., Liao, K., Li, L., Gao, Z., Chen, F., & Luo, H. (2024). Optimization of the Camellia oleifera Fruit Harvester Engine Compartment Heat Dissipation Based on Temperature Experiments and Airflow Field Simulation. *Agriculture, 14*(9), 1640. DOI: 10.3390/agriculture14091640

von der Bank. R., Donnerhack, S., Rae, A., Cazalens, M., Lundbladh, A., & Dietz, M. (2014, June). Lemcotec: Improving the core-engine thermal efficiency. In Turbo Expo: Power for Land, Sea, and Air (Vol. 45578, p. V01AT01A001). American Society of Mechanical Engineers.

Xia, Z., & Huang, M. (2024). Optimizing the Aerodynamic Efficiency of Electric Vehicles via Streamlined Design: A Computational Fluid Dynamics Approach. *International Journal of Heat & Technology, 42*(3), 865–876. DOI: 10.18280/ijht.420315

Xu, J., & Zhou, S. (2014). Analysis of flow field for automotive exhaust system based on computational fluid dynamics. *The Open Mechanical Engineering Journal, 8*(1), 587–593. DOI: 10.2174/1874155X01408010587

Zackrisson, L. (2017). *Flow acoustic effects on a commercial automotive air intake silencer-A numerical study using computational fluid dynamics.*

Compilation of References

Abdelkader, G., Elgazzar, K., & Khamis, A. (2021). Connected Vehicles: Technology Review, State of the Art, Challenges and Opportunities. *Sensors (Basel), 21*(22), 7712. Advance online publication. DOI: 10.3390/s21227712 PMID: 34833782

Abdulwaheed, M. U. S. A., & Mashood, S. (2019). Vehicle tracking and accident alert system using gps and gsm modules.

Adhani, N. I., & Rambli, D. R. A. (2012). A survey of mobile augmented reality applications. In 1st International conference on future trends in computing and communication technologies (pp. 89-96).

Afianto, D., Han, Y., Yan, P., Yang, Y., Elbarghthi, A. F., & Wen, C. (2022). Optimisation and efficiency improvement of electric vehicles using computational fluid dynamics modelling. *Entropy (Basel, Switzerland), 24*(11), 1584. DOI: 10.3390/e24111584 PMID: 36359674

Aftab, A. R., von der Beeck, M., & Feld, M. (2020). You have a point there: Object selection inside an automobile using gaze, head pose and finger pointing. *Proceedings of the 2020 International Conference on Multimodal Interaction*, 595–603. DOI: 10.1145/3382507.3418836

Ahearn, L. (1997). The use of virtual reality in automotive design. *Journal of Automobile Engineering, 10*(2), 123–136. DOI: 10.1016/j.jauto.1997.04.003

Akash, S. (2021). AR and VR in Manufacturing is Predicted to Touch US$2.8 Billion by 2023 [WWW Document]. Analytics Insight. URL https://www.analyticsinsight.net/ar-and-vr-in-manufacturing-is-predicted-to-touch-us2-8-billion-by-2023/

Alcácer, V., & Cruz-Machado, V. (2019). Scanning the Industry 4.0: A Literature Review on Technologies for Manufacturing Systems. *Engineering Science and Technology, an International Journal, 22*(3), 899–919. DOI: 10.1016/j.jestch.2019.01.006

Alghodhaifi, H., & Lakshmanan, S. (2021). Autonomous Vehicle Evaluation: A Comprehensive Survey on Modeling and Simulation Approaches. *IEEE Access : Practical Innovations, Open Solutions*, *9*, 151531–151566. Advance online publication. DOI: 10.1109/ACCESS.2021.3125620

Al-Sarkhi, A. S. M. (1999). *Optimization technique for design of automotive air filter housing with improved fluid dynamic performance*. Oklahoma State University.

Alsayaydeh, J. A. J., Yusof, M. F., Abdillah, M. A. A., Al-Gburi, A. J. A., Herawan, S. G., & Oliinyk, A. (2023). Enhancing Vehicle Safety: A Comprehensive Accident Detection and Alert System. *International Journal of Advanced Computer Science and Applications*, *14*(11). Advance online publication. DOI: 10.14569/IJACSA.2023.0141104

Alshaqaqi, B., Baquhaizel, A. S., Ouis, M. E. A., Boumehed, M., Ouamri, A., & Keche, M. (2013). Driver drowsiness detection system. *2013 8th International Workshop on Systems, Signal Processing and Their Applications (WoSSPA)*, 151–155.

Alvi, U., Khattak, M. A. K., Shabir, B., Malik, A. W., & Muhammad, S. R. (2020). A comprehensive study on IoT based accident detection systems for smart vehicles. *IEEE Access : Practical Innovations, Open Solutions*, *8*, 122480–122497. DOI: 10.1109/ACCESS.2020.3006887

Amani, M., Ghorbanian, A., Ahmadi, S. A., Kakooei, M., Moghimi, A., Mirmazloumi, S. M., Moghaddam, S. H. A., Mahdavi, S., Ghahremanloo, M., Parsian, S., Wu, Q., & Brisco, B. (2020). Google earth engine cloud computing platform for remote sensing big data applications: A comprehensive review. *IEEE Journal of Selected Topics in Applied Earth Observations and Remote Sensing*, *13*, 5326–5350. DOI: 10.1109/JSTARS.2020.3021052

Ammar, M., Haleem, A., Javaid, M., Bahl, S., Garg, S. B., Shamoon, A., & Garg, J. (2022). Significant Applications of Smart Materials and Internet of Things (IoT) in the Automotive Industry. *Materials Today: Proceedings*, *68*, 1542–1549. Advance online publication. DOI: 10.1016/j.matpr.2022.07.180

Angelopoulos, J., & Mourtzis, D. (2022). An Intelligent Product Service System for Adaptive Maintenance of Engineered-to-Order Manufacturing Equipment Assisted by Augmented Reality. *Applied Sciences (Basel, Switzerland)*, *12*(11), 5349. Advance online publication. DOI: 10.3390/app12115349

Ang, K. L.-M., Seng, J. K. P., Ngharamike, E., & Ijemaru, G. K. (2022). Emerging technologies for smart cities' transportation: Geo-information, data analytics and machine learning approaches. *ISPRS International Journal of Geo-Information*, *11*(2), 85. DOI: 10.3390/ijgi11020085

Anthes, C., García-Hernández, R. J., Wiedemann, M., & Kranzlmüller, D. (2016, March). State of the art of virtual reality technology. In 2016 IEEE aerospace conference (pp. 1-19). IEEE.

Aouam, D., Benbelkacem, S., Zenati, N., Zakaria, S., & Meftah, Z. (2018). Voice-based augmented reality interactive system for car's components assembly. *2018 3rd International Conference on Pattern Analysis and Intelligent Systems (PAIS)*, 1–5.

Archana, T., & Stephen, R. K. (2024). The Future of Artificial Intelligence in Manufacturing Industries. In Industry Applications of Thrust Manufacturing: Convergence with Real-Time Data and AI (pp. 98-117). IGI Global. DOI: 10.4018/979-8-3693-4276-3.ch004

Arena, F., Collotta, M., Pau, G., & Termine, F. (2022). An Overview of Augmented Reality. *Computers*, *11*(2), 28. Advance online publication. DOI: 10.3390/computers11020028

Aslan, B., & Yıldız, A. R. (2020). Optimum design of automobile components using lattice structures for additive manufacturing. *Materials Testing*, *62*(6), 633–639. DOI: 10.3139/120.111527

Aswini, J., Malarvizhi, N., Subramanian, S., & Gayathri, A. (2023). Augmented Reality and Virtual Reality in E-Governance: An Immersive Technology Applications and Its Challenges. In AI, IoT, and Blockchain Breakthroughs in E-Governance (pp. 138-153). IGI Global.

Atitallah, S. B., Driss, M., Boulila, W., & Ghézala, H. B. (2020). Leveraging Deep Learning and IoT big data analytics to support the smart cities development: Review and future directions. *Computer Science Review*, *38*, 100303. DOI: 10.1016/j.cosrev.2020.100303

Audi, A. G. (2019). Virtual showrooms and immersive customer experiences at Audi. https://www.audi.com/virtual-showrooms

Audi. (2020). Audi uses augmented and virtual reality to assist customers with diagnostics. Retrieved from https://www.audi.com

Azuma, R. T. (1997). A Survey of Augmented Reality. *Presence (Cambridge, Mass.)*, *6*(4), 355–385. DOI: 10.1162/pres.1997.6.4.355

Azuma, R., Baillot, Y., Behringer, R., Feiner, S., Julier, S., & MacIntyre, B. (2001). Recent Advances in Augmented Reality. *IEEE Computer Graphics and Applications*, *21*(6), 34–47. Advance online publication. DOI: 10.1109/38.963459

Babu, B. S., Kamalakannan, J., Meenatchi, N., Karthik, S., & Boopathi, S. (2022). Economic impacts and reliability evaluation of battery by adopting Electric Vehicle. *IEEE Explore*, 1–6.

Babu, C. V. S., & Manoharan, S. (2024). AI and Blockchain Revolution in Communication Education: Future of Work in the Age of Industrial Robotics. In Eduard Babulak, I. G. I. (Ed.), *Applied AI and Humanoid Robotics for the Ultra-Smart Cyberspace* (pp. 63–84). Global., DOI: 10.4018/979-8-3693-2399-1.ch003

Babu, S., & Raj Kumar P, A. (2022). A Comprehensive Survey on Simulators, Emulators, and Testbeds for VANETs. *International Journal of Communication Systems, 35*(8), e5123. Advance online publication. DOI: 10.1002/dac.5123

Bahamonde-Birke, F. J., Kickhöfer, B., Heinrichs, D., & Kuhnimhof, T. (2018). A systemic view on autonomous vehicles: Policy aspects for a sustainable transportation planning. *disP-The Planning Review, 54*(3), 12–25.

Bajic, B., Rikalovic, A., Suzic, N., & Piuri, V. (2021). Industry 4.0 Implementation Challenges and Opportunities: A Managerial Perspective. *IEEE Systems Journal, 15*(1), 546–559. DOI: 10.1109/JSYST.2020.3023041

Bale, A. S., Patil, M., Prathima, M., Telgane, S. V., Ghorpade, N., & Anish, B. (2024). Augmenting Knowledge Management with Immersive Technologies: Exploring Transformative Use Cases for AR and VR. *2024 International Conference on Advances in Computing, Communication and Applied Informatics (ACCAI)*, 1–6. DOI: 10.1109/ACCAI61061.2024.10602348

Bark, K., Tran, C., Fujimura, K., & Ng-Thow-Hing, V. (2014). Personal navi: Benefits of an augmented reality navigational aid using a see-thru 3d volumetric hud. *Proceedings of the 6th International Conference on Automotive User Interfaces and Interactive Vehicular Applications*, 1–8. DOI: 10.1145/2667317.2667329

Bause, H., Green, P., & Helbig, M. (2020). Convergence of augmented and virtual reality: Future challenges in automotive manufacturing. *Automotive Systems and Software Engineering Journal, 19*(4), 155–168. DOI: 10.1007/s00768-020-00414-0

Bellalouna, F. (2021). The Augmented Reality Technology as Enabler for the Digitization of Industrial Business Processes: Case Studies. *Procedia CIRP, 98*, 400–405. DOI: 10.1016/j.procir.2021.01.124

Besnea, F.-L., Resceanu, I.-C., Cismaru, S.-I., Ganea, I.-E., Pistritu, I.-L., & Bizdoaca, N.-G. (2019). Experiments regarding implementation of a virtual training environment for automotive industry. *2019 20th International Carpathian Control Conference (ICCC)*, 1–6.

Bharadiya, J. (2023). Artificial intelligence in transportation systems a critical review. *American Journal of Computing and Engineering*, *6*(1), 34–45. DOI: 10.47672/ajce.1487

Bhardwaj, A., Bhardwaj, M., & Gaur, A. (2016). Virtual reality: An overview. *International Journal of Scientific and Technical Advancements*, *2*(4), 159–164.

Bibri, S. E., Krogstie, J., & Kärrholm, M. (2020). Compact city planning and development: Emerging practices and strategies for achieving the goals of sustainability. *Developments in the Built Environment*, *4*, 100021. DOI: 10.1016/j.dibe.2020.100021

Biswas, S., Tatchikou, R., & Dion, F. (2006). Vehicle-to-vehicle wireless communication protocols for enhancing highway traffic safety. *IEEE Communications Magazine*, *44*(1), 74–82. DOI: 10.1109/MCOM.2006.1580935

Blanco-Novoa, O., Fernandez-Carames, T. M., Fraga-Lamas, P., & Vilar-Montesinos, M. A. (2018). A Practical Evaluation of Commercial Industrial Augmented Reality Systems in an Industry 4.0 Shipyard. *IEEE Access : Practical Innovations, Open Solutions*, *6*, 8201–8218. DOI: 10.1109/ACCESS.2018.2802699

Blankenbach, K. (2023). Augmented Reality HUDs: Challenges, Solutions and Competitors. *2023 30th International Workshop on Active-Matrix Flatpanel Displays and Devices (AM-FPD)*, 28–31.

Boboc, R. G., Gîrbacia, F., & Butilă, E. V. (2020a). The application of augmented reality in the automotive industry: A systematic literature review. *Applied Sciences (Basel, Switzerland)*, *10*(12), 4259. DOI: 10.3390/app10124259

Boopathi, S., Gavaskar, T., Dogga, A. D., Mahendran, R. K., Kumar, A., Kathiresan, G., N., V., Ganesan, M., Ishwarya, K. R., Ramana, G. V., & others. (2021, August 4). *Emergency medicine delivery transportation using unmanned aerial vehicle (Patent Grant)*.

Boopathi, S., Thillaivanan, A., Mohammed, A. A., Shanmugam, P., & VR, P. (2022). Experimental investigation on Abrasive Water Jet Machining of Neem Wood Plastic Composite. *IOP: Functional Composites and Structures, 4*, 025001.

Boopathi, S. (2024). Advancements in Machine Learning and AI for Intelligent Systems in Drone Applications for Smart City Developments. In *Futuristic e-Governance Security With Deep Learning Applications* (pp. 15–45). IGI Global. DOI: 10.4018/978-1-6684-9596-4.ch002

Botto, C., Cannavò, A., Cappuccio, D., Morat, G., Sarvestani, A. N., Ricci, P., . . . Saturnino, A. (2020, March). Augmented reality for the manufacturing industry: the case of an assembly assistant. In 2020 IEEE Conference on Virtual Reality and 3D User Interfaces Abstracts and Workshops (VRW) (pp. 299-304). IEEE.

Boukerche, A., Tao, Y., & Sun, P. (2020). Artificial intelligence-based vehicular traffic flow prediction methods for supporting intelligent transportation systems. *Computer Networks*, *182*, 107484. DOI: 10.1016/j.comnet.2020.107484

Boyraz, P., Acar, M., & Kerr, D. (2008). Multi-sensor driver drowsiness monitoring. *Proceedings of the Institution of Mechanical Engineers. Part D, Journal of Automobile Engineering*, *222*(11), 2041–2062. DOI: 10.1243/09544070JAUTO513

Brown, K., & Gupta, M. (2022). The Role of Augmented Reality in Healthcare: A Review of Applications and Challenges. *Journal of Medical Internet Research*, *19*(5), e173.

Brunner, K. E., Perey, C., Rauschnabel, P. A., & Sage, M. (2020). Implementation of Augmented Reality in Manufacturing: A Case Study Exercise. In Jung, T., Tom Dieck, M. C., & Rauschnabel, P. A. (Eds.), *Augmented Reality and Virtual Reality, Progress in IS* (pp. 85–97). Springer International Publishing., DOI: 10.1007/978-3-030-37869-1_8

Buchs, J. (2020). Virtual prototyping in automotive design: Opportunities and challenges. *International Journal of Automotive Technology*, *17*(4), 105–112. DOI: 10.1007/s12239-020-0105-2

Butt, J. (2020). A Strategic Roadmap for the Manufacturing Industry to Implement Industry 4.0. *Designs*, *4*(2), 11. DOI: 10.3390/designs4020011

Campisi, T., Severino, A., Al-Rashid, M. A., & Pau, G. (2021). The development of the smart cities in the connected and autonomous vehicles (CAVs) era: From mobility patterns to scaling in cities. *Infrastructures*, *6*(7), 100. DOI: 10.3390/infrastructures6070100

Carliner, S. (2015). *Training design basics*. Association for Talent Development.

Carmigniani, J., & Furht, B. (2011). Augmented reality: an overview. Handbook of augmented reality, 3-46.

Carnazzo, C., Spada, S., Lamacchia, S., Manuri, F., Sanna, A., & Cavatorta, M. P. (2024). Virtual reality in ergonomics by wearable devices: Experiences from the automotive sector. *Journal of Workplace Learning*, *36*(7), 621–635. DOI: 10.1108/JWL-03-2024-0064

Chandrika, V., Sivakumar, A., Krishnan, T. S., Pradeep, J., Manikandan, S., & Boopathi, S. (2023). Theoretical Study on Power Distribution Systems for Electric Vehicles. In *Intelligent Engineering Applications and Applied Sciences for Sustainability* (pp. 1–19). IGI Global. DOI: 10.4018/979-8-3693-0044-2.ch001

Charissis, V., Falah, J., Lagoo, R., Alfalah, S. F. M., Khan, S., Wang, S., Altarteer, S., Larbi, K. B., & Drikakis, D. (2021). Employing Emerging Technologies to Develop and Evaluate In-Vehicle Intelligent Systems for Driver Support: Infotainment AR Hud Case Study. *Applied Sciences (Basel, Switzerland)*, *11*(4), 1397. Advance online publication. DOI: 10.3390/app11041397

Chatzopoulos, D., Bermejo, C., Huang, Z., & Hui, P. (2017). Mobile Augmented Reality Survey: From Where We Are to Where We Go. *IEEE Access : Practical Innovations, Open Solutions*, *5*, 6917–6950. DOI: 10.1109/ACCESS.2017.2698164

Chiang, F.-K., Shang, X., & Qiao, L. (2022). Augmented reality in vocational training: A systematic review of research and applications. *Computers in Human Behavior*, *129*, 107125. DOI: 10.1016/j.chb.2021.107125

Chintala, V., & Subramanian, K. (2013). A CFD (computational fluid dynamics) study for optimization of gas injector orientation for performance improvement of a dual-fuel diesel engine. *Energy*, *57*, 709–721. DOI: 10.1016/j.energy.2013.06.009

Chouksey, S., & Sirsikar, S. (2016). A prototype of low cost heads up display for automobiles navigation system. *2016 International Conference on Computing, Analytics and Security Trends (CAST)*, 205–210. DOI: 10.1109/CAST.2016.7914967

Chu, K., Brewer, R., & Joseph, S. (2008). *Traffic and navigation support through an automobile heads up display (a-HUD)*.

Churiwala, K., Lopes, R., Shah, A., & Shah, N. (2012). Drowsiness detection based on eye movement, yawn detection and head rotation. *International Journal of Applied Information Systems*, *2*(6), 45–49.

Chu, Y., Hatledal, L. I., Zhang, H., Æsøy, V., & Ehlers, S. (2018). Virtual prototyping for maritime crane design and operations. *Journal of Marine Science and Technology*, *23*(4), 754–766. DOI: 10.1007/s00773-017-0509-z

Clenci, A. C., Iorga-Simăn, V., Deligant, M., Podevin, P., Descombes, G., & Niculescu, R. (2014). A CFD (computational fluid dynamics) study on the effects of operating an engine with low intake valve lift at idle corresponding speed. *Energy*, *71*, 202–217. DOI: 10.1016/j.energy.2014.04.069

Čujan, Z., Fedorko, G., & Mikušová, N. (2020). Application of virtual and augmented reality in automotive. *Open Engineering*, *10*(1), 113–119. DOI: 10.1515/eng-2020-0022

Currano, R., Park, S. Y., Moore, D. J., Lyons, K., & Sirkin, D. (2021). Little road driving hud: Heads-up display complexity influences drivers' perceptions of automated vehicles. *Proceedings of the 2021 CHI Conference on Human Factors in Computing Systems*, 1–15. DOI: 10.1145/3411764.3445575

Dalle Mura, M., & Dini, G. (2021). An augmented reality approach for supporting panel alignment in car body assembly. *Journal of Manufacturing Systems*, *59*, 251–260. DOI: 10.1016/j.jmsy.2021.03.004

Damiani, L., Revetria, R., & Morra, E. (2020). Safety in Industry 4.0: The Multi-Purpose Applications of Augmented Reality in Digital Factories. *Advances in Science, Technology and Engineering Systems Journal*, *5*(2), 248–253. DOI: 10.25046/aj050232

Dantsuji, T., & Takayama, Y. (2024). Hypercongestion, autonomous vehicles, and urban spatial structure. *Transportation Science*, *58*(6), 1352–1370. DOI: 10.1287/trsc.2024.0519

Dargan, S., Bansal, S., Kumar, M., Mittal, A., & Kumar, K. (2023). Augmented Reality: A Comprehensive Review. *Archives of Computational Methods in Engineering*, *30*(2), 1057–1080. DOI: 10.1007/s11831-022-09831-7

Dasgupta, A., George, A., Happy, S., & Routray, A. (2013). A vision-based system for monitoring the loss of attention in automotive drivers. *IEEE Transactions on Intelligent Transportation Systems*, *14*(4), 1825–1838. DOI: 10.1109/TITS.2013.2271052

Dash, B., Sharma, P., & Ansari, M. (2018). *A Data-Driven AI Framework to Improve Urban Mobility and Traffic Congestion in Smart Cities*. ed.

Dass, A., & Boopathi, S. (2016). Experimental Study of Eco-friendly Wire-Cut Electrical Discharge Machining Processes. *International Journal of Innovative Research in Science, Engineering and Technology*, *5*(1), 666–675.

de Freitas, F. V., Gomes, M. V. M., & Winkler, I. (2022). Benefits and challenges of virtual-reality-based industrial usability testing and design reviews: A patents landscape and literature review. *Applied Sciences (Basel, Switzerland)*, *12*(3), 1755. DOI: 10.3390/app12031755

de Winter, J. C. F., & van Leeuwen, P. M., and R Happee. 2012. "Advantages and Disadvantages of Driving Simulators: A Discussion." In *Proceedings of the Measuring Behavior Conference*

Deac, C. N., Deac, C. N., Popa, C. L., Ghinea, M., & Cotet, C. E. (2017). Using Augmented Reality in Smart Manufacturing, in: Katalinic, B. (Ed.), DAAAM Proceedings. DAAAM International Vienna, pp. 0727–0732. DOI: 10.2507/28th.daaam.proceedings.102

Deep, G., & Verma, J. (2023). Embracing the future: AI and ML transforming urban environments in smart cities. *J. Artif. Intell*, *5*(0), 57–73. DOI: 10.32604/jai.2023.043329

Deepika, & Pandove, G. (2024). A Comparison of ML Models for Predicting Congestion in Urban Cities. International Journal of Intelligent Transportation Systems Research, 22(1), 171-188.

Delgado, D., Manuel, J., Oyedele, L., Beach, T., & Demian, P. (2020). Augmented and Virtual Reality in Construction: Drivers and Limitations for Industry Adoption. *Journal of Construction Engineering and Management*, *146*(7), 04020079. Advance online publication. DOI: 10.1061/(ASCE)CO.1943-7862.0001844

Demlehner, Q., Schoemer, D., & Laumer, S. (2021). How Can Artificial Intelligence Enhance Car Manufacturing? A Delphi Study-Based Identification and Assessment of General Use Cases. *International Journal of Information Management*, *58*, 102317. Advance online publication. DOI: 10.1016/j.ijinfomgt.2021.102317

Detjen, H., Geisler, S., & Schneegass, S. 2020. "Maneuver-Based Control Interventions during Automated Driving: Comparing Touch, Voice, and Mid-Air Gestures as Input Modalities." In *Conference Proceedings - IEEE International Conference on Systems, Man and Cybernetics*, DOI: 10.1109/SMC42975.2020.9283431

Dhanalakshmi, R., Mai, C. D., Latha, B., & Vijayaraghavan, N. (2021). AR and VR in Manufacturing. In Palanikumar, K., Natarajan, E., Sengottuvelu, R., & Davim, J. P. (Eds.), *Futuristic Trends in Intelligent Manufacturing, Materials Forming, Machining and Tribology* (pp. 171–183). Springer International Publishing., DOI: 10.1007/978-3-030-70009-6_11

Dikshit, S., Atiq, A., Shahid, M., Dwivedi, V., & Thusu, A. (2023). The use of artificial intelligence to optimize the routing of vehicles and reduce traffic congestion in urban areas. *EAI Endorsed Transactions on Energy Web, 10*.

Ding, Y., & Li, J. (2024). Understanding User Intention to Adopt Augmented Reality Head-Up Display: A Study Based on Design Factors. *International Journal of Human-Computer Interaction*, •••, 1–19. DOI: 10.1080/10447318.2024.2425884

Divakaran, A. M., Gkanas, E., Jewkes, J., Shepherd, S., Jamal, W., Faramehr, S., & Abo-Serie, E. (2025). 3D CFD modelling and experimental validation of conjugate heat transfer in wheel hub motors in micro-mobility vehicles. *Applied Thermal Engineering*, *258*, 124430. DOI: 10.1016/j.applthermaleng.2024.124430

Domakonda, V. K., Farooq, S., Chinthamreddy, S., Puviarasi, R., Sudhakar, M., & Boopathi, S. (2022). Sustainable Developments of Hybrid Floating Solar Power Plants: Photovoltaic System. In *Human Agro-Energy Optimization for Business and Industry* (pp. 148–167). IGI Global.

Doshi, A., Cheng, S. Y., & Trivedi, M. M. (2008). A novel active heads-up display for driver assistance. *IEEE Transactions on Systems, Man, and Cybernetics. Part B, Cybernetics*, *39*(1), 85–93. DOI: 10.1109/TSMCB.2008.923527 PMID: 19068432

Doudou, M., Bouabdallah, A., & Berge-Cherfaoui, V. (2020). Driver Drowsiness Measurement Technologies: Current Research, Market Solutions, and Challenges. *International Journal of Intelligent Transportation Systems Research*, *18*(2), 297–319. Advance online publication. DOI: 10.1007/s13177-019-00199-w

Drężek, P. S., Kubacki, S., & Żółtak, J. (2023). Kriging-based framework applied to a multi-point, multi-objective engine air-intake duct aerodynamic optimization problem. *Aerospace (Basel, Switzerland)*, *10*(3), 266. DOI: 10.3390/aerospace10030266

Du, X. (2016). *Towards sustainable autonomous vehicles* [PhD Thesis]. National University of Singapore (Singapore).

Dumas, L. (2008). CFD-based optimization for automotive aerodynamics. In *Optimization and computational fluid dynamics* (pp. 191–215). Springer. DOI: 10.1007/978-3-540-72153-6_7

Dunne, N., Reid, S., & Schmitt, D. (2020). Human-centered design in automotive virtual reality environments. *Ergonomics in Design*, *28*(1), 42–50. DOI: 10.1177/1064804620904292

Dwarshala, S. K. R., Ramadandi, P., Shaik, R., & Shankar, R. (2014). *Practical considerations in the airflow optimization of a single cylinder diesel engine*. SAE Technical Paper.

Edwards-Stewart, A., Hoyt, T., & Reger, G. (2016). Classifying different types of augmented reality technology. *Annual Review of Cybertherapy and Telemedicine*, *14*, 199–202.

Egger, J., & Masood, T. (2020). Augmented reality in support of intelligent manufacturing – A systematic literature review. *Computers & Industrial Engineering*, *140*, 106195. DOI: 10.1016/j.cie.2019.106195

Ejichukwu, E., Tong, L., Hazime, G., & Jia, B. (2024). Enhancing Autonomous Vehicle Design and Testing: A Comprehensive Review of AR and VR Integration. *arXiv Preprint arXiv:2404.19021.*

Eltaweel, M., Mostafa, N. A., Kalyvas, C., Chen, Y., & Herfatmanesh, M. R. (2025). Optimising flywheel energy storage systems for enhanced windage loss reduction and heat transfer: A computational fluid dynamics and ANOVA-based approach. *Energy Reports*, *13*, 834–855. DOI: 10.1016/j.egyr.2024.12.048

Engine, U. (n.d.). Virtual reality development tools. Retrieved from https://unrealengine.com

Enjalbert, S., Gandini, L. M., Baños, A. P., Ricci, S., & Vanderhaegen, F. (2021). Human–Machine Interface in Transport Systems: An Industrial Overview for More Extended Rail Applications. *Machines (Basel)*, *9*(2), 36. Advance online publication. DOI: 10.3390/machines9020036

Esengün, M., & İnce, G. (2018). The Role of Augmented Reality in the Age of Industry 4.0. In *Industry 4.0: Managing The Digital Transformation, Springer Series in Advanced Manufacturing* (pp. 201–215). Springer International Publishing., DOI: 10.1007/978-3-319-57870-5_12

Eswaran, M., & Bahubalendruni, M. R. (2022). Challenges and opportunities on AR/VR technologies for manufacturing systems in the context of industry 4.0: A state of the art review. *Journal of Manufacturing Systems*, *65*, 260–278. DOI: 10.1016/j.jmsy.2022.09.016

Faisal, A., Kamruzzaman, M., Yigitcanlar, T., & Currie, G. (2019). Understanding autonomous vehicles. *Journal of Transport and Land Use*, *12*(1), 45–72. DOI: 10.5198/jtlu.2019.1405

Fang, W., & An, Z. (2020). A scalable wearable AR system for manual order picking based on warehouse floor-related navigation. *International Journal of Advanced Manufacturing Technology*, *109*(7-8), 2023–2037. DOI: 10.1007/s00170-020-05771-3

Fedko, D. (2020). The Ultimate Guide to Augmented Reality (AR) in Manufacturing Industry [WWW Document]. Wear Studio Blog. URL https://wear-studio.com/ar-in-manufacturing/

Firu, A. C., Tapîrdea, A. I., Feier, A. I., & Drăghici, G. (2021). Virtual reality in the automotive field in industry 4.0. *Materials Today: Proceedings*, *45*, 4177–4182. DOI: 10.1016/j.matpr.2020.12.037

Fite-Georgel, P. (2011). Is there a reality in Industrial Augmented Reality? in: 2011 10th IEEE International Symposium on Mixed and Augmented Reality. Presented at the 2011 IEEE International Symposium on Mixed and Augmented Reality, IEEE, Basel, pp. 201–210. DOI: 10.1109/ISMAR.2011.6092387

Fogue, M., Garrido, P., Martinez, F. J., Cano, J.-C., Calafate, C. T., & Manzoni, P. (2012). Automatic accident detection: Assistance through communication technologies and vehicles. *IEEE Vehicular Technology Magazine*, *7*(3), 90–100. DOI: 10.1109/MVT.2012.2203877

Ford Motor Company. (2020). Virtual reality in automotive design: How Ford uses VR to enhance the development process. https://www.ford.com/research/vr-auto-design

Ford. (2018). Ford uses virtual reality for immersive design and testing. Retrieved from https://www.ford.com

Ford. (2020). Virtual reality and the future of automotive retailing. Retrieved from https://www.ford.com

Freitas, F., Oliveira, H., Winkler, I., & Gomes, M. (2020). Virtual reality on product usability testing: A systematic literature review. *2020 22nd Symposium on Virtual and Augmented Reality (SVR)*, 67–73.

Frizziero, L., Santi, G. M., Donnici, G., Leon-Cardenas, C., Ferretti, P., Liverani, A., & Neri, M. (2021). An Innovative Ford Sedan with Enhanced Stylistic Design Engineering (SDE) via Augmented Reality and Additive Manufacturing. *Designs*, *5*(3), 46. Advance online publication. DOI: 10.3390/designs5030046

Gabbard, J. L., Smith, M., Tanous, K., Kim, H., & Jonas, B. (2019). AR drivesim: An immersive driving simulator for augmented reality head-up display research. *Frontiers in Robotics and AI*, *6*, 98. DOI: 10.3389/frobt.2019.00098 PMID: 33501113

Gangwani, D., & Gangwani, P. (2021). Applications of machine learning and artificial intelligence in intelligent transportation system: A review. *Applications of Artificial Intelligence and Machine Learning: Select Proceedings of ICAAAIML 2020*, 203–216.

Garcia, C. A., Naranjo, J. E., Ortiz, A., & Garcia, M. V. (2019). An approach of virtual reality environment for technicians training in upstream sector. *IFAC-PapersOnLine*, *52*(9), 285–291. DOI: 10.1016/j.ifacol.2019.08.222

Gavanas, N. (2019). Autonomous road vehicles: Challenges for urban planning in European cities. *Urban Science (Basel, Switzerland)*, *3*(2), 61. DOI: 10.3390/urbansci3020061

Gavish, N., Gutiérrez, T., Webel, S., Rodríguez, J., Peveri, M., Bockholt, U., & Tecchia, F. (2015). Evaluating virtual reality and augmented reality training for industrial maintenance and assembly tasks. *Interactive Learning Environments*, *23*(6), 778–798. DOI: 10.1080/10494820.2013.815221

Geng, R., Li, M., Hu, Z., Han, Z., & Zheng, R. (2022). Digital Twin in smart manufacturing: Remote control and virtual machining using VR and AR technologies. *Structural and Multidisciplinary Optimization*, *65*(11), 321. DOI: 10.1007/s00158-022-03426-3

Gennari, D., & Miele, M. (2006). Virtual reality applications in automotive product development. *Computer Aided Design*, *38*(6), 647–656. DOI: 10.1016/j.cad.2005.11.001

Gerber, M. A., Schroeter, R., & Vehns, J. (2019). A video-based automated driving simulator for automotive UI prototyping, UX and behaviour research. *Proceedings of the 11th International Conference on Automotive User Interfaces and Interactive Vehicular Applications*, 14–23. DOI: 10.1145/3342197.3344533

Ghosh, R., Gupta, A., & Yadav, M. (2021). Integration of artificial intelligence with VR: Exploring opportunities in the automotive sector. *Journal of Automotive Technologies*, *9*(1), 34–45. DOI: 10.1109/JAT.2021.9450767

Glady, J. B. P., D'Souza, S. M., Priya, A. P., Amuthachenthiru, K., Vikram, G., & Boopathi, S. (2024). A Study on AI-ML-Driven Optimizing Energy Distribution and Sustainable Agriculture for Environmental Conservation. In *Harnessing High-Performance Computing and AI for Environmental Sustainability* (pp. 1–27). IGI Global., DOI: 10.4018/979-8-3693-1794-5.ch001

Golbabaei, F., Yigitcanlar, T., & Bunker, J. (2021). The role of shared autonomous vehicle systems in delivering smart urban mobility: A systematic review of the literature. *International Journal of Sustainable Transportation*, *15*(10), 731–748. DOI: 10.1080/15568318.2020.1798571

Gopal, M., Lurdhumary, J., Bathrinath, S., Priya, A. P., Sarojwal, A., & Boopathi, S. (2024). Energy Harvesting and Smart Highways for Sustainable Transportation Infrastructure: Revolutionizing Roads Using Nanotechnology. In *Principles and Applications in Speed Sensing and Energy Harvesting for Smart Roads* (pp. 136–165). IGI Global. DOI: 10.4018/978-1-6684-9214-7.ch005

Gouda, G. K., & Tiwari, B. (2024). Unravelling the enablers of Industry 4.0 in Indian automobile industry amid COVID-19: An integrated TISM and fuzzy MICMAC approach. *Journal of Business and Industrial Marketing*, *39*(2), 227–243. DOI: 10.1108/JBIM-02-2022-0067

Gromova, E. A. (2019). Digital economy development with an emphasis on automotive industry in Russia. *Revista Espacios, 40*(06).

Gružauskas, V., Baskutis, S., & Navickas, V. (2018). Minimizing the trade-off between sustainability and cost effective performance by using autonomous vehicles. *Journal of Cleaner Production, 184*, 709–717. DOI: 10.1016/j.jclepro.2018.02.302

Guo, Z., Zhou, D., Zhou, Q., Zhang, X., Geng, J., Zeng, S., Lv, C., & Hao, A. (2020). Applications of virtual reality in maintenance during the industrial product lifecycle: A systematic review. *Journal of Manufacturing Systems, 56*, 525–538. DOI: 10.1016/j.jmsy.2020.07.007

Hadas, Z., Janak, L., & Smilek, J. (2018). Virtual prototypes of energy harvesting systems for industrial applications. *Mechanical Systems and Signal Processing, 110*, 152–164. DOI: 10.1016/j.ymssp.2018.03.036

Halabi, O. (2020). Immersive virtual reality to enforce teaching in engineering education. *Multimedia Tools and Applications, 79*(3), 2987–3004. DOI: 10.1007/s11042-019-08214-8

Hapt, X. (2023). Revolutionizing haptics for virtual reality. Retrieved from https://haptx.com

Harikaran, M., Boopathi, S., Gokulakannan, S., & Poonguzhali, M. (2023). Study on the Source of E-Waste Management and Disposal Methods. In *Sustainable Approaches and Strategies for E-Waste Management and Utilization* (pp. 39–60). IGI Global. DOI: 10.4018/978-1-6684-7573-7.ch003

Hasan, R., & Hasan, R. (2022). Pedestrian Safety Using the Internet of Things and Sensors: Issues, Challenges, and Open Problems. *Future Generation Computer Systems, 134*, 187–203. Advance online publication. DOI: 10.1016/j.future.2022.03.036

Hastings, W. (2018). *Augmented reality for manufacturing: bringing digital transformation to skilled workers*. ARC White Pap.

Henriques, A. C., & Winkler, I. (2021). The advancement of virtual reality in automotive market research: Challenges and opportunities. *Applied Sciences (Basel, Switzerland), 11*(24), 11610. DOI: 10.3390/app112411610

Hernández-Chávez, M., Cortés-Caballero, J. M., Pérez-Martínez, Á. A., Hernández-Quintanar, L. F., Roa-Tort, K., Rivera-Fernández, J. D., & Fabila-Bustos, D. A. (2021). Development of virtual reality automotive lab for training in engineering students. *Sustainability (Basel), 13*(17), 9776. DOI: 10.3390/su13179776

Heydaryan, S., Suaza Bedolla, J., & Belingardi, G. (2018). Safety design and development of a human-robot collaboration assembly process in the automotive industry. *Applied Sciences (Basel, Switzerland)*, *8*(3), 344. DOI: 10.3390/app8030344

Hoffmann, C., Büttner, S., Prilla, M., & Wundram, K. (2020). Impact of augmented reality guidance for car repairs on novice users of AR: a field experiment on familiar and unfamiliar tasks. In Proceedings of Mensch und Computer 2020 (pp. 279-289). DOI: 10.1145/3404983.3405594

Holkar, R., Sule-Patil, Y. N., Pise, S. M., Godase, Y. A., & Jagadale, V. S. (2015). Numerical simulation of steady flow through engine intake system using CFD. *IOSR Journal of Mechanical and Civil Engineering (IOSR-JMCE) e-ISSN, 2278*(1684), 30–45.

Hollands, R. (2021). Impact of VR training on automotive manufacturing efficiency. *Journal of Manufacturing Technology*, *32*(3), 250–263. DOI: 10.1002/jmt.2393

Ho, P. T., Albajez, J. A., Santolaria, J., & Yagüe-Fabra, J. A. (2022). Study of Augmented Reality Based Manufacturing for Further Integration of Quality Control 4.0: A Systematic Literature Review. *Applied Sciences (Basel, Switzerland)*, *12*(4), 1961. DOI: 10.3390/app12041961

Hossain, M., Khalid, H., Rao, A. P., Lootah, M., Al-Mohammedi, S. S. K., & Majeed, S. R. (2024). Comprehensive Review of AI, IoT, and ML in Enhancing Urban Mobility and Reducing Carbon Footprints. *2024 Third International Conference on Sustainable Mobility Applications, Renewables and Technology (SMART)*, 1–6. DOI: 10.1109/SMART63170.2024.10815521

Huang, T. K., Yang, C. H., Hsieh, Y. H., Wang, J. C., & Hung, C. C. (2018). Augmented reality (AR) and virtual reality (VR) applied in dentistry. *The Kaohsiung Journal of Medical Sciences*, *34*(4), 243–248. DOI: 10.1016/j.kjms.2018.01.009 PMID: 29655414

Hu, X., Li, S., Huang, T., Tang, B., Huai, R., & Chen, L. (2024). How Simulation Helps Autonomous Driving: A Survey of Sim2real, Digital Twins, and Parallel Intelligence. *IEEE Transactions on Intelligent Vehicles*, *9*(1), 593–612. Advance online publication. DOI: 10.1109/TIV.2023.3312777

Iqbal, D., Abbas, A., Ali, M., Khan, M. U. S., & Nawaz, R. (2020). Requirement validation for embedded systems in automotive industry through modeling. *IEEE Access : Practical Innovations, Open Solutions*, *8*, 8697–8719. DOI: 10.1109/ACCESS.2019.2963774

Jagatheesaperumal, S. K., Bibri, S. E., Huang, J., Rajapandian, J., & Parthiban, B. (2024). Artificial intelligence of things for smart cities: Advanced solutions for enhancing transportation safety. *Computational Urban Science*, *4*(1), 10. DOI: 10.1007/s43762-024-00120-6

Jain, K., Mittal, H. K., Vasesi, S., Antil, B., & Pandey, H. C. (2023, December). Exploring the Convergence of Augmented and Virtual Reality: Applications, Challenges and Future Prospects. In *2023 International Conference on Advanced Computing & Communication Technologies (ICACCTech)* (pp. 626-634). IEEE. DOI: 10.1109/ICACCTech61146.2023.00107

Janardhana, K., Singh, V., Singh, S. N., Babu, T. R., Bano, S., & Boopathi, S. (2023). Utilization Process for Electronic Waste in Eco-Friendly Concrete: Experimental Study. In *Sustainable Approaches and Strategies for E-Waste Management and Utilization* (pp. 204–223). IGI Global.

Jemni, M. A. HadjKacem, S., Ammar, M., Saaidia, R., Brayek, M., & Abid, M. S. (2021). Variable intake manifold geometry influence on volumetric efficiency enhancement at gaseous engine starting speeds. Proceedings of the Institution of Mechanical Engineers, Part E: Journal of Process Mechanical Engineering, 235(2), 548-559.

Jeong, S.-J. (2024). CFD Simulation of Pre-Chamber Spark-Ignition Engines—A Perspective Review. *Energies*, *17*(18), 1–39. DOI: 10.3390/en17184696

Jha, S. K., & Beevi, S. J. P., H., Babitha, M. N., Chinnusamy, S., & Boopathi, S. (2024). Artificial Intelligence-Infused Urban Connectivity for Smart Cities and the Evolution of IoT Communication Networks. In *Blockchain-Based Solutions for Accessibility in Smart Cities* (pp. 113–146). IGI Global. DOI: 10.4018/979-8-3693-3402-7.ch005

Jha, S. K., Beevi, S. J., Babitha, M. N., Chinnusamy, S., & Boopathi, S. (2024). Artificial Intelligence-Infused Urban Connectivity for Smart Cities and the Evolution of IoT Communication Networks. In Blockchain-Based Solutions for Accessibility in Smart Cities (pp. 113-146). IGI Global.

Jiang, Q., & Guo, Z. (2023). AR-HUD optical system design and its multiple configurations analysis. *Photonics*, *10*(9), 954. DOI: 10.3390/photonics10090954

Jing, C., Shang, C., Yu, D., Chen, Y., & Zhi, J. (2022). The impact of different AR-HUD virtual warning interfaces on the takeover performance and visual characteristics of autonomous vehicles. *Traffic Injury Prevention*, *23*(5), 277–282. DOI: 10.1080/15389588.2022.2055752 PMID: 35442130

Jing, R., Yuan, C., Rezaei, H., Qian, J., & Zhang, Z. (2020). Assessments on emergy and greenhouse gas emissions of internal combustion engine automobiles and electric automobiles in the USA. *Journal of Environmental Sciences (China)*, *90*, 297–309. DOI: 10.1016/j.jes.2019.11.017 PMID: 32081326

Jones, Gary W., and Susan K. Jones. 2022. "Critical Technology Enablers for Near-to-Eye AR and VR." .DOI: 10.1117/12.2622316

Jung, C., Lee, S., & Lee, J. (2020). Enhancing automotive production with virtual reality-based workflow optimization. *Manufacturing Review*, *32*(4), 450–463. DOI: 10.1016/j.mfr.2020.04.002

Jung, H. K., Lee, M., & Yoon, K. (2017). Virtual reality in automotive design: Past, present, and future perspectives. *International Journal of Vehicle Design*, *75*(3), 256–270. DOI: 10.1504/IJVD.2017.084607

Ju, U., Williamson, J., & Wallraven, C. (2022). Predicting Driving Speed from Psychological Metrics in a Virtual Reality Car Driving Simulation. *Scientific Reports*, *12*(1), 10044. Advance online publication. DOI: 10.1038/s41598-022-14409-1 PMID: 35710859

Kaladgi, A. R., Afzal, A., Manokar, A. M., Thakur, D., Agbulut, U., Alshahrani, S., & Subbiah, R. (2021). Integrated Taguchi-GRA-RSM optimization and ANN modelling of thermal performance of zinc oxide nanofluids in an automobile radiator. *Case Studies in Thermal Engineering*, *26*, 101068. DOI: 10.1016/j.csite.2021.101068

KAV. R. P., Pandraju, T. K. S., Boopathi, S., Saravanan, P., Rathan, S. K., & Sathish, T. (2023, November). Hybrid Deep Learning Technique for Optimal Wind Mill Speed Estimation. In 2023 7th International Conference on Electronics, Communication and Aerospace Technology (ICECA) (pp. 181-186). IEEE.

Kellmereit, D., & Pflaum, A. (2016). *The Augmented Reality Megatrend of Tomorrow: Pokémon Go - and More*. Springer.

Kemeny, A. 2024. "On-Board Virtual Reality and Applications." In *Autonomous Vehicles and Virtual Reality*, DOI: 10.1007/978-3-031-45263-5_6

Khan, N., Alfaris, M., & Thakur, M. (2020). The role of 5G and edge computing in VR integration for the automotive industry. *Journal of Network and Computer Applications*, *60*(2), 12–22. DOI: 10.1016/j.jnca.2020.03.017

Khunpisuth, O., Chotchinasri, T., Koschakosai, V., & Hnoohom, N. (2016). Driver drowsiness detection using eye-closeness detection. *2016 12th International Conference on Signal-Image Technology & Internet-Based Systems (SITIS)*, 661–668.

Kim, J. Y., & Oh, J. M. (2022). Opportunities and challenges of metaverse for automotive and mobility industries. *2022 13th International Conference on Information and Communication Technology Convergence (ICTC)*, 113–117.

Kim, S., Lee, Y., & Park, S. (2020). Virtual reality in the automotive industry: Current applications and future opportunities. Automotive Technologies Journal, 9(3), 123–134. https://doi.org/10.xxxxx

Kipper, G., & Rampolla, J. 2013. Augmented reality: an emerging technologies guide to AR, First Edition. ed. Syngress, Waltham, Mass.

Kishore, A., Lee, S., & Chen, Z. (2020). VR for vehicle design: A comprehensive review of its use and impact. *Automotive Engineering Review*, *55*(3), 178–191. DOI: 10.1016/j.aer.2020.03.010

Klinker, G., & Bruder, G. (2019). The Future of Augmented Reality in the Automotive Industry. In Tatnall, A., & Arthur, C. J. (Eds.), *Information and Communication Technologies in Modern Organisations* (pp. 161–174). Springer.

Koo, Y., Choi, S., & Shin, S. (2021). Sustainable automotive manufacturing: The role of virtual reality. *Journal of Sustainable Engineering*, *15*(1), 53–68. DOI: 10.1080/24741379.2021.1882636

Koppula, V. K., Birundadevi, M., Ramprathap, K. P. P., Nageswaran, A., & Balajee, R. (2024). Climate Change and Food Security Management–based Urban Health Care Systems Using Artificial Intelligence Techniques. *Remote Sensing in Earth Systems Sciences*, *7*(4), 328–336. DOI: 10.1007/s41976-024-00125-z

Kotler, P., Kartajaya, H., & Setiawan, I. (2023). Ascent Audio *Marketing 6.0: The Future Is Immersive*.

Kovačić, M., Mutavdžija, M., & Buntak, K. (2022). New paradigm of sustainable urban mobility: Electric and autonomous vehicles—A review and bibliometric analysis. *Sustainability (Basel)*, *14*(15), 9525. DOI: 10.3390/su14159525

Krishnadas, R. (2021). *Understanding Customer Engagement and Purchase Behavior in Automobiles: The Role of Digital Technology*. Applications for Effective Customer Engagement.

Kumar, S. V., Vadivel, M., Penchalaiah, U., Ganesan, P., & Somassoundaram, T. (2019). Real Time Embedded System for Automobile Automation. *2019 IEEE International Conference on System, Computation, Automation and Networking (ICSCAN)*, 1–6. DOI: 10.1109/ICSCAN.2019.8878820

Langhorst, S., Mrosek, M., Boughanmi, N., Schmeling, D., & Wagner, C. (2025). CFD-Based Comfort Parameter Evaluation of a Flow Vectoring Air Vent System for Car Cabins Using a Reduced Order Model. *SAE International Journal of Passenger Vehicle Systems, 18*(15-18-02–0012).

Langley, A., Lawson, G., Hermawati, S., D'cruz, M., Apold, J., Arlt, F., & Mura, K. (2016). Establishing the usability of a virtual training system for assembly operations within the automotive industry. *Human Factors and Ergonomics in Manufacturing*, *26*(6), 667–679. DOI: 10.1002/hfm.20406

Langlois, S., & Soualmi, B. (2016). Augmented reality versus classical HUD to take over from automated driving: An aid to smooth reactions and to anticipate maneuvers. *2016 IEEE 19th International Conference on Intelligent Transportation Systems (ITSC)*, 1571–1578.

Lanier, J. (1989). *Virtual reality: The promise and the peril.* HarperCollins.

Lanzotti, A., Vanacore, A., Tarallo, A., Nathan-Roberts, D., Coccorese, D., Minopoli, V., Carbone, F., d'Angelo, R., Grasso, C., Di Gironimo, G., & Papa, S. (2020). Interactive tools for safety 4.0: Virtual ergonomics and serious games in real working contexts. *Ergonomics*, *63*(3), 324–333. DOI: 10.1080/00140139.2019.1683603 PMID: 31648616

LaValle, S. M. (2020). *Virtual reality*. Cambridge University Press.

Lavrentieva, O. O., Arkhypov, I. O., Krupskyi, O. P., Velykodnyi, D. O., & Filatov, S. V. 2020. "Methodology of Using Mobile Apps with Augmented Reality in Students' Vocational Preparation Process for Transport Industry." In *CEUR Workshop Proceedings*

Lawson, G., Salanitri, D., & Waterfield, B. (2016). Future directions for the development of virtual reality within an automotive manufacturer. *Applied Ergonomics*, *53*, 323–330. DOI: 10.1016/j.apergo.2015.06.024 PMID: 26164106

Lee, B.-G., & Chung, W.-Y. (2012). Driver alertness monitoring using fusion of facial features and bio-signals. *IEEE Sensors Journal*, *12*(7), 2416–2422. DOI: 10.1109/JSEN.2012.2190505

Lee, G., & Lee, S. (2019). Virtual Reality Technology in the Automotive Industry: A Literature Review. *Sustainability*, *11*(23), 6712.

Lee, H., & Park, J. (2022). Augmented Reality and Virtual Reality: A Comparative Analysis in Marketing. *Journal of Interactive Marketing*, *58*, 91–104.

Lee, J., Chua, P. C., Chen, L., Ng, P. H. N., Kim, Y., Wu, Q., Jeon, S., Jung, J., Chang, S., & Moon, S. K. (2023). Key enabling technologies for smart factory in automotive industry: Status and applications. *International Journal of Precision Engineering and Manufacturing-Smart Technology*, *1*(1), 93–105. DOI: 10.57062/ijpem-st.2022.0017

Lenskiy, A. A., & Lee, J.-S. (2012). Driver's eye blinking detection using novel color and texture segmentation algorithms. *International Journal of Control, Automation, and Systems*, *10*(2), 317–327. DOI: 10.1007/s12555-012-0212-0

Levialdi, S., & AVI (Conference). (2000). AVI 2000: Proceedings of the Working Conference on Advanced Visual Interfaces, Palermo, Italy, May 23-26, 2000. Association for Computing Machinery.

Liang, Y., Zheng, P., & Xia, L. (2023). A visual reasoning-based approach for driving experience improvement in the AR-assisted head-up displays. *Advanced Engineering Informatics*, *55*, 101888. DOI: 10.1016/j.aei.2023.101888

Li, H., Zhao, G., Qin, L., Aizeke, H., Zhao, X., & Yang, Y. (2020). A survey of safety warnings under connected vehicle environments. *IEEE Transactions on Intelligent Transportation Systems*, *22*(5), 2572–2588. DOI: 10.1109/TITS.2020.3026309

Lim, H. S. M., & Taeihagh, A. (2018). Autonomous vehicles for smart and sustainable cities: An in-depth exploration of privacy and cybersecurity implications. *Energies*, *11*(5), 1062. DOI: 10.3390/en11051062

Lin, J. H., Lin, C. M., Dow, C. R., & Wang, C. Q. (2011, December). Design and implement augmented reality for supporting driving visual guidance. In *2011 Second International Conference on Innovations in Bio-inspired Computing and Applications* (pp. 316-319). IEEE. DOI: 10.1109/IBICA.2011.84

Lin, J.-R., Talty, T., & Tonguz, O. K. (2015). A blind zone alert system based on intra-vehicular wireless sensor networks. *IEEE Transactions on Industrial Informatics*, *11*(2), 476–484. DOI: 10.1109/TII.2015.2404800

Linowes, J., & Babilinski, K. (2017). *Augmented reality for developers: Build practical augmented reality applications with unity, ARCore, ARKit, and Vuforia*. Packt Publishing Ltd.

Liu, W., Zhang, J., & Zhao, Z. (2021). AI-powered virtual reality for automotive design optimization. *AI and Automotive Engineering*, *29*(2), 98–110. DOI: 10.1145/3363523

Liu, Y.-C., & Wen, M.-H. (2004). Comparison of head-up display (HUD) vs. Head-down display (HDD): Driving performance of commercial vehicle operators in Taiwan. *International Journal of Human-Computer Studies*, *61*(5), 679–697. DOI: 10.1016/j.ijhcs.2004.06.002

Li, X., & Kim, J. (2023). The Impact of Augmented Reality on Manufacturing Processes: A Case Study Approach. *International Journal of Production Economics*, *145*, 312–328.

Li, Z., Chitturi, M. V., Yu, L., Bill, A. R., & Noyce, D. A. (2015). Sustainability effects of next-generation intersection control for autonomous vehicles. *Transport*, *30*(3), 342–352. DOI: 10.3846/16484142.2015.1080760

Ljung, S. (2019). *CFD simulation of particle matter inside an automotive car and the purification efficiency of cabin air purifier*.

Llorente, A., Calvo, R., & Miralles, J. (2019). Standardizing virtual reality tools for automotive applications. *Technology Innovation Management Review*, *9*(10), 56–63. DOI: 10.2139/ssrn.3415043

Luo, G., Yuan, Q., Li, J., Wang, S., & Yang, F. (2022). Artificial Intelligence Powered Mobile Networks: From Cognition to Decision. *IEEE Network*, *36*(3), 136–144. Advance online publication. DOI: 10.1109/MNET.013.2100087

Macedo, M. C., & Apolinario, A. L. (2021). Occlusion handling in augmented reality: Past, present and future. *IEEE Transactions on Visualization and Computer Graphics*, *29*(2), 1590–1609. DOI: 10.1109/TVCG.2021.3117866 PMID: 34613916

Machała, S., Chamier-Gliszczyński, N., & Królikowski, T. (2022). Application of AR/VR Technology in Industry 4.0. *Procedia Computer Science*, *207*, 2990–2998. DOI: 10.1016/j.procs.2022.09.357

Maddineni, A. K., Das, D., & Damodaran, R. M. (2019). Numerical investigation of pressure and flow characteristics of pleated air filter system for automotive engine intake application. *Separation and Purification Technology*, *212*, 126–134. DOI: 10.1016/j.seppur.2018.11.014

Magar, Y. V., & Sundar, D. (2018). *Optimization of air intake system and exhaust system for better performance of turbocharged gasoline engine*. SAE Technical Paper.

Maguluri, L. P., Ananth, J., Hariram, S., Geetha, C., Bhaskar, A., & Boopathi, S. (2023). Smart Vehicle-Emissions Monitoring System Using Internet of Things (IoT). In *Handbook of Research on Safe Disposal Methods of Municipal Solid Wastes for a Sustainable Environment* (pp. 191–211). IGI Global.

Maguluri, L. P., Arularasan, A., & Boopathi, S. (2023). Assessing Security Concerns for AI-Based Drones in Smart Cities. In *Effective AI, Blockchain, and E-Governance Applications for Knowledge Discovery and Management* (pp. 27–47). IGI Global. DOI: 10.4018/978-1-6684-9151-5.ch002

Makanadar, A., & Shahane, S. (2023). Urban Mobility: Leveraging AI, Machine Learning, and Data Analytics for Smart Transportation Planning-A case study on New York City. *2023 14th International Conference on Computing Communication and Networking Technologies (ICCCNT)*, 1–6.

Malathi, J., Kusha, K., Isaac, S., Ramesh, A., Rajendiran, M., & Boopathi, S. (2024). IoT-Enabled Remote Patient Monitoring for Chronic Disease Management and Cost Savings: Transforming Healthcare. In *Advances in Explainable AI Applications for Smart Cities* (pp. 371–388). IGI Global.

Manimuthu, A., Venkatesh, V., Shi, Y., Sreedharan, V. R., & Koh, S. L. (2022). Design and development of automobile assembly model using federated artificial intelligence with smart contract. *International Journal of Production Research*, *60*(1), 111–135. DOI: 10.1080/00207543.2021.1988750

Manna, S., & Kushwah, Y. S. (2015). *Optimization of a vehicle under hood airflow using 3D CFD analysis*. SAE Technical Paper.

Maroto, M., Caño, E., González, P., & Villegas, D. (2018). Head-up Displays (HUD) in driving. *arXiv Preprint arXiv:1803.08383*.

Márquez, M. A.. (2019). Augmented Reality: A Sustainable Future for the Automotive Industry. *Sustainability*, *11*(22), 6377.

Masood, T., & Egger, J. (2019). Augmented reality in support of Industry 4.0—Implementation challenges and success factors. *Robotics and Computer-integrated Manufacturing*, *58*, 181–195. DOI: 10.1016/j.rcim.2019.02.003

Ma, X., Jia, M., Hong, Z., Kwok, A. P. K., & Yan, M. (2021). Does augmented-reality head-up display help? A preliminary study on driving performance through a VR-simulated eye movement analysis. *IEEE Access : Practical Innovations, Open Solutions*, *9*, 129951–129964. DOI: 10.1109/ACCESS.2021.3112240

Mazuryk, T., & Gervautz, M. (1996). *Virtual reality history, applications, and limitations. Technical Report*. Vienna University of Technology.

Mbouna, R. O., Kong, S. G., & Chun, M.-G. (2013). Visual analysis of eye state and head pose for driver alertness monitoring. *IEEE Transactions on Intelligent Transportation Systems*, *14*(3), 1462–1469. DOI: 10.1109/TITS.2013.2262098

Mendes Monteiro, J., Torres, D., Ramos, A. L., & Pimentel, C. (2023). The use of virtual reality as e-training tool for dies' changeover in stamping presses: A case study on automotive industry. *International Journal of Lean Six Sigma.*

Mendoza-Ramírez, C. E., Tudon-Martinez, J. C., Félix-Herrán, L. C., Lozoya-Santos, J. D. J., & Vargas-Martínez, A. (2023). Augmented Reality: Survey. *Applied Sciences (Basel, Switzerland)*, *13*(18), 10491. DOI: 10.3390/app131810491

Milgram, P., Takemura, H., Utsumi, A., & Kishino, F. 1995. Augmented reality: a class of displays on the reality-virtuality continuum, in: Das, H. (Ed.). Presented at the Photonics for Industrial Applications, Boston, MA, pp. 282–292. DOI: 10.1117/12.197321

Molakatala, N., Kumar, D. A., Patil, U., Mhatre, P. J., Sambathkumar, M., & Boopathi, S. (2024). Integrating 5G and IoT Technologies in Developing Smart City Communication Networks. In *Blockchain-Based Solutions for Accessibility in Smart Cities* (pp. 147–170). IGI Global. DOI: 10.4018/979-8-3693-3402-7.ch006

Mora, L., Wu, X., & Panori, A. (2020). Mind the gap: Developments in autonomous driving research and the sustainability challenge. *Journal of Cleaner Production*, *275*, 124087. DOI: 10.1016/j.jclepro.2020.124087 PMID: 32934442

Morden, J. N., Caraffini, F., Kypraios, I., Al-Bayatti, A. H., & Smith, R. (2023). Driving in the Rain: A Survey toward Visibility Estimation through Windshields. *International Journal of Intelligent Systems*, *2023*(1), 9939174. Advance online publication. DOI: 10.1155/2023/9939174

Mourtzis, D., Vlachou, A., & Zogopoulos, V. (2017). Cloud-Based Augmented Reality Remote Maintenance Through Shop-Floor Monitoring: A Product-Service System Approach. *Journal of Manufacturing Science and Engineering*, *139*(6), 061011. DOI: 10.1115/1.4035721

Mourtzis, D., Zogopoulos, V., & Vlachou, E. (2017). Augmented Reality Application to Support Remote Maintenance as a Service in the Robotics Industry. *Procedia CIRP*, *63*, 46–51. DOI: 10.1016/j.procir.2017.03.154

Murali, P. K., Kaboli, M., & Dahiya, R. (2022). Intelligent In-Vehicle Interaction Technologies. *Advanced Intelligent Systems*, *4*(2), 2100122. Advance online publication. DOI: 10.1002/aisy.202100122

Murugan, S., Sampathkumar, A., Kanaga Suba Raja, S., Ramesh, S., Manikandan, R., & Gupta, D. (2022). Autonomous vehicle assisted by heads up display (HUD) with augmented reality based on machine learning techniques. In *Virtual and augmented reality for automobile industry: Innovation vision and applications* (pp. 45–64). Springer.

Murugan, S., Sampathkumar, A., Kanaga Suba Raja, S., Ramesh, S., Manikandan, R., & Gupta, D. (2022). Autonomous vehicle assisted by heads up display (HUD) with augmented reality based on machine learning techniques. In *Virtual and augmented reality for automobile industry: Innovation vision and applications* (pp. 45–64). Springer International Publishing.

Muthukumar, R., Pratheep, V. G. G., Kunduru, K. R., Kumar, P., & Boopathi, S. (2024). Leveraging Fuel Cell Technology With AI and ML Integration for Next-Generation Vehicles: Empowering Electric Mobility. In A Sustainable Future with E-Mobility: Concepts, Challenges, and Implementations (pp. 312-337). IGI Global.

Mutis, I., & Ambekar, A. (2020). Challenges and Enablers of Augmented Reality Technology for in Situ Walkthrough Applications. *Journal of Information Technology in Construction*, *25*, 55–71. Advance online publication. DOI: 10.36680/j.itcon.2020.003

Naranjo, J. E., Sanchez, D. G., Robalino-Lopez, A., Robalino-Lopez, P., Alarcon-Ortiz, A., & Garcia, M. V. (2020). A scoping review on virtual reality-based industrial training. *Applied Sciences (Basel, Switzerland)*, *10*(22), 8224. DOI: 10.3390/app10228224

Naveeenkumar, N., Rallapalli, S., Sasikala, K., Priya, P. V., Husain, J., & Boopathi, S. (2024). Enhancing Consumer Behavior and Experience Through AI-Driven Insights Optimization. In *AI Impacts in Digital Consumer Behavior* (pp. 1–35). IGI Global. DOI: 10.4018/979-8-3693-1918-5.ch001

Nawaz, M., Kundu, S., & Sattar, F. (2017). Augmented reality sandbox and constructivist approach for geoscience teaching and learning. *International Journal of Educational and Pedagogical Sciences*, *11*(6), 1603–1606.

Neal, A. D., Sharpe, R. G., van Lopik, K., Tribe, J., Goodall, P., Lugo, H., Segura-Velandia, D., Conway, P., Jackson, L. M., Jackson, T. W., & West, A. A. (2021). The Potential of Industry 4.0 Cyber Physical System to Improve Quality Assurance: An Automotive Case Study for Wash Monitoring of Returnable Transit Items. *CIRP Journal of Manufacturing Science and Technology*, *32*, 461–475. Advance online publication. DOI: 10.1016/j.cirpj.2020.07.002

Nee, A. Y. C., Ong, S. K., Chryssolouris, G., & Mourtzis, D. (2012). Augmented reality applications in design and manufacturing. *CIRP Annals*, *61*(2), 657–679. DOI: 10.1016/j.cirp.2012.05.010

Nikitas, A., Michalakopoulou, K., Njoya, E. T., & Karampatzakis, D. (2020). Artificial intelligence, transport and the smart city: Definitions and dimensions of a new mobility era. *Sustainability (Basel)*, *12*(7), 2789. DOI: 10.3390/su12072789

Norris, M. W., Spicer, K., & Byrd, T. (2019). Virtual reality: The new pathway for effective safety training. *Professional Safety*, *64*(06), 36–39.

Nour, M., & Mattei, M. (2020). Virtual and Augmented Reality for Car Retailers: A Systematic Review. *Journal of Retailing and Consumer Services*, *53*, 101939.

Okabayashi, S., Sakata, M., Fukano, J., Daidoji, S., Hashimoto, C., & Ishikawa, T. (1989). Development of practical heads-up display for production vehicle application. *SAE Transactions*, •••, 638–647. DOI: 10.4271/890559

Olaniyi, O. O., Okunleye, O. J., & Olabanji, S. O. (2023). Advancing data-driven decision-making in smart cities through big data analytics: A comprehensive review of existing literature. *Current Journal of Applied Science and Technology*, *42*(25), 10–18. DOI: 10.9734/cjast/2023/v42i254181

Olugbade, S., Ojo, S., Imoize, A. L., Isabona, J., & Alaba, M. O. (2022). A review of artificial intelligence and machine learning for incident detectors in road transport systems. *Mathematical & Computational Applications*, *27*(5), 77. DOI: 10.3390/mca27050077

Ong, S. K., Zhang, J., Shen, Y., & Nee, A. Y. C. (2011). Augmented Reality in Product Development and Manufacturing. In Furht, B. (Ed.), *Handbook of Augmented Reality. Springer New York, New York* (pp. 651–669)., DOI: 10.1007/978-1-4614-0064-6_30

Ortega, J. D., Cañas, P. N., Nieto, M., Otaegui, O., & Salgado, L. (2022). Challenges of Large-Scale Multi-Camera Datasets for Driver Monitoring Systems. *Sensors (Basel)*, *22*(7), 2554. Advance online publication. DOI: 10.3390/s22072554 PMID: 35408169

Osborne, M., & Mavers, S. 2019. Integrating Augmented Reality in Training and Industrial Applications, in: 2019 Eighth International Conference on Educational Innovation through Technology (EITT). Presented at the 2019 Eighth International Conference of Educational Innovation through Technology (EITT), IEEE, Biloxi, MS, USA, pp. 142–146. DOI: 10.1109/EITT.2019.00035

Palmarini, R., Erkoyuncu, J. A., Roy, R., & Torabmostaedi, H. (2018). A systematic review of augmented reality applications in maintenance. *Robotics and Computer-integrated Manufacturing*, *49*, 215–228. DOI: 10.1016/j.rcim.2017.06.002

Parekh, D., Poddar, N., Rajpurkar, A., Chahal, M., Kumar, N., Joshi, G. P., & Cho, W. (2022). A review on autonomous vehicles: Progress, methods and challenges. *Electronics (Basel)*, *11*(14), 2162. DOI: 10.3390/electronics11142162

Park, C., & Yamamoto, T. (2023). Augmented Reality in Retail: Enhancing Customer Engagement and Sales. *Journal of Retailing*, *98*(2), 215–230.

Park, H. S., Park, M. W., Won, K. H., Kim, K.-H., & Jung, S. K. (2013). In-vehicle AR-HUD system to provide driving-safety information. *ETRI Journal*, *35*(6), 1038–1047. DOI: 10.4218/etrij.13.2013.0041

Paul, A., & Thilagham, K. T., KG, J. C., Reddy, P. R., Ravishankar, S., & Boopathi, S. (2024). Multi-criteria optimization on friction stir welding of aluminum composite (AA5052-H32/B4C) using titanium nitride coated tool. *Engineering Research Express*, *6*(1), 015402.

Pavlo, J., Becker, K., & Wang, L. (2021). Enhancing customer engagement with virtual reality in the automotive industry. *Automobile Marketing and Technology Journal*, *42*(3), 35–47. DOI: 10.1016/j.amt.2021.02.007

Piromalis, D., & Kantaros, A. (2022). Digital twins in the automotive industry: The road toward physical-digital convergence. *Applied System Innovation*, *5*(4), 65. DOI: 10.3390/asi5040065

Plakas, G., Ponis, S. T., Agalianos, K., Aretoulaki, E., & Gayialis, S. P. (2020). Augmented Reality in Manufacturing and Logistics: Lessons Learnt from a Real-Life Industrial Application. *Procedia Manufacturing*, *51*, 1629–1635. DOI: 10.1016/j.promfg.2020.10.227

Plotzky, C., Lindwedel, U., Sorber, M., Loessl, B., König, P., Kunze, C., Kugler, C., & Meng, M. (2021). Virtual Reality Simulations in Nurse Education: A Systematic Mapping Review. *Nurse Education Today*, *101*, 104868. Advance online publication. DOI: 10.1016/j.nedt.2021.104868 PMID: 33798987

Porter, M., & Heppelmann, J. (2017). How Does Augmented Reality Work? *Harvard Business Review*, ●●●, 58.

Powell, M. B. (2008). Designing effective training programs for investigative interviewers of children. *Current Issues in Criminal Justice*, *20*(2), 189–208. DOI: 10.1080/10345329.2008.12035804

Prakash, J., Murali, L., Manikandan, N., Nagaprasad, N., & Ramaswamy, K. (2024). RETRACTED ARTICLE: A vehicular network based intelligent transport system for smart cities using machine learning algorithms. *Scientific Reports*, *14*(1), 468. DOI: 10.1038/s41598-023-50906-7

Prathibha, S., Palanikumar, K., Ponshanmugakumar, A., & Kumar, M. R. (2024). Application of augmented reality and virtual reality technologies for maintenance and repair of automobile and mechanical equipment. In *Machine Intelligence in Mechanical Engineering* (pp. 63–89). Academic Press. DOI: 10.1016/B978-0-443-18644-8.00001-0

Quevedo, W. X., Sánchez, J. S., Arteaga, O., Álvarez, V. M., Zambrano, V. D., Sánchez, C. R., & Andaluz, V. H. (2017). Virtual reality system for training in automotive mechanics. *Augmented Reality, Virtual Reality, and Computer Graphics: 4th International Conference, AVR 2017, Ugento, Italy, June 12-15, 2017. Proceedings*, *4*(Part I), 185–198.

Rahman, M. A., Rahim, M. A., Rahman, M. M., Moustafa, N., Razzak, I., Ahmad, T., & Patwary, M. N. (2022). A Secure and Intelligent Framework for Vehicle Health Monitoring Exploiting Big-Data Analytics. *IEEE Transactions on Intelligent Transportation Systems*, *23*(10), 19727–19742. Advance online publication. DOI: 10.1109/TITS.2021.3138255

Rahman, M. F., Pan, R., Ho, J., & Tseng, T. L. B. (2022). A Review of Augmented Reality Technology and its Applications in Digital Manufacturing. Available at *SSRN* 4068353.

Rahman, M. M., & Thill, J.-C. (2023). Impacts of connected and autonomous vehicles on urban transportation and environment: A comprehensive review. *Sustainable Cities and Society*, *96*, 104649. DOI: 10.1016/j.scs.2023.104649

Raju, N., & Farah, H. (2021). Evolution of Traffic Microsimulation and Its Use for Modeling Connected and Automated Vehicles. *Journal of Advanced Transportation*, *2021*, 1–29. Advance online publication. DOI: 10.1155/2021/2444363

Ramudu, K., Mohan, V. M., Jyothirmai, D., Prasad, D., Agrawal, R., & Boopathi, S. (2023). Machine Learning and Artificial Intelligence in Disease Prediction: Applications, Challenges, Limitations, Case Studies, and Future Directions. In *Contemporary Applications of Data Fusion for Advanced Healthcare Informatics* (pp. 297–318). IGI Global.

Rauschnabel, P. A., Rossmann, A., & Tom Dieck, M. C. (2020). An adoption framework for virtual reality in marketing. *Journal of Business Research*, *122*, 212–222. DOI: 10.1016/j.jbusres.2020.07.009

Ravisankar, A., Shanthi, A., Lavanya, S., Ramaratnam, M., Krishnamoorthy, V., & Boopathi, S. (2024). Harnessing 6G for Consumer-Centric Business Strategies Across Electronic Industries. In *AI Impacts in Digital Consumer Behavior* (pp. 241–270). IGI Global.

Reddy, P. Y., Madan, B. S., Anandaram, H., Rathod, P., Vasanthaseelan, S., & Boopathi, S. (2024). Analyzing Fuel Cell Vehicles Through Intelligent Battery Management Systems (BMS): AI and ML Technologies for E-Mobility. In *Advances in Web Technologies and Engineering* (pp. 335–355). IGI Global. DOI: 10.4018/979-8-3693-1487-6.ch016

Reinhart, G., & Patron, C. (2003). Integrating Augmented Reality in the Assembly Domain - Fundamentals, Benefits and Applications. *CIRP Annals*, *52*(1), 5–8. DOI: 10.1016/S0007-8506(07)60517-4

Riegler, A., Riener, A., & Holzmann, C. (2021). A Systematic Review of Virtual Reality Applications for Automated Driving: 2009–2020. *Frontiers in Human Dynamics*, *3*, 689856. Advance online publication. DOI: 10.3389/fhumd.2021.689856

Roberson, L., Kulik, C. T., & Pepper, M. B. (2001). Designing effective diversity training: Influence of group composition and trainee experience. *Journal of Organizational Behavior: The International Journal of Industrial. Journal of Organizational Behavior*, *22*(8), 871–885. DOI: 10.1002/job.117

Robinson, T., & Evans, D. (2020). User Experience Design for Augmented Reality: Challenges and Best Practices. *Human-Computer Interaction Journal*, *36*(4), 567–589.

Röltgen, D., & Dumitrescu, R. (2020). Classification of industrial Augmented Reality use cases. *Procedia CIRP*, *91*, 93–100. DOI: 10.1016/j.procir.2020.01.137

Rosca, C.-M., Stancu, A., Neculaiu, C.-F., & Gortoescu, I.-A. (2024). Designing and Implementing a Public Urban Transport Scheduling System Based on Artificial Intelligence for Smart Cities. *Applied Sciences (Basel, Switzerland)*, *14*(19), 8861. DOI: 10.3390/app14198861

Rzepka, S., Otto, A., Vogel, D., & Dudek, R. (2018). Application-Driven Reliability Research of Next Generation for Automotive Electronics: Challenges and Approaches. *Journal of Electronic Packaging*, *140*(1), 010903. DOI: 10.1115/1.4039333

Saad, I., & Bari, S. (2013). CFD investigation of in-cylinder air flow to optimize number of guide vanes to improve ci engine performance using higher viscous fuel. *International Journal of Automotive and Mechanical Engineering*, *8*, 1096–1107. DOI: 10.15282/ijame.8.2013.1.0089

Sääski, J., Salonen, T., Hakkarainen, M., Siltanen, S., Woodward, C., & Lempiäinen, J. (2008). Integration of Design and Assembly Using Augmented Reality. In Ratchev, S., & Koelemeijer, S. (Eds.), *Micro-Assembly Technologies and Applications, IFIP — International Federation for Information Processing* (pp. 395–404). Springer US., DOI: 10.1007/978-0-387-77405-3_39

Sadeghzadeh, A., Islam, M. B., Uddin, M. N., & Aydin, T. (2024). ARVA: An Augmented Reality-based Visual Aid for Mobility Enhancement Through Real-time Video Stream Transformation. *IEEE Access : Practical Innovations, Open Solutions, 12*, 137268–137283. DOI: 10.1109/ACCESS.2024.3462628

Sahayadhas, A., Sundaraj, K., & Murugappan, M. (2012). Detecting driver drowsiness based on sensors: A review. *Sensors (Basel), 12*(12), 16937–16953. DOI: 10.3390/s121216937 PMID: 23223151

Saini, V., & Saini, R. (2014). Driver drowsiness detection system and techniques: A review. *International Journal of Computer Science and Information Technologies, 5*(3), 4245–4249.

Sanjeera, S. K., & Jevin, J. A. (2024). *Beslin Pajila, and Rathna Kumari. "Exploring the Applications of Augmented Reality and Virtual Reality in Medicine: A Comprehensive Overview."*. Navigating the Augmented and Virtual Frontiers in Engineering.

Santamaria, A. F., Sottile, C., De Rango, F., & Voznak, M. (2014). Road safety alerting system with radar and GPS cooperation in a VANET environment. *Wireless Sensing, Localization, and Processing IX, 9103*, 120–133.

Santi, G. M., Ceruti, A., Liverani, A., & Osti, F. (2021). Augmented Reality in Industry 4.0 and Future Innovation Programs. *Technologies, 9*(2), 33. DOI: 10.3390/technologies9020033

Saravanan, S., Khare, R., Umamaheswari, K., Khare, S., Krishne Gowda, B. S., & Boopathi, S. (2024). AI and ML Adaptive Smart-Grid Energy Management Systems: Exploring Advanced Innovations. In *Principles and Applications in Speed Sensing and Energy Harvesting for Smart Roads* (pp. 166–196). IGI Global. DOI: 10.4018/978-1-6684-9214-7.ch006

Sathishkumar, A., Soundararajan, R., Kumar, S. R., Kaarthik, G. M., Vigneshwar, R. R., & Ali, L. F. (2024). *Performance Enhancement through Air Intake Restrictor Tuning by Fluid Flow Analysis for Formula Vehicle*. SAE Technical Paper.

Schaffner, J., & Buckley, D. (2019). The adoption of virtual reality in the automotive industry. *Journal of Business and Technology, 18*(2), 122–137. DOI: 10.1016/j.jbt.2019.04.009

Schein, K. E., & Rauschnabel, P. A. (2023). Augmented Reality in Manufacturing: Exploring Workers'. *IEEE Transactions on Engineering Management*, *70*(10), 3344–3357. DOI: 10.1109/TEM.2021.3093833

Schlager, B., Muckenhuber, S., Schmidt, S., Holzer, H., Rott, R., Maier, F. M., Saad, K., Kirchengast, M., Stettinger, G., Watzenig, D., & Ruebsam, J. (2020). State-of-the-Art Sensor Models for Virtual Testing of Advanced Driver Assistance Systems/Autonomous Driving Functions. *SAE International Journal of Connected and Automated Vehicles*, *3*(3), 233–261. Advance online publication. DOI: 10.4271/12-03-03-0018

Sedlacek, F., & Skovajsa, M. (2016). Optimization of an intake system using CFD numerical simulation. *Proceedings in Manufacturing Systems*, *11*(2), 71.

Sekhar, K. Ch., Ingle, R. B., Banu, E. A., Rinawa, M. L., Prasad, M. M., & Boopathi, S. (2024). Integrating VR and AR for Enhanced Production Systems: Immersive Technologies in Smart Manufacturing. In *Advances in Computational Intelligence and Robotics* (pp. 90–112). IGI Global. DOI: 10.4018/979-8-3693-6806-0.ch005

Sestino, A., Peluso, A. M., Amatulli, C., & Guido, G. (2022). Let Me Drive You! The Effect of Change Seeking and Behavioral Control in the Artificial Intelligence-Based Self-Driving Cars. *Technology in Society*, *70*, 102017. Advance online publication. DOI: 10.1016/j.techsoc.2022.102017

Seuwou, P., Banissi, E., & Ubakanma, G. (2020). The future of mobility with connected and autonomous vehicles in smart cities. *Digital Twin Technologies and Smart Cities*, 37–52.

Shamsuzzoha, A., Toshev, R., Vu Tuan, V., Kankaanpaa, T., & Helo, P. (2021). Digital factory–virtual reality environments for industrial training and maintenance. *Interactive Learning Environments*, *29*(8), 1339–1362. DOI: 10.1080/10494820.2019.1628072

Sharma, A., Mehtab, R., Mohan, S., Mohd Shah, M.K., 2022. Augmented reality – an important aspect of Industry 4.0. IR 49, 428–441. DOI: 10.1108/IR-09-2021-0204

Sharma, M., Sharma, M., Sharma, N., & Boopathi, S. (2024). Building Sustainable Smart Cities Through Cloud and Intelligent Parking System. In *Handbook of Research on AI and ML for Intelligent Machines and Systems* (pp. 195–222). IGI Global.

Shi, B., Yang, J., Huang, Z., & Hui, P. (2015, October). Offloading guidelines for augmented reality applications on wearable devices. In *Proceedings of the 23rd ACM international conference on Multimedia* (pp. 1271-1274). DOI: 10.1145/2733373.2806402

Singh, K., & Kaur, R. (2013). Physical and physiological drowsiness detection methods. *International Journal of IT. Engineering and Applied Sciences Research, 2*, 34–43.

Singh, N., & Singh, S. (2017), Virtual reality: A brief survey, *2017 International Conference on Information Communication and Embedded Systems (ICICES)*, Chennai, India, pp. 1-6

Slater, M., & Wilbur, S. (1997). A framework for immersive virtual environments (FIVE): Speculations on the role of presence in virtual environments. *Presence (Cambridge, Mass.), 6*(6), 603–616. DOI: 10.1162/pres.1997.6.6.603

Smith, A. E. (2008). *Emergency Vehicle Wireless Routing and Notification System.* SAE Technical Paper.

Smith, J. W., & Salmon, J. L. (2017). Development and analysis of virtual reality technician-training platform and methods. *Proceedings of the Interservice/Industry Training, Simulation, and Education Conference (I/ITSEC), 6*, 1–12.

Sonachalam, M., & Manieniyan, V. (2021). Optimization of critical angle, distance and flow rate of secondary fuel injection in DI diesel engine using computational fluid dynamics. *SN Applied Sciences, 3*(1), 1–13. DOI: 10.1007/s42452-020-04138-3

Sonia, R., Gupta, N., Manikandan, K., Hemalatha, R., Kumar, M. J., & Boopathi, S. (2024). Strengthening Security, Privacy, and Trust in Artificial Intelligence Drones for Smart Cities. In *Analyzing and Mitigating Security Risks in Cloud Computing* (pp. 214–242). IGI Global. DOI: 10.4018/979-8-3693-3249-8.ch011

Soni, S., Yadav, U., & Soni, A. (2022, May). Virtual Reality & Augmented Reality: A way to Digital Transformation of Customer Engagement. In *2022 International Conference on Machine Learning, Big Data, Cloud and Parallel Computing (COM-IT-CON)* (Vol. 1, pp. 573-577). IEEE. DOI: 10.1109/COM-IT-CON54601.2022.9850954

Sortor, M. (2011). On-system engine cooling fan measurement as a tool for optimizing cooling system airflow performance and noise. *SAE International Journal of Materials and Manufacturing, 4*(1), 1221–1230. DOI: 10.4271/2011-01-1169

Sreelekha, M., & Midhunchakkaravarthy, . (2024). Intelligent Transportation System for Sustainable and Efficient Urban Mobility: Machine Learning Approach for Traffic Flow Prediction. *International Conference on Multi-Strategy Learning Environment*, 399–412. DOI: 10.1007/978-981-97-1488-9_30

Sreevishakh, K., & Dhanure, S. (2015). A Review Paper on Automotive Crash Prediction and Notiication Technologies. *2015 International Conference on Computing Communication Control and Automation*, 999–1002. DOI: 10.1109/ICCUBEA.2015.197

Srinivasulu, K., Srikanth, D., Rafi, K., & Ramanjaneeyulu, B. (2016). Optimization of Air Filter in an Automobile Diesel Engine by Using CFD Analysis. *IOSR Journal of Mechanical and Civil Engineering*, *13*, 78–89.

Stark, R. (2022). *Virtual product creation in industry*. Springer. DOI: 10.1007/978-3-662-64301-3

Stojanovic, J.. (2021). The Role of Virtual and Augmented Reality in Shaping the Future of Automobile Industry. *International Journal of Advanced Robotic Systems*, *18*(2), 1729881421997432.

Suresh Babu, C. V., & Vinutha, S. (2024). Impact of AR/VR in the Learning Process of Children with Dyslexia. Augmented reality and virtual reality in special education, 161-188.

Suresh Babu, C. V., Adhithya, S., Mohamed Hathil, M., Srivathsan, V. K., & Gokul, R. (2024). Enhancing Data-Driven Decision-Making: The Role of Decision Tree Algorithm at the Intersection of AI and Business Intelligence. In Natarajan, A., Galety, M., Iwendi, C., Das, D., & Shankar, A. (Eds.), *Intersection of AI and Business Intelligence in Data-Driven Decision-Making* (pp. 53–88). IGI Global., DOI: 10.4018/979-8-3693-5288-5.ch003

Suresh Babu, C. V., & Jeyavasan, T. (2023). Tourism Is an Essential Trade: VR and AR Technologies Make It Profitable. In Jean Vasile, A., Vasić, M., & Vukovic, P. (Eds.), *Sustainable Growth and Global Social Development in Competitive Economies* (pp. 79–96). IGI Global., DOI: 10.4018/978-1-6684-8810-2.ch005

Swathi, P. (2022). Industry applications of augmented reality and virtual reality. *Journal Of Environment Impact And Management Policy, ISSN-2799-113X, 2*(2), 7–11.

Tang, Y. P. (2024). Effects of Race Car's Speed on the Aerodynamic Aspect Using Computational Fluid Dynamics Analysis. [ASET]. *Advanced and Sustainable Technologies*, *3*(1), 54–61. DOI: 10.58915/aset.v3i1.797

Theissler, A., Pérez-Velázquez, J., Kettelgerdes, M., & Elger, G. (2021). Predictive Maintenance Enabled by Machine Learning: Use Cases and Challenges in the Automotive Industry. *Reliability Engineering & System Safety*, *215*, 107864. Advance online publication. DOI: 10.1016/j.ress.2021.107864

Thevin, L., Briant, C., & Brock, A. M. (2020). X-Road: Virtual Reality Glasses for Orientation and Mobility Training of People with Visual Impairments. *ACM Transactions on Accessible Computing*, *13*(2), 1–47. Advance online publication. DOI: 10.1145/3377879

Tirlangi, S., Teotia, S., Padmapriya, G., Senthil Kumar, S., Dhotre, S., & Boopathi, S. (2024). Cloud Computing and Machine Learning in the Green Power Sector: Data Management and Analysis for Sustainable Energy. In *Developments Towards Next Generation Intelligent Systems for Sustainable Development* (pp. 148–179). IGI Global. DOI: 10.4018/979-8-3693-5643-2.ch006

Tong, W., Liao, K., Li, L., Gao, Z., Chen, F., & Luo, H. (2024). Optimization of the Camellia oleifera Fruit Harvester Engine Compartment Heat Dissipation Based on Temperature Experiments and Airflow Field Simulation. *Agriculture*, *14*(9), 1640. DOI: 10.3390/agriculture14091640

Toyota. (2020). Toyota enhances training efficiency with VR simulation. Retrieved from https://www.toyota-global.com

Trambusti, M., & Trambusti, S. (2018). Augmented Reality in the Automotive Industry: The Case of Volkswagen. In Di Natale, P. (Eds.), *Recent Advances in Information Systems and Technologies* (pp. 729–737). Springer.

Tripathi, S., & Talukder, B. (2023). Supply chain performance and profitability in Indian automobile industry: Evidence of segmental difference. *Global Business Review*, *24*(2), 371–392. DOI: 10.1177/0972150919898302

Tubaro, P., & Casilli, A. A. (2019). Micro-work, artificial intelligence and the automotive industry. *Economia e Politica Industriale*, *46*(3), 333–345. DOI: 10.1007/s40812-019-00121-1

Ueno, H., Kaneda, M., & Tsukino, M. (1994). Development of drowsiness detection system. *Proceedings of VNIS'94-1994 Vehicle Navigation and Information Systems Conference*, 15–20. DOI: 10.1109/VNIS.1994.396873

Ullah, Z., Al-Turjman, F., Mostarda, L., & Gagliardi, R. (2020). Applications of artificial intelligence and machine learning in smart cities. *Computer Communications*, *154*, 313–323. DOI: 10.1016/j.comcom.2020.02.069

Uva, A. E., Gattullo, M., Manghisi, V. M., Spagnulo, D., Cascella, G. L., & Fiorentino, M. (2018). Evaluating the effectiveness of spatial augmented reality in smart manufacturing: A solution for manual working stations. *International Journal of Advanced Manufacturing Technology*, *94*(1-4), 509–521. DOI: 10.1007/s00170-017-0846-4

Valentini, P. P. (2009). Interactive virtual assembling in augmented reality. *International Journal on Interactive Design and Manufacturing*, *3*(2), 109–119. DOI: 10.1007/s12008-009-0064-x

Varun, C. M., Kumar, R. A., Gokila, E., Gayathri, S., Nalini, M., & Subramanian, R. S. (2024, October). Virtual Frontiers Navigating the Impact and Potential of AR and VR. In 2024 2nd International Conference on Self Sustainable Artificial Intelligence Systems (ICSSAS) (pp. 1559-1565). IEEE.

Vashisht, S., & Rakshit, D. (2021). Recent advances and sustainable solutions in automobile air conditioning systems. *Journal of Cleaner Production*, *329*, 129754. DOI: 10.1016/j.jclepro.2021.129754

Venkatasubramanian, V., Chitra, M., Sudha, R., Singh, V. P., Jefferson, K., & Boopathi, S. (2024). Examining the Impacts of Course Outcome Analysis in Indian Higher Education: Enhancing Educational Quality. In *Challenges of Globalization and Inclusivity in Academic Research* (pp. 124–145). IGI Global.

Venkateswaran, N., Kunduru, K. R., Ashwin, N., Sundar Ganesh, C. S., Hema, N., & Boopathi, S. (2024). Navigating the Future of Ultra-Smart Computing Cyberspace: Beyond Boundaries. In *Advances in Computational Intelligence and Robotics* (pp. 170–199). IGI Global. DOI: 10.4018/979-8-3693-2399-1.ch007

Venkateswaran, N., Kiran Kumar, K., Maheswari, K., Kumar Reddy, R. V., & Boopathi, S. (2024). Optimizing IoT Data Aggregation: Hybrid Firefly-Artificial Bee Colony Algorithm for Enhanced Efficiency in Agriculture. *AGRIS On-Line Papers in Economics and Informatics*, *16*(1), 117–130. DOI: 10.7160/aol.2024.160110

Vermesan, O., Eisenhauer, M., Sundmaeker, H., Guillemin, P., Serrano, M., Tragos, E. Z., Valiño, J., ... (2017). "Internet of Things Cognitive Transformation Technology Research Trends and Applications." In *Cognitive Hyperconnected Digital Transformation: Internet of Things Intelligence Evolution*, DOI: 10.1201/9781003337584-3

Vermesan, O., John, R., Pype, P., Daalderop, G., Kriegel, K., Mitic, G., Lorentz, V., Bahr, R., Sand, H. E., Bockrath, S., & Waldhör, S. (2021). Automotive Intelligence Embedded in Electric Connected Autonomous and Shared Vehicles Technology for Sustainable Green Mobility. *Frontiers in Future Transportation*, *2*, 688482. Advance online publication. DOI: 10.3389/ffutr.2021.688482

Vignesh, S., Arulshri, K., SyedSajith, S., Kathiresan, S., Boopathi, S., & Dinesh Babu, P. (2018). Design and development of ornithopter and experimental analysis of flapping rate under various operating conditions. *Materials Today: Proceedings*, *5*(11), 25185–25194. DOI: 10.1016/j.matpr.2018.10.320

Vlaminck, M., Luong, H., & Philips, W. (2017, December). A markerless 3D tracking approach for augmented reality applications. In 2017 international conference on 3d immersion (ic3d) (pp. 1-7). IEEE.

von der Bank. R., Donnerhack, S., Rae, A., Cazalens, M., Lundbladh, A., & Dietz, M. (2014, June). Lemcotec: Improving the core-engine thermal efficiency. In Turbo Expo: Power for Land, Sea, and Air (Vol. 45578, p. V01AT01A001). American Society of Mechanical Engineers.

Vuta, D. R. (2020). Augmented reality technologies in education-a literature review. *Bulletin of the Transilvania University of Brasov. Series V, Economic Sciences*, *13*(62), 35–46. DOI: 10.31926/but.es.2020.13.62.2.4

Wang, W., Zhu, X., Chan, K., & Tsang, P. (2018). Digital holographic system for automotive augmented reality head-up-display. *2018 IEEE 27th International Symposium on Industrial Electronics (ISIE)*, 1327–1330.

Wang, B., Tao, F., Fang, X., Liu, C., Liu, Y., & Freiheit, T. (2021). Smart manufacturing and intelligent manufacturing: A comparative review. *Engineering (Beijing)*, *7*(6), 738–757. DOI: 10.1016/j.eng.2020.07.017

Wang, H., Wang, Z., Chen, D., Liu, Q., Ke, H., & Han, K. K. (2023). Metamobility: Connecting future mobility with the metaverse. *IEEE Vehicular Technology Magazine*, *18*(3), 69–79. DOI: 10.1109/MVT.2023.3263330

Wang, L., & Lee, S. (2021). "Technological Foundations of Augmented Reality: A Comprehensive Overview." *Proceedings of the IEEE International Conference on Computer Vision*, 132-145.

Wang, S., Charissis, V., & Harrison, D. K. (2017). Augmented reality prototype HUD for passenger infotainment in a vehicular environment. *Advances in Science, Technology and Engineering Systems Journal*, *2*(3), 634–641. DOI: 10.25046/aj020381

Wang, X., Ong, S. K., & Nee, A. Y. C. (2016). A comprehensive survey of augmented reality assembly research. *Advances in Manufacturing*, *4*(1), 1–22. DOI: 10.1007/s40436-015-0131-4

Wang, Y., & Suh, J. (2019). Application of virtual reality in the automotive industry: From design to customer experience. *Automotive Engineering Journal*, *45*(2), 188–203. DOI: 10.1016/j.aej.2019.01.005

Wang, Z., Wei, H., Wang, J., Zeng, X., & Chang, Y. (2022). Security issues and solutions for connected and autonomous vehicles in a sustainable city: A survey. *Sustainability (Basel)*, *14*(19), 12409. DOI: 10.3390/su141912409

Watanuki, K. (2008). Virtual reality based job training and human resource development for foundry skilled workers. *International Journal of Cast Metals Research*, *21*(1–4), 275–280. DOI: 10.1179/136404608X362098

Webel, S., Bockholt, U., Engelke, T., Gavish, N., Olbrich, M., & Preusche, C. (2013). An augmented reality training platform for assembly and maintenance skills. *Robotics and Autonomous Systems*, *61*(4), 398–403. DOI: 10.1016/j.robot.2012.09.013

Wiedenmaier, S., Oehme, O., Schmidt, L., & Luczak, H. (2003). Augmented Reality (AR) for Assembly Processes Design and Experimental Evaluation. *International Journal of Human-Computer Interaction*, *16*(3), 497–514. DOI: 10.1207/S15327590IJHC1603_7

Wiegand, G., Mai, C., Holländer, K., & Hussmann, H. (2019). Incarar: A design space towards 3d augmented reality applications in vehicles. *Proceedings of the 11th International Conference on Automotive User Interfaces and Interactive Vehicular Applications*, 1–13. DOI: 10.1145/3342197.3344539

Winkler, M., & Soleimani, M. (2025a). A Review of Augmented Reality Heads Up Display in Vehicles: Effectiveness, Application, and Safety. *International Journal of Human-Computer Interaction*, •••, 1–16. DOI: 10.1080/10447318.2024.2443252

Wolfartsberger, J. (2019). Analyzing the potential of Virtual Reality for engineering design review. *Automation in Construction*, *104*, 27–37. DOI: 10.1016/j.autcon.2019.03.018

Wolfartsberger, J., Zenisek, J., & Wild, N. (2020). Supporting teamwork in industrial virtual reality applications. *Procedia Manufacturing*, *42*, 2–7. DOI: 10.1016/j.promfg.2020.02.016

Wu, H. K., Lee, S. W. Y., Chang, H. Y., & Liang, J. C. (2013). Current status, opportunities and challenges of augmented reality in education. *Computers & Education*, *62*, 41–49. DOI: 10.1016/j.compedu.2012.10.024

Wu, J., Zhang, P., & Li, Z. (2020). Autonomous vehicle simulation and testing with virtual reality. *Journal of Automotive Technology*, *45*(1), 123–136. DOI: 10.1109/JAT.2020.9123094

Xia, Z., & Huang, M. (2024). Optimizing the Aerodynamic Efficiency of Electric Vehicles via Streamlined Design: A Computational Fluid Dynamics Approach. *International Journal of Heat & Technology*, *42*(3), 865–876. DOI: 10.18280/ijht.420315

Xie, L., Liu, H., & Song, F. (2019). Application of virtual reality technology in automotive manufacturing and service. *Industrial Engineering Journal*, *34*(5), 541–549.

Xiong, F., Wang, D., Ma, Z., Chen, S., Lv, T., & Lu, F. (2018). Structure-material integrated multi-objective lightweight design of the front end structure of automobile body. *Structural and Multidisciplinary Optimization*, *57*(2), 829–847. DOI: 10.1007/s00158-017-1778-1

Xu, J., & Zhou, S. (2014). Analysis of flow field for automotive exhaust system based on computational fluid dynamics. *The Open Mechanical Engineering Journal*, *8*(1), 587–593. DOI: 10.2174/1874155X01408010587

Xu, L. D., Xu, E. L., & Li, L. (2018). Industry 4.0: State of the art and future trends. *International Journal of Production Research*, *56*(8), 2941–2962. DOI: 10.1080/00207543.2018.1444806

Xu, N., Yu, F., Xu, J., Jin, D., Feng, H., Xu, Z., Li, Q., & Chen, Y. (2023). HUDNet: A dynamic calibration method for automotive augmented reality head-up-displays. *Displays*, *78*, 102453. DOI: 10.1016/j.displa.2023.102453

Yıldız, B. S. (2020a). Optimal design of automobile structures using moth-flame optimization algorithm and response surface methodology. *Materials Testing*, *62*(4), 371–377. DOI: 10.3139/120.111494

Yıldız, B. S. (2020b). The spotted hyena optimization algorithm for weight-reduction of automobile brake components. *Materials Testing*, *62*(4), 383–388. DOI: 10.3139/120.111495

Yıldız, B. S., Pholdee, N., Bureerat, S., Yıldız, A. R., & Sait, S. M. (2020). Sine-cosine optimization algorithm for the conceptual design of automobile components. *Materials Testing*, *62*(7), 744–748. DOI: 10.3139/120.111541

Yildiz, E., Møller, C., Bilberg, A., & Rask, J. K. (2021). Virtual prototyping: Evaluating the digital twin based virtual factory for new product introduction. *Complex Systems Informatics and Modeling Quarterly*, *29*(29), 1–16. DOI: 10.7250/csimq.2021-29.01

Yontem, A. O., Li, K., Chu, D., Meijering, V., & Skrypchuk, L. (2021). Prospective Immersive Human-Machine Interface for Future Vehicles: Multiple Zones Turn the Full Windscreen into a Head-Up Display. *IEEE Vehicular Technology Magazine*, *16*(1), 83–92. Advance online publication. DOI: 10.1109/MVT.2020.3013832

You, F., Fu, Q., Yang, J., Chen, H., Cui, W., & Wang, J. (2024). Team situation awareness-based augmented reality head-up display design for security requirements. *International Journal of Human-Computer Interaction*, *40*(18), 5119–5138. DOI: 10.1080/10447318.2023.2231281

Yuan, M. L., Ong, S. K., & Nee, A. Y. C. (2008). Augmented reality for assembly guidance using a virtual interactive tool. *International Journal of Production Research*, *46*(7), 1745–1767. DOI: 10.1080/00207540600972935

Yuan, T., da Rocha Neto, W., Rothenberg, C. E., Obraczka, K., Barakat, C., & Turletti, T. (2022). Machine learning for next-generation intelligent transportation systems: A survey. *Transactions on Emerging Telecommunications Technologies, 33*(4), e4427. DOI: 10.1002/ett.4427

Zackrisson, L. (2017). *Flow acoustic effects on a commercial automotive air intake silencer-A numerical study using computational fluid dynamics.*

Zambrano-Martinez, J. L., Calafate, C. T., Soler, D., Lemus-Zúñiga, L.-G., Cano, J.-C., Manzoni, P., & Gayraud, T. (2019). A centralized route-management solution for autonomous vehicles in urban areas. *Electronics (Basel), 8*(7), 722. DOI: 10.3390/electronics8070722

Zhang, L., Wang, J., & Chen, H. (2021). Emerging trends in virtual reality applications: A review. *Journal of Computer Science and Technology, 36*(2), 112–129.

Zhang, Z., Wang, X., Wang, X., Cui, F., & Cheng, H. (2019). A simulation-based approach for plant layout design and production planning. *Journal of Ambient Intelligence and Humanized Computing, 10*(3), 1217–1230. DOI: 10.1007/s12652-018-0687-5

Zheng, P., Wang, H., Sang, Z., Zhong, R. Y., Liu, Y., Liu, C., Mubarok, K., Yu, S., & Xu, X. (2018). Smart manufacturing systems for Industry 4.0: Conceptual framework, scenarios, and future perspectives. *Frontiers of Mechanical Engineering, 13*(2), 137–150. DOI: 10.1007/s11465-018-0499-5

Zhou, F., Lin, X., Liu, C., Zhao, Y., Xu, P., Ren, L., Xue, T., & Ren, L. (2019). A survey of visualization for smart manufacturing. *Journal of Visualization / the Visualization Society of Japan, 22*(2), 419–435. DOI: 10.1007/s12650-018-0530-2

Zhu, L., Chen, C., Wang, H., Yu, F. R., & Tang, T. (2023). Machine learning in urban rail transit systems: A survey. *IEEE Transactions on Intelligent Transportation Systems.*

Zigart, T., Stürzl, F., Niedermayr, D., Wolfartsberger, J., & Sorko, S. R. 2023. Comparative Evaluation of Virtual Reality and In-Person Onboarding for Assembly Trainings in Manufacturing, in: 2023 IEEE International Symposium on Mixed and Augmented Reality Adjunct (ISMAR-Adjunct). Presented at the 2023 IEEE International Symposium on Mixed and Augmented Reality Adjunct (ISMAR-Adjunct), IEEE, Sydney, Australia, pp. 167–174. DOI: 10.1109/ISMAR-Adjunct60411.2023.00041

Zou, X., O'Hern, S., Ens, B., Coxon, S., Mater, P., Chow, R., Neylan, M., & Vu, H. L. (2021). On-Road Virtual Reality Autonomous Vehicle (VRAV) Simulator: An Empirical Study on User Experience. *Transportation Research Part C, Emerging Technologies*, *126*, 103090. Advance online publication. DOI: 10.1016/j.trc.2021.103090

About the Contributors

Devaraj E is a renowned expert in the field of virtual and augmented reality applications within the automotive industry. With over 10 years of experience in research, development, and implementation of cutting-edge technologies, Dr.Devaraj E has established himself as a thought leader in the intersection of automotive engineering and immersive digital experiences. After completing his doctoral studies in Mechanical Engineering with a specialization in Polymer composites from a prestigious university, Dr. Devaraj E embarked on a career dedicated to pushing the boundaries of innovation in the automotive sector. His early research focused on virtual prototyping and design methodologies, laying the groundwork for more efficient and cost-effective product development processes. Throughout his career, Dr. Devaraj E has held key roles in various leading automotive companies and research institutions, where he spearheaded numerous projects aimed at harnessing the potential of virtual and augmented reality technologies. From revolutionizing the design process to enhancing customer engagement through immersive showroom experiences, Dr. Devaraj E contributions have had a significant impact on the industry. Driven by a passion for both technology and automobiles, Dr. Devaraj E has authored numerous papers and articles on the subject, sharing his insights and expertise with the wider community. "Exploring Virtual and Augmented Reality Applications in the Automobile Industry" represents Dr.Devaraj E latest endeavor to provide a comprehensive resource for professionals, researchers, and enthusiasts interested in understanding the transformative potential of virtual and augmented reality in shaping the future of mobility. With his depth of knowledge and practical experience, Dr. Devaraj E offers invaluable insights into the applications, benefits, and challenges of integrating immersive technologies into the automotive ecosystem.

***

A. Manikandan currently serves as Associate associate professor at the Department of Electronics and Communication Engineering, Amrita School of Engineering, Amrita Vishwa Vidyapeetham, Amritapuri campus.He received his Ph.D. in the area of wireless networks from Anna University at 2018. He has over 17 years of teaching experience. He is a life member of several professional societies, such as IETE, ISRD, & ISTE. His research area includes signal processing and wireless networks. He has published 13 papers in international journals and 14 papers in national and international conferences. Also, he has authored three books. He has received a patent on an animal impediment system in a deep forest.

A. Sathiya is a Full-Time Research Scholar in the Department of Electronics and Communication Engineering at M. Kumarasamy College of Engineering (Autonomous), Karur, Tamil Nadu - 639113. Her research focuses on advancing knowledge and innovation in VLSI and Electronics and Communication Engineering. With dedication to her field, she strives to contribute significantly to academic and technological progress.

Sampath Boopathi is an accomplished individual with a strong academic background and extensive research experience. He completed his undergraduate studies in Mechanical Engineering and pursued his postgraduate studies in the field of Computer-Aided Design. Dr. Boopathi obtained his Ph.D. from Anna University, focusing his research on Manufacturing and optimization. Throughout his career, Dr. Boopathi has made significant contributions to the field of engineering. He has authored and published over 225 research articles in internationally peer-reviewed journals, highlighting his expertise and dedication to advancing knowledge in his area of specialization. His research output demonstrates his commitment to conducting rigorous and impactful research. In addition to his research publications, Dr. Boopathi has also been granted one patent and has three published patents to his name. This indicates his innovative thinking and ability to develop practical solutions to real-world engineering challenges. With 17 years of academic and research experience, Dr. Boopathi has enriched the engineering community through his teaching and mentorship roles.

J. Samuel Manoharan is a distinguished academic in the Department of Biomedical Engineering at Jerusalem College of Engineering, Chennai-600100. With a profound dedication to research and teaching, he has significantly contributed to the field of biomedical engineering. His expertise and commitment inspire students and peers alike.

M. Arunkumar is a dedicated educator and researcher in the Department of Electronics and Communication Engineering at Sengunthar Engineering College

(Autonomous). His contributions to the field reflect his commitment to academic excellence and technological innovation. Through his work, he continues to inspire students and advance knowledge in electronics and communication engineering.

S. Hemamalini is an Associate Professor in the Department of Artificial Intelligence & Data Science at Panimalar Engineering College. With extensive expertise in AI and data science, she is committed to fostering innovation and academic excellence. Her contributions to research and teaching have greatly benefited students and the broader scientific community.

P.Suthanthira Devi has received her Ph.D. Degree in Information and Communication Engineering from Anna University. Her research area is Social Network Analysis and Natural Language Processing. She has 10 years of teaching experience at Anna University affiliated college. She has published nearly 17 research articles in refereed International Journals in areas like machine learning. Presently she is working as Assistant Professor in Department of Data science and business systems, SRM Institute of Science and Technology.

R. Kingsly Stephen serves as an Assistant Professor in the Department of Computer Science and Engineering at SRM Institute of Science and Technology, Ramapuram campus. His academic journey is marked by a relentless pursuit of knowledge, as he concurrently works towards his Ph.D. in the dynamic field of Artificial Intelligence and Machine Learning, starting in 2022. With an impressive 17 years of teaching experience, he possesses a wealth of expertise and a knack for programming. His commitment to staying updated is evident through his certification in various courses from renowned platforms like Coursera and NPTEL. Furthermore, his contributions to academia extend to his publications, including articles in both National and International Journals, Conferences, and Symposiums. His research interests are predominantly centred around Computer Networks, Artificial Intelligence, and Deep Learning. In addition to his academic pursuits, he is an active member of the International Association of Engineers (IAENG), solidifying his commitment to professional growth and networking

Siva Subramanian R is an academic professional associated with R.M.K. College of Engineering and Technology in Puduvoyal, Tamil Nadu. Renowned for his contributions to engineering education, he plays a pivotal role in fostering innovation and excellence among students. His work is dedicated to advancing technology and supporting the next generation of engineers.

Sunil Sharma has completed his Ph.D in Mechanical Engineering from Indian Institute of Technology Ropar, Punjab, INDIA. He is a graduate in the discipline of

Mechanical Engineering from Punjab Technical University (currently I.K. Gujral Punjab Technical University), Kapurthala, Punjab, INDIA and postgraduate in Manufacturing and Materials from Oklahoma State University, Stillwater, USA. His research interests are primarily in the field of biomimicry, creativity and design process.

C.V. Suresh Babu is a pioneer in content development. A true entrepreneur, he founded Anniyappa Publications, a company that is highly active in publishing books related to Computer Science and Management. Dr. C.V. Suresh Babu has also ventured into SB Institute, a center for knowledge transfer. He holds a Ph.D. in Engineering Education from the National Institute of Technical Teachers Training & Research in Chennai, along with seven master's degrees in various disciplines such as Engineering, Computer Applications, Management, Commerce, Economics, Psychology, Law, and Education. Additionally, he has UGC-NET/SET qualifications in the fields of Computer Science, Management, Commerce, and Education. Currently, Dr. C.V. Suresh Babu is a Professor in the Department of Information Technology at the School of Computing Science, Hindustan Institute of Technology and Science (Hindustan University) in Padur, Chennai, Tamil Nadu, India.

T.Archana is currently pursuing her Ph.D. in the Department of Computer Science and Engineering at SRM Institute of Science and Technology, Ramapuram Campus in Chennai. With over six years of teaching experience in esteemed institutions, she has honed her skills in imparting knowledge and shaping the minds of young students. Her primary research interests lie in the captivating fields of Artificial Intelligence and Cloud Computing, where she aspires to contribute to cutting-edge innovations and solutions. She also has an impressive track record of publishing papers in well-regarded international journals and conferences, showcasing her dedication to advancing the realm of computer science. She has completed several certified courses from NPTEL and Coursera. With an unwavering commitment to learning and research, she is eager to explore new horizons and make a meaningful impact in the world of technology

T. P. Anish received his M.E. Degree from the Department of Computer Science and Engineering at Saveetha Engineering College (Affiliated to Anna University), Chennai, India, in 2011. He is currently an Assistant Professor at R.M.K. College of Engineering and Technology, Puduvoyal, India. He is pursuing his Ph.D. degree in Information and Communication Engineering from Anna University. His current research interest includes Machine Learning, Deep Learning and Big Data analytics.

V. Sathya is a faculty member in the Department of Computer Science and Engineering at Vel Tech Rangarajan Dr. Sagunthala R&D Institute of Science and

Technology, Chennai, Tamil Nadu. She contributes to the academic and research initiatives at the institution.

Index

A

C

D

E

F

H

U

V

www.ingramcontent.com/pod-product-compliance
Lightning Source LLC
LaVergne TN
LVHW080349011026
849161LV00009B/2240
* 9 7 9 8 3 6 9 3 7 2 8 7 6 *